T0074328

Lecture Notes in Social Networks

For further volumes:
http://www.springer.com/series/8768

Alessandro Aldini · Alessandro Bogliolo
Editors

User-Centric Networking

Future Perspectives

 Springer

Editors
Alessandro Aldini
Alessandro Bogliolo
Dipartimento di Scienze di Base e
    Fondamenti
University of Urbino
Urbino
Italy

ISSN 2190-5428   ISSN 2190-5436 (electronic)
ISBN 978-3-319-05217-5   ISBN 978-3-319-05218-2 (eBook)
DOI 10.1007/978-3-319-05218-2
Springer Cham Heidelberg New York Dordrecht London

Library of Congress Control Number: 2014938220

Printed on acid-free paper

Springer is part of Springer Science+Business Media (www.springer.com)

Preface

User centricity promises to become one of the main driving principles of the Internet value chain. The flexibility and the widespread diffusion of Wi-Fi technologies, combined with the cooperative attitude typical of online community members, have determined the emergence of user-centric initiatives aimed at sharing services and resources to extend Internet access networks beyond the coverage of the infrastructures managed by established operators. In *user-centric networks* (UCNs), individuals share subscribed Internet access and network resources in exchange of specific incentives from social, economic, and technical viewpoints. Community members make use of their Wi-Fi-enabled equipment to cover the so-called *last mile* and allow other members to establish a local loop and take advantage of services and Internet backhauling. The success of such new models of Internet connectivity would not be the same without users involved actively in the provision of contents, applications, and services, rather than just being the recipients of those made available by established providers.

The user-centric character of a user-friendly and plug and play connectivity model and the need to assist the autonomic deployment of user-centric wireless local loops were the subject of study for the EU ULOOP (User-centric Wireless Local Loop) project, funded by the EU IST Seventh Framework Programme (FP7/ 2007–2013). The main aspects under investigation were classified into the following themes: general aspects of user-centric networking, trust management and cooperation incentives, resource management, mobility aspects, and marketplace issues. The structure of this book reflects such a classification into five parts and presents the main problems faced and the solutions developed within the ULOOP framework.

Part I of this book provides a general introduction to *User-Centric Networking*. Architectural models and technological aspects are presented to provide the basis to discuss the potential, the features, the feasibility, and the applicability of user-centric environments. Living examples are used to illustrate the key concepts and to point out the benefits that they bring and the challenges that they raise.

In Part II, trust management is introduced for the specification of trust relationships assisting users in the selection of trustworthy entities to cooperate with. Cooperation is considered as an enabling condition for a successful transformation of the end-user from the typical consumer of Internet services to an active hop (i.e., a *prosumer*) of the connectivity and service distribution chain. In spite of the widespread cooperative attitude that has determined the success of, e.g., social networks, content sharing systems, and peer-to-peer applications, the willingness to cooperate cannot be taken for granted in terms of sharing network functionality. The typical constraints of Wi-Fi technologies in terms of bandwidth, energy consumption, and computation resources, contribute to keep users from prosocial behaviors. Last but not least, cooperation involves also economic dynamics in which Internet stakeholders and operators play a role essential to guarantee socioeconomic sustainability. Hence, cooperation incentives are fundamental to avoid selfish behaviors, provide motivations to sharing, and extend the benefits to external actors. In the setting of ULOOP, incentives of different nature are combined that include reputation-based management of trust on the fly as well as virtual currency-based reward management.

Part III is focused on resource management as a basis for an efficient maintenance of ad-hoc wireless infrastructures. A fair and self-organizing use of network resources depends on several factors, and solutions ranging from the adoption of cooperative trust-based strategies to the implementation of resource allocation algorithms and workload/congestion control mechanisms. In this context, it is essential to take into account the growth of the community, which can be highly dynamic because of traffic fluctuations due to mobile stations that leave and join the community continuously.

Mobility aspects are treated in Part IV, which surveys on challenges and requirements for mobility management in user-centric networks. In the case of ULOOP, the main difficulties are related to the dynamic, erratic behavior of users and to the interoperability to other systems. Mobility management entails issues like, e.g., tracking, which is supported by estimation models based on the analysis of human social behaviors, and handover, which pursues the ideal of the *always best connection*. The challenge consists of predicting the right gateway to which any ULOOP node shall connect to, while moving, in order to maintain as transparently as possible connectivity and quality of experience.

To conclude the survey, Part V is dedicated to a perspective on market analysis and exploitation in the framework of user-centric wireless local-loop networks, with a particular emphasis on the analysis of the ULOOP case study.

In order to provide an adequate analysis of the ULOOP impact and sustainability from the socioeconomic standpoint, a deep investigation of Wi-Fi-related regulatory frameworks was performed at the beginning of the project. A prominent result is a collection of definitions taken from European Community Directives that are relevant in the framework of ULOOP and that we report in the glossary following this preface. Based on a logical, concept-driven order, such definitions make it clear the general notions of Service and Universal Service, and then introduce the specific notions related to the service deployment through electronic communication means, i.e., Electronic Communications Network, Public Communications Network, Network Termination Point, Electronic Communications Service, Publicly Available Electronic Communications Services, Associated Services, Associated Facilities, Access, Local Loop, Interconnection Conditional Access System, Provision of an Electronic Communications Network. Afterwards, regulatory issues are clarified through the notions of National Regulatory Authority, General Authorisation, Exclusive Rights, Special Rights, while all the involved actors are defined via the terms Operator, Service Provider, Established Service Provider, Commercial Communication, Recipient of the Service, User, Consumer, Subscriber, End-User. Application-specific terms complete the list, i.e., Enhanced Digital Television Equipment, Application Program Interface, Traffic Data, Location Data, Communication, Call, Consent, Value Added Service, Electronic Mail. For the sake of accessibility, the glossary is proposed in alphabetical order. Moreover, it is integrated with ULOOP specific definitions that will be helpful for the reader prior to, or while, reading the book.

January 2014   Alessandro Aldini
                                                          Alessandro Bogliolo

Contents

Part I
User-Centric Environments

User-Centric Networking: Bringing the Home Network to the Core

Rute Sofia

Abstract This paper goes over the concept of User-centric networking, as a paradigm for networking architectures usually located in the Customer Premises, and which is steadily changing the way the Internet has been devised. Such change is due to the fact that the Internet end-user is empowered due to novel approaches such as software defined networking, thus being in control of functionality that so far was restricted to be placed in the core and access regions of networks. Such change introduces the need to revisit home networking, and from an end-to-end perspective, to introduce new concepts and technology into the networking functionality. The paper presents a user-centric model and its functional blocks, and describes the architectural example that has been conceived, implemented, and validated in the context of the European project ULOOP—User-centric Wireless Local Loop.

Keywords User-centric networks · Wireless · Home networks · Internet architectures

1 Introduction

Today's end-user is connected to the Internet by means of a variety of broadband access technologies that usually do not directly reach the *end-user equipment* (*UE*). Rather, this final segment of the local-loop (last mile) is provided by a number of short-range technologies, among which *Wireless Fidelity* (*Wi-Fi*) is the de-facto solution. The growing popularity of Wi-Fi as a complementary technology relates to its low-cost and worldwide availability, to the ease of use, and to the high interoperability

R. Sofia (✉)
COPELABS, University Lusófona, Building U First Floor, Campo Grande 388, 1749-024
Lisbon, Portugal
e-mail: rute.sofia@ulusofona.pt

A. Aldini and A. Bogliolo (eds.), *User-Centric Networking*,
Lecture Notes in Social Networks, DOI: 10.1007/978-3-319-05218-2_1,
© Springer International Publishing Switzerland 2014

3

that it is capable of sustaining, when interfacing with Internet broadband access technologies.

Having the last-hop of the Internet (towards the user) based on Wi-Fi often creates bottlenecks. Nevertheless, there are clear advantages in terms of Internet wholesale models, as each residential household becomes a Wi-Fi hotspot that becomes, most of the time, underused.

Due to such deployment as well as due to the introduction of new paradigms such as *Software Defined Networking (SDN)*, Wi-Fi is giving rise to new models of Internet connectivity. In these new Internet access models, the end-user is one of the stakeholders thus ceasing the plain role of consumer of Internet services (be it connectivity or content), to become an active hop in the connectivity distribution chain.

User-centric networking (UCN) [8, 19] explores concepts to allow user-centric wireless local-loops to form autonomously. The term user-centric in this context is meant to express a community model that extends the reach of a high debit, multi-access broadband backbone from different perspectives (technical, business model). Such a model is expected to be beneficial both from an end-to-end and from an access perspective, given that it allows expanding high debit reach in a seamless, cooperative, and low-cost manner. Moreover, UCN follows an evolutionary path in the context of future Internet architectural design, by building on existing work related to the recent trend of *Do-it-Yourself Networks (DIYN)*. Hence, a fundamental difference between such work and previous contributions on ad-hoc or mesh networking relates to the fact that UCN assumes that an infrastructure providing Internet access to specific locations is widely available, and users are simply willing to expand such infrastructure by exchanging some resources (networking resources, services). UCN is based on the notion that trust circles can assist in cooperative behavior involving both the access and the end-user.

These aspects empower the user role as an active element of networking given that (i) the user becomes a producer/provider of specific services; (ii) the user device is part of the network. The aforementioned aspects go against the Internet end-to-end principle, which describes a clear functional splitting between end-systems and the network. UCN therefore introduces challenges that are worth to be analyzed from an operational network perspective, as UCN is supported both by access devices as well as by devices residing in the *Customer Premises (CP)* and for the largest majority, in residential (home network) scenarios. Hence, future Internet models have to integrate properties that allow nomadic end-user experience for any application across multi-access or single-access networks, assuming that one or more operators are involved. UCNs subvert the current notion of Internet architectural design, as they impact the Internet wholesale model chain, by empowering the end-user as an active stakeholder in the core network. Functionality that usually resides in the core network becomes more useful if placed closer to the Internet end-user.

From a business perspective, wireless local-loops that are built mainly based upon an Internet stakeholder willingness to cooperate should be a starting point to revisit current business models for broadband access and to analyze new business models. Similarly to what occurs in the energy sector in micro-generation models, in UCNs, the end-user becomes a micro-provider of a specific community by sharing his/her subscribed broadband access within his/her community, as well as by providing specific Internet services, according to specific incentives. Such incentives may simply relate with a well defined human trust (social) network, or even with some form of reward, e.g., gain coverage and Internet access beyond the end-user's premises. They may be user-based; access-provider based; a mix of both cases. Moreover, from a network access provider perspective and at a first glance, the motivation to invest on such models could just seem related to the possibility to expand capillarity in a low-cost way, as well as to the exploration of new services, which the users can help to define (community-based services). However, UCN advocates cooperation between access providers and end-users in terms of user-centric networking services, as such cooperation opens up new possibilities in terms of business models, based upon a clear separation between the service and network layers, as well as between the network manager and the infrastructure owner. New types of operators that would act as organizers will appear therefore fostering competition and clearly addressing the goal of an open and competitive digital economy.

Technical advantages must be explored from an access perspective and are one of the main aspects to pursue in UCNs. For instance, by deploying UCN it is possible to keep traffic local, namely, to take advantage of the physical proximity of sources and destinations and therefore, to prevent traffic from crossing the full access backbone when sources and destinations are "close" (according to previously defined criteria). Traffic locality rules can be applied in a wireless local-loop and will have as consequence a reduction in the access OPEX as well as an optimization of spectrum. Another intuitive advantage is the fact that the subscription relation between the end-user and the access operator can be strengthened by having the access operator empowering the end-user with partial networking functionality, in a way that is completely transparent to the end-user. In other words: such cooperative model (based upon Internet service micro-generation) gives the means for the access operator to provide value-added services that are more appealing to the end-user and that go beyond regular (Triple Play) Internet subscriptions, common today both in the bundled and in Service Provider centric models . For instance, models such as the one embodied today e.g., by FON, when used in strong cooperation with access providers, give the means to access providers to offer Internet access subscriptions with worldwide wireless roaming included, which by itself differentiates such service towards competitors.

This paper contributes to a better understanding of UCN, to its impact in Internet architectural design and operation. For such purpose, the rest of this document is organized as follows. Section 2 gives insight concerning UCN notions and background, while Sect. 3 covers motivation and details concerning the functional blocks that we believe are essential to consider in any UCN model. Section 4 provides an example of a UCN model that has been conceived and implemented, as well as validated in the context of the European project *ULOOP—User-centric Wireless Local Loop*. The paper concludes in Sect. 5, which also provides pointers for future research in UCN.

2 UCN Background

UCN relate to a recent trend in spontaneous wireless deployments where individual users or communities of users share subscribed access in exchange of specific incentives. In literature, names that relate with the UCN concept are *personal hotspot, spontaneous user-centric networks*.

In UCN, there are two fundamental networking roles: *node* and *gateway*. A UCN node concerns a role (software functionality) that a wireless capable device takes. Concrete examples of nodes can be specific user equipment, access points, or even some management server. A UCN gateway is a role (software functionality) that reflects an operational behavior making a UCN node capable of acting as a mediator between UCN systems and non-UCN systems—the outside world. The gateway role may or may not be owned and controlled by a UCN user; it may also or only be controlled by an access operator. The key differentiating factor of the role of gateway, in contrast to a regular UCN node, is the operational intelligence and mediation capability. Similarly to UCN nodes, the UCN gateway functionality may reside in the user-equipment, in APs, or even in the access network. Hence, they exhibit a feature that is key in user-centric environments: their behavior as part of the network is expected to be highly variable. Gateways will be active or inactive based on several conditions such as users' wishes and network load.

As previously mentioned, each UCN node has a unique *owner, micro-provider (MP)*, assigned. An owner is an entity (end-user, operator, virtual operator) that is responsible for any actions concerning his/her device. The term "responsible" reflects liability, i.e., from an operator's perspective the owner is the single responsible for the adequate/inadequate usage of the user's device within a specific, trust-bounded community.

A *community* in UCN is a set of UCN nodes that hold common interests (such as sharing connectivity or resources/peripherals) at some instant in time and space. In other words, nodes exhibit a space and time correlation that is the basis to establish a robust connectivity model. This is expected to be extrapolated by adequately modeling trust associations between nodes. We highlight that the notion of community does not have any relation whatsoever to an *Online Social Network (OSN)*, nor even to some specific OSN subset.

An interest is here defined as a parameter capable of providing a measure (cost) of the "attention" of a node towards a specific location in a specific time instant. In other words, an interest is a parameter that provides a node with a measure of a specific time and space correlation. For instance, assuming that a user goes each Saturday morning to the coffee-shop on the neighborhood corner, an interest here could be "having a coffee". Other users in the same location (exhibiting a similar time and space correlation) are in the same place during an overlapping period of time. They all share an interest as they are all collocated in the same location for a specific period of time. The shared interest here is: attending the same coffee-shop. Therefore, owners may be complete strangers and yet, connectivity may be

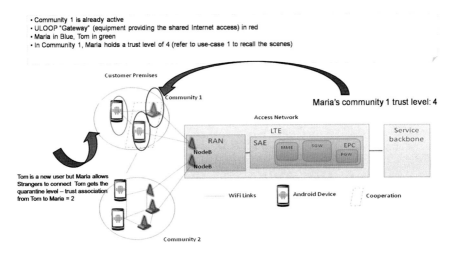

Fig. 1 Expanded capillarity and 3G offloading UCN applicability example

set across the devices, based on parameters such as specific *Quality of Experience* (*QoE*) metrics; node movement history; roaming and service sharing patterns.

Figure 1 illustrates a UCN where two different communities are represented, Community 1 and Community 2. The term community here is simply representative and identifies a set of users within the same WLAN. It could be, for instance, a mesh network in a city, or a hotspot at a coffee-shop.

Community 1 represents an example of a dense wireless network, infrastructure mode (e.g., shopping-mall, football stadium, indoor spaces in a school campus). By dense it is meant that several users may activate devices in AP mode and therefore, there is a strong signal overlap. Hence, the result of this is that despite the fact that spectrum abounds, *Signal to Noise Ratio* (*SNR*) can be very low in some areas (known as grey areas).

As for Community 2, it stands for a mesh network also interconnected to the same LTE provider. There is no strict relation between a community and a geographic location.

Maria is a user in Community 1 carrying her Android smart-phone (UE). Maria's UE selects a specific gateway (AP or UE) to be associated with in a certain location. After the reception of her association request, the gateway broadcasts a query both via the Wi-Fi interface and the LTE interface (to reach the backend) in order to figure out whether Maria is or not an authorized and trusted user. At the same time, the chosen gateway also triggers an adequate gateway selection mechanism that takes into consideration not only Maria's expectations, but also the potential overlap and electromagnetic noise in the area, as well as the optimization of the load across the entire network. While roaming in Community 1, the gateway onto which Maria's UE is currently associated detects that she is on the move (e.g., due to SNR variations) and immediately attempts to estimate/anticipate a potential new anchor for connectivity

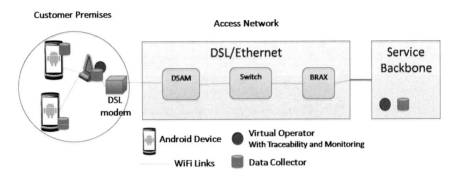

Fig. 2 Traceability and collective monitoring UCN applicability

(new gateway). Upon agreement between the gateways, Maria's UE is automatically attached to a new gateway which can fullfill Maria's service expectations.

Tom, another user of Community 1, is in a gray area. His device realizes that Maria's device allows connectivity relaying and therefore Tom's device triggers a request to connect to Maria's device. Maria allows other users with whom her device does not yet have a trust association established to interconnect by providing them a small amount of resources based on specific QoE requirements (e.g., only if her UE has enough battery level and up to 20 % of Maria's link capacity). Therefore, Maria and Tom's UE automatically negotiate connectivity and Tom goes online through Maria's device. Michael, another Community 1 user, is a subscriber of a network operator different than the one Maria is subscribed to and also belongs to Community 1. Given that they share interests in the context of Community 1, Michael and Maria are able to connect and exchange data directly, without going through their respective operators. Moreover, provided that there is such an available device, Michael can also profit from the Internet access while in Community 1. Every time Michael is within the coverage of Community 1 devices, his UE handovers from the 3G network to Community 1 (through Maria's UE). As one of the services provided by the network, all the communication between Michael and other users inside Community 1 is performed locally, including voice and video calls, and thus, for Michael, this means that his traffic is offloaded from the 3G network to the cloud. Whenever Michael leaves Community 1 area, his UE handovers back to the 3G network. Thanks to the resource optimization and load-balancing features of gateways within Community 1 continuously exchange data and thus offload/transfer some UE's to other elected gateways. Besides Community 1, a second group of users belonging to Community 2 gets information about data being shared in Community 1 (e.g., through the backend system). The second group is located in Paris, at Bob's place. Bob is using a tethered Android powered smart phone to connect to the LTE network and then uses the UCN functionality on the phone, via Wi-Fi, to share Internet access.

A second applicability case is provided in Fig. 2, where from an end-to-end perspective a single community is illustrated. Community 4 is connected to the Internet by means of a fixed operator (carrier-grade Ethernet/DSL).

The last hop to the user is Wi-Fi based. We also highlight that the number of UCN communities can profit from services that other users share. In this example the ultimate goal is not to expand coverage but instead to consider ULOOP functionality as an enabling technology platform for cooperative data dissemination. In regular deployments, such data cannot be available, as it is simply the result of a cooperative effort based on a self-organizing system. Moreover, end-user devices that are UCN enabled may be able to gather open data (data collected from the users' surrounding environment). This is, for instance, the case of Maria, who needs to print her boarding pass at an airport with Wi-Fi coverage. Due to the UCN functionality implemented in gateways and also provided directly by other users, Maria can print her boarding pass through John's device, a user that Maria's device trusts through a bi-directional trust association.

In Community 4, UCN functionality tracks user expectations and service response. Therefore, users providing expanded coverage have feedback about their resource usage on-the-fly. Moreover, the users are provided with incentives for sharing, e.g., more bandwidth in exchange of receiving some advertisement. Such tracking/monitoring can be performed based on the CPE (UE and gateway) or directly via UE. Moreover, such tracking relates to information that is not personal and that the user always acknowledges to provide beforehand. In this use-case, tracked data does impose neither any confidentiality nor privacy risk for the user. The UE serves the purpose of being part of the data dissemination towards users that share some form of interest, or for which there is some interconnection due to a dynamically established trust circle.

3 UCN Functional Blocks

UCN envisions increasing the potential of the Internet by devising communication and networking technologies which support the creation of techno-social communities, providing a combination of information, communication and human elements, by relying on adequate modeling of trust associations and trust levels. Communication opportunities due to sharing of Internet access as well as due to relaying across multiple hops provide a way to reduce costs, thus creating opportunities for new Internet wholesale models. Hence, new services provided by communities as well as new business models for end-users and access operators are expected to emerge due to novel features, e.g., an increase in spectrum and energy efficiency in managing wireless communications.

UCNs assume that an existing infrastructure is available and that Internet users are willing to expand such infrastructure in a way that is user-friendly and self-organizing. UCNs assumes also that within specific trust circles some form of cooperation incentives can be provided in order for both the access and the end-user to cooperate and assist in further developing Internet architectures. In order for that to happen, UCNs consider four main functional blocks as illustrated in Fig. 3: trust management and cooperation incentives; resource management; mobility aspects; and backward compatibility.

Fig. 3 UCN functional blocks

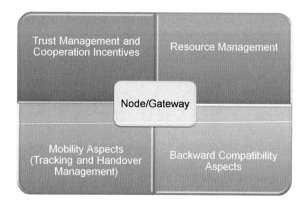

3.1 Trust Management and Cooperation Incentives

Trust management and cooperation incentives relate with understanding how to define and build circles of trust on-the-fly. Such circles of trust are capable of sustaining an environment where stakeholders share some form of Internet resources in order to support the dynamic behavior of UCN. Trust management is based on reputation mechanisms able to identify end-user misbehaviour and to address social aspects, e.g., the different types of levels of trust users may have in different communities (e.g., family, affiliation). In situations where the created network of trust is not enough to allow resources to be shared, devices are able to use a cooperation incentive scheme based on the transfer of credits directly proportional to the amount of shared resources.

3.1.1 Motivating Usage via Cooperation

Cooperation incentives in UCN are considered both from a specific technology perspective, as well as from a business perspective. Technical incentives may relate to natural features of the technology that result in a win-win match when cooperation is applied. A concrete example of a technical incentive relates to potential improvements of the 802.11 MAC layer. UCN engineers the MAC layer in a way that mitigates problems related with low data rate stations. Hence, when low data rate stations and high data rate stations cooperate, all elements are expected to take some advantage of such cooperation. While as for a business incentive, we can think of a specific peering scheme that may assist the access operators in understanding how to obtain revenue based on UCN architectures.

As part of communities and also as individual nodes, cooperation must consider the willingness of owners/nodes in participating in communication. Willingness can be driven by different facts such as energy saving, low processing power, and/or lack of storage room. Although a node is not willing to share resources due to one of the aforementioned facts, the cooperation functionality should encourage such user in doing it so, as he/she can get an immediate return (e.g., more processing) while sharing that resource it has the most (e.g., storage). Instead of simply paying users with the same "currency", e.g., you get more bandwidth if you give more bandwidth, the cooperation functionality should reward involved entities with the type of resource the user wants and at the moment the user needs (i.e., immediately or later on).

3.1.2 The User Perspective

The lack of trust between users can influence their level of willingness and our belief is that motivation should be based on shared interests. Users sharing the same interest (e.g., movies), although being completely unknown to one another, can be easily encouraged in carrying information on behalf of others. A user interested in comedy movies surely won't mind to carry a copy of a movie destined to some other user if he/she is able to get a copy also. At this point, cooperation not only helps users disseminate information quickly and seamlessly (as the movie will reach different interested users other than the destination) as it also contributes to sparing resources from users who are not interested in that specific content. Cooperation shall be easily encouraged if users share some social relationship. Thus, social ties have an important role in making cooperation among users even more reliable. Software functionality in UCN nodes is expected to track user expectations and service response. In this case, users are expected to cooperate in order to provide surrounding UCN nodes with information that not only can improve their own but also the other users' network experience. Users can exchange: (i) SNR information, e.g., to aid in the handover process; (ii) behavior information, to strictly penalize malicious/greedy users; (iii) connectivity quality levels, to aid in load balancing and interference reduction.

3.1.3 The Provider Perspective

UCN is a perfect solution for operators who are looking for higher density at limited cost, letting them to rely on created communities, in order to provide the required resources to demanding users at specific instants in time. This will offer an energy-efficient and cost optimized solution to increase density of the operators' networks. Moreover, the subscription relation between the end-user and the access operator can be strengthened by having the access operator empowering the end-user with partial networking functionality, in a way that is completely transparent to the end-user. In other words: such cooperative model (based upon Internet service micro-generation) gives the means for the access operator to provide value-added services that are more appealing to the end-user and that go beyond regular subscriptions, common today both in the bundled and in Service Provider centric models such as the one embodied today by e.g., FON, when used in strong cooperation with access providers, give the means to access providers to offer Internet access. For instance, access operators can take advantage of UCN capabilities to further expand its control towards the customer premises devices.

Some reasons for UCN adoption (and hence for the relevancy of operator-based incentives) are: to provide adequate feedback to customers; to ensure an optimal network operation, where expanded coverage is also offered; to be able to deal with interference in dense areas; to provide residential areas with the same authentication/authorization model used in UCN coverage and thus, reduce CAPEX; to gain in reputation by supporting communities, following what is today common practice in open-source business models.

An "Open Source" model with "some limitation" can favorite a win-win equilibrium between UCN and operator's competitiveness goals. Hence, operator's based incentives are expected to improve the potential of interoperability and of business opportunity for access and service stakeholders.

3.2 Augmented Resource Management

As UCN relies on wireless infrastructures that are often deployed in an ad-hoc way, resource management optimization is a key aspect to pursuit. UCN has as purpose to assist in developing robust and high debit wireless local-loops in a way that meets current broadband access technologies debit as possible, and in a way that reduces the chances for bottlenecks to occur. Throughput maximization is to be addressed across more than one hop by means of cooperative networking techniques of which one possibility is relaying. In regards to resource management, and to achieve a fair and self-organizing network operation, there are aspects to be looked for such as the need to adequately and dynamically be able to control growth of UCN communities; dynamic fluctuations of the network both in terms of traffic due to stations joining and leaving frequently, as well as due to the movement of stations. Another aspect that is considered crucial to look for is to develop cooperative and distributed mechanisms that assist the network in adequately selecting nodes that are willing to be micro-providers. Such selection is to be performed in a way that considers not only throughput maximization, but also the lowest-cost in terms of energy-efficiency.

3.3 Dealing with Frequent Roaming: Anchor Point Control and Movement Estimation

UCN is based on the notion of users carrying (or owning) low-cost and limited capacity portable devices which are cooperative in nature and which extend the network in a user-centric way, not necessarily implying the support for networking services such as multi-hop routing. For instance, in UCNs transmission may simply be relayed based on simple mechanisms already existing in end-user devices. These emerging architectures therefore represent networks where the nodes that integrate the network are in fact end-user devices which may have additional storage capability and which may or may not sustain networking services. Such nodes, being carried by end-users exhibit a highly dynamic behavior. Nodes move frequently following social patterns and based on their carriers interests; inter-contact exchange is the basis for the definition of connectivity models as well as data transmission. The network is also expected to frequently change (and even to experience frequent partitions) due to the fact that such nodes, being portable, are limited in terms of energy resources.

In terms of mobility and adding to the currently available solutions, UCN is focused on two main aspects: mobility tracking and estimation, as well as handover support. Ways of addressing patterns of node movement to estimate mobility patterns based on existing or novel social models is one aspect to be addressed [17]. The purpose is to assist in improving the underlying connectivity model, and hence overall network operation. Social mobility modeling is an aspect that assists in deriving algorithms and functionality that can anticipate the way nodes move based on analysis and tracking of node movement through time.

A final aspect to consider is to ensure that the functionality to be developed can assist in dealing with the unmanaged aspects of UCN architecture and should get rid of anchors in the network. This may be required, for instance, if a UCN community is not capable of providing a node with adequate mobility management e.g., due to trust aspects.

Being focused in Wi-Fi, the regular 250 m range is small compared to the geographic distances that users are expected to travel in UCN scenarios. Hence, performing a complete handover would impose strong requirements on speed of message exchange. In UCN this is to be tackled by considering the regularity (routine) present in users' movement, which may assist in determining the place and type of resources that may be required to set up to assist seamless handovers. For example, based on movement analysis, the system may determine with a high probability that the user will handover towards the range of UCN gateway A or UCN gateway B. In this case specific functions may assist in defining the adequate next target, and how to handover to it. The challenge here is to identify with enough accuracy and reliability the gateways that a UCN node should connect to, while moving.

3.4 Backward Compatibility

When dealing with unaware systems, i.e., systems that do not support UCN functionality, it is important to consider that a UCN device needs to be available to connect to a "legacy" network, and therefore, we must ensure proper backward-compatibility. This implies handing the interoperability aspects that deal with existing networking models and paradigms. Interoperability with legacy networks is especially important in the case where nodes roam through different types of networks, such as mobility between UCN and other types of networking architectures. Interoperability also relates to backward-compatibility, i.e., assisting devices outside of UCN to be able to participate in UCN communities in a regular way. Regardless of who supports the community information for the agnostic nodes, the connection, access must remain transparent (limitations can apply). Therefore, by allowing legacy nodes to connect to a UCN supported community, the trust and resource management implications must be considered. Within these aspects, the solution greatly depends on how much interoperability is required, and at which phase should it be considered (either as design constraint or proxy/adaptation/add-on functionality). The impact on different functionality can be substantial, depending on the type of integration/interoperability required.

4 An Architectural UCN Model: The ULOOP Example

In this section we provide a model for UCN that has been conceived, validated, as well as implemented in the context of the European project ULOOP, being currently available to the community as open-source LGPLv3.0 software [14, 15].

ULOOP considers a software-defined approach to implement the UCN architecture. A unique software suite is provided to devices, which then take the role of node or gateway, depending on a series of external conditions. The next sections provide input into the ULOOP architecture, which is composed of three main entities: the Trust Manager entity; the Resource Manager entity; the Mobility Aspects Entity. Each of this entities comprises specific sub-modules that are dynamically activated depending on the role assumed by a device implementing ULOOP.

4.1 The Trust Manager Entity

In ULOOP, trust management and incentives for cooperation are related to understanding how to define and build circles of trust on-the-fly to provide the user with liability [6]. Trust management is based on reputation mechanisms able to identify end-user misbehavior and to address social aspects, e.g., the different types of levels of trust users may have in different communities (e.g., family, affiliation). In situations where the created network of trust is not enough to allow resources to be shared, ULOOP devices are able to use a cooperation incentive scheme based on the transfer of credits directly proportional to the amount of shared resources [2, 3, 7].

4.1.1 Trust Setup

Trust setup in ULOOP is a one-time process that a user (owner) executes on one of its devices. This process does not need to be repeated on other devices of the user. After the setup procedure, the trust value may be updated based on a new value for the initial trust value, which in ULOOP is known as the *dispositional trust level* (*DT*)

[5], which can be always adjusted in each of the devices owned by the same user as a first step.

Trust setup is triggered in any ULOOP node and comprises a series of steps which result in: (i) a unique identifier, the crypto-id; (ii) a wallet with an initial set of credits; (iii) an initial trust value towards any new neighbor, familiar or not—the dispositional trust value.

The first step towards building trust references in communities, i.e., from a ULOOP node to others, is to be able to uniquely identify owners of ULOOP nodes. Ideally, the recognition must be attack-proof. Hence, the end-user must be able to authenticate her/him. However, it also important to protect the privacy of this end-user, so this building block contains both identity management and *Privacy-Enhancing Technologies* (*PETs*).

ULOOP reuses the concept of crypto-identifiers (*crypto-ids*) based on asymmetric cryptography. With such crypto-ids, the end-users can prove in a decentralized way and with cryptographic strength that they really own the secret linked to the crypto-id. Concerning privacy, creation and proof of ownership of crypto-ids does not require a centralized identity authority. Thus, end-users in ULOOP will protect their privacy through crypto-ids that they generate themselves and act as their pseudonyms not linked to their real world identity. The crypto-ids are based on a set of information provided to the user by an authorized entity (e.g., the personal identification number embedded in a citizen identity card provided by a government to any citizen, or a mobile phone number associated to a unique SIM card). Such personal identification number will be used to generate a unique crypto-id based on a hash function that is implemented in any ULOOP node or gateway. The local generated crypto-id will need to be verified by an authorized entity in order to allow the ULOOP node/gateway to gain full access to the ULOOP community.

When such verification cannot happen, the ULOOP device gets a minimum trust level in the community, allowing it to use a predefined set of minimum resources.

In ULOOP, owners are likely to be responsible for more than one active device. One would be a primary device, and the remainder equipment will share the same crypto-id generated by the first personal device, as well as the reputation level and trust associations associated to the unique crypto-id. This is possible by using secure in range wireless or wired communications. Synchronizing the reputation levels and trust associations among personal devices will allow the user to always make use of the earned reputation level, trust associations and credits that resulted from the usage of the unique crypto-id in another personal device. Synchronization of trust information can be done by using prior-art on file and data synchronization. The validation of the unique crypto-id can be done by making use of any opportunity to access the Internet (limited Internet access should be allowed by the minimum trust level). This may create some problem in extreme cases, in which Internet access is not possible for a long time. However, such scenarios are more related to delay-tolerant networks and not to ULOOP, in which it is expected that trust management and cooperation incentives will create the conditions to make Internet access more pervasive than today. Nevertheless, it is clear that the usage of a unique crypto-id may limit the usage of ULOOP in fully decentralized environments, namely in the presence of isolated ULOOP networks (without any Internet access whatsoever) and new users (that still need their crypto-ids to be validated).

UCNs such as ULOOP are supported both by static, fully dedicated nodes as well as by nodes provided by end-users on-the-fly. Since some nodes are carried by Internet end-users, their networking composition, surrounding environment and organization can rapidly change. As such, the dispositional trust level on a given node might not be appropriate in all circumstances and should be able to be adapted and changed over time, in order to protect the node's integrity. The process of dispositional trust adaptation might occur in two different cases: (i) the node has a dispositional trust level that is inappropriate and leaves it too open to attacks; (ii) The node joins a different community than the initial one in which the dispositional trust level had been setup.

An untrustworthy node in ULOOP goes through a boot-up procedure where the node may be the first one an owner is responsible for, or one of several nodes. In the former case the owner is prompted to set its DT, e.g., being able to select from a list of predefined values, which range from 0, being 0 "paranoid" which means that a priori the node will not trust anyone, and being 1 "blind trust" which means that the node will trust no matter what. In the second case, the user is presented with two options: (i) to clone the dispositional trust level assigned to other devices that are already in ULOOP and that she/he owns, for the usage of unique crypto-ids in different personal devices: (ii) to assign a new DT level for the node being introduced, as explained in the previous paragraph.

After the one-time step of trust setup, any node starts in background a trust management process.

4.1.2 Trust Management

Trust management is performed in two different phases of ULOOP: (i) when connectivity is attempted; (ii) during data transmission. When a node attempts to connect to a wireless network (e.g., via a captive portal), this triggers a request for resources, an aspect that is tackled by the Trust Manager entity in ULOOP.

The Trust Manager entity is in charge of executing the three sub-blocks it integrates (setup; management; cooperation control and rewarding), as well as in charge of establishing and maintaining the external interfaces (communication via TCP sockets, for the sake of proof-of-concept) with the Trust Manager of other ULOOP nodes (requester to requestee and vice-versa), as well as the internal interfaces with other operational modules (Resource Manager and Mobility Aspects) within the same node. When first instantiated, the Trust Manager performs a series of initial setup procedures, such as the virtual crypto-id generation and validation, as well as the dispositional trust setup. After this, and before going to the main operational mode, it starts a set of periodic activities from the reward manager that have to be executed in the background in order to ensure the proper operation and update of the bank account and the wallet of the node. Finally, the main functionality allows the node to perform its main operation, such as exchanging crypto-ids with other ULOOP nodes in order to start a cooperation process, performing social trust computation of those nodes and carrying out the control of cooperation, fundamental to decide if a service is obtained or allowed from/to another node.

4.1.3 Cooperation and Rewarding

The computation of trust is provided by a dynamic cost function, implemented via the sub-module social trust computation. Then, ULOOP contains an entity, the *Cooperation Manager* (*CM*), that is responsible for coordinating the cooperation. On the control of cooperation, if the device is a requester and requires a service, it must compute the amount of credits that will convince the prospective requestee in engaging in cooperation. Then, a Reward Manager entity takes care of controlling promises of payments, and payments itself, in coordination with the Cooperation Manager entity.

At bootstrap, the Cooperation Manager entity starts by assigning an initial amount of cooperation credits to the user. This initial amount takes into consideration the node trust level and established minimum and maximum amount of credits thresholds for ULOOP devices. As the Reward Manager handles credits, the Cooperation Manager informs it of this amount in order to make the Reward Manager aware of how much cooperation credits the device has. During negotiation, credits are used by the requestee to express the cost of the service/resource he/she provides. The negotiation phase is positively concluded if and only if an agreement is reached both in terms of service level and in terms of credits between requester and requestee. In ULOOP trust can also be used as a parameter to affect the cost of the negotiated service. The ULOOP incentive framework is open to the implementation of any functional relation between cost and trust [1].

On the control of cooperation, if the device is a requester and requires a service, it must compute the amount of credits that will convince the prospective requestee in engaging in cooperation [4]. Additionally, as the tokens are the common language among the different managers, a number of tokens is computed by means of Social Trust Computation and a promise of payment is done by means of Reward Manager. Then, the CM sends (by means of external interface made available to all modules of the Trust Manager) a service request to the potential requestee, which in turn replies specifying whether or not it will engage in cooperation. In the case the device is a requestee, it receives the service request and evaluates whether the received credits are enough to provide the requested service. Then, a check on the amount of resources is done is order to assess whether the requestee can answer the service request. If so, the requestee (i.e., Reward Manager) accepts the received amount of credits, Social trust Computation updates the trust level, and issues a service reply informing the Cooperation and Trust Managers that requestee is ready to engage in cooperation.

The CM is expected to run in the both ULOOP node and gateway. The role of a device will be set by detecting the conditions around and feeding that data to the respective daemon. For instance, a device may become a gateway because it is connected to the Internet and if it has the required trust level. So, if by some reason the trust level changes, that node may be automatically prevented from becoming a gateway.

The Reward Manager is the ULOOP software module that handles payments and credit transfers, used as additional rewarding incentives for cooperation. Credit transfers revolve around credit units, which are a form of virtual currency. The Reward Manager software module has been characterized in a way to ensure that the transmission of credits is validated and secure, by preventing the creation of fake credits and the forging or duplicating of payments. The resulting virtual currency model is secure and, while being centralized in nature, allows the nodes to exchange credits when offline. The Reward Manager is a software module running in each ULOOP node. The module does not require any additional external interfaces and it provides a set of APIs (in the form of function calls) that can be directly used by any other ULOOP module on the same node. Communication between nodes and the central authority managing credit exchanges and ownership (also known as the "Bank") requires HTTP connectivity. Software in need of exchanging credits must use the Reward Manager's APIs. The system allows users, uniquely identified and registered with the central authority (Bank), to generate credits when registering into the system for the first time and to exchange such credits between registered users at any time. Each payment is uniquely identified. Payments may be made and exchanged even while disconnected from the Internet, but they must eventually be acknowledged by the Bank in order to be processed. The Reward Manager is expected to run in both a ULOOP node (end-user equipment) and on a ULOOP gateway (e.g., Access Point).

4.1.4 Social Trust Computation

The computation of trust is provided by a function implemented in ULOOP nodes and gateways. Trust computation is a dynamic cost function that has to be sufficiently strong to provide, based on a local perspective, attack resistance. It comprises therefore the dispositional trust of a node, as well as evidence concerning contacts with other nodes. To explain the notions behind social trust computation we provide an example based on three nodes: node A, the node that is about to compute a trust level towards a node B, and node C representing a node in the same community as node A. Node A has a dispositional trust level e.g., 0.5. In order for node A to compute the trust association cost towards node B, it takes into consideration recommendations sent by nodes belonging to the community. Such recommendation may be direct, i.e., the node has a direct trust association to node B, or indirect, i.e., a node has an indirect trust association to node B with the association being established through some other node, e.g., C.

Direct trust associations are more relevant (have more weight on the trust cost function) than indirect recommendations. Recommendations provide with a trust cost that nodes in the community have towards a new node. A direct recommendation received by node represents an answer from a node in the community, and contains the computed cost of one or several trust associations between and the target node. An indirect recommendation received by a node represents an answer from a node in the community which contains the computed cost of one or several trust associations between and the target node, but is not yet in the trust table of that node. ULOOP proposes a specific trust computation function which considers both direct and indirect recommendation values, as well as the owner's own beliefs—dispositional trust. Moreover, the more stable acquaintances are, the more trusted their recommendations become. Other functions may be applied easily via the interfaces created for such purpose.

The operational behavior of this module is as follows. After boot up the nodes check for their dispositional trust D and activate a trust table. The trust table is a structure where each row is a tuple with the following structure: <*Node Id, trust level, aging*>. When activated, the node provides each of its neighbors with an equal trust level of D. In other words, in environments where relations were not yet established, ULOOP nodes trust equally all nodes around. Then, the trust table can be periodically updated via recommendations by neighbors to assist in computing periodically the trust table of each node. Requests for social trust computation come from the trust manager, cooperation manager and are provided via a look up to the trust table.

4.2 The Resource Management Entity

In ULOOP the management of network resources takes advantage of the willingness that users have in cooperating, based on the mentioned two types of incentives: trust-based and reward-based.

The resource management operation takes place for nodes that have credits and are trusted in the community. The resource management operation itself starts when a Gateway gets a request for resources. This request is mainly from trust management block on a ULOOP gateway. If the resources are available in the Gateway, the resource management block provides a positive feedback to trust management and the new node can then join the network. The resource management block also provides updates about the channel to the mobility aspects functional block.

4.3 Call Admission Control Based on Trust

Call Admission Control (CAC) is responsible for checking if there are available resources, on the gateway, to accept or deny a request from RM. The CAC is only enabled on the requestee side (gateway). After the RM initializes the CAC function, the CAC stays in an idle state until a RM calls CAC or when the thread, scheduled to run before, wakes up to check the priority queue (pqueue). When the RM calls CAC, CAC handles the incoming request, prioritizes it, and puts it on a virtual queue, *pqueue*. After this, CAC schedules the thread to run. When the thread wakes up, it checks if the pqueue is empty or not. If not, it enqueues the request with highest priority and then checks if the gateway can accept it or not. Acceptance is provided by another sub-module, responsible for resource allocation, the *Elastic Spectrum Management (ESM)* functional block.

4.3.1 Resource Allocation

ULOOP is envisioned to be applicable to dense area networks, which face, among other problems, the issue of interference. Moreover, in these environments, spectrum abounds and is underused. In ULOOP and in addition to augmented call admission control and self-organizing mechanisms to elect and to select gateways, a key aspect to be developed relates to considering mechanisms that allow the MAC layer to become more elastic in multi-user environments in a way that is fully backward compatible with current IEEE 802.11 standards. This is achieved by just working with the current MAC frame format, and with the interpretation of such frames by ULOOP nodes [16].

Resource allocation in ULOOP follows the recent trend concerning frequency assignment and sub-division which argues that the channel width of nodes should be adaptive, in particular by considering an alternative way of arranging wireless channel assignments, based on *Orthogonal Frequency Division Multiplexing (OFDM)*. OFDM is supported by IEEE 802.11a/g/n standards, which are the ones that are dealt with in ULOOP. To achieve this purpose—which ULOOP named *Elastic Spectrum Management (ESM)*—resource allocation integrates a new mechanism that employs adaptive multi-user access, modulation, error coding and power allocation techniques to judge the tradeoff between costs versus performance gain [9, 10, 21].

ESM is initialized by the RM, indicating which mode the ESM must work, as a gateway or as a station. For the gateway mode, the ESM is responsible to assign a number of bits to a station depending on the number of tokens. Each station connected to the gateway has its own number of bits. This number will be used to create a super-frame, containing multiple parts of different payloads of each station. A station with a higher number of tokens, compared to the others stations, has the right to write more data inside the super-frame. A station with the least tokens may get to write less data or even nothing inside the payload, in case the 6 slots, each corresponding to 8 bits, have already been assigned to others stations with higher number of tokens. After the slots have all been assigned to the stations, the ESM will create the super-frame and send it to the driver for transmission.

Still in the context of resource allocation ULOOP researched cooperative diversity techniques at the MAC Layer, mechanism entitled RelaySpot [11–13]. Cooperative relaying is quite helpful in IEEE 802.11 networks, since terminals end-up with different data rates due to the rate adaptation mechanism. Terminals far away from an AP may grape the medium for long time to complete their slow transmission. Relaying data over a node that has better data rate towards source and AP will release the medium earlier and providing better throughput and reduced delay.

4.3.2 Cooperative Load-Balancing

Due to the dynamic behavior of ULOOP, nodes willing to share resources are more prone to be exposed to interference due to associations of other nodes. One of the aspects that is required to consider based on a self-organizing behavior that is inherent to ULOOP gateways is to assist in preventing excessive resource consumption, i.e., by performing network load optimization [20]. Part of this mechanism relates to being able to shift in an optimal way stations across different gateways and also to be able to adequately perform load-balancing among gateways. The aggregation of resource utilization, QoS and QoE measurements in a semantic form is the reasoning mechanism of the decision-making engines having the responsibility of load balancing trigger. Resource consumption monitoring that works in a passive way provides ULOOP gateways with the ability of classifying its clients according to bandwidth usage. The gateway arranges the client Id's with respect to their bandwidth consumption and marks the most consuming stations as "resource hungry". With this categorization, gateways can be aware of nodes that are less beneficial to the system, and if required assist their handover to other gateways while balancing load in the network in a fairer way.

4.4 The Mobility Aspects Entity

From a mobility perspective UCNs exhibit a highly dynamic behavior where the selection of the "best" mobility anchor points requires the pursuit of two main aspects: adequate selection and redundancy. This has to be achieved by always weighting user expectations and the support each user is willing to give as well as the network support (access sharing) each user can in fact provide to its counter-peers in the network. Mobility anchor point location and selection optimization is therefore a crucial requirement of UCNs. Mobility anchor points may be part of the SP equipment, of the NAP equipment (edge node) or in fact be part of the equipment of the MP and this can increase heavily a UCN complexity.

4.4.1 Achieving a Better Control of Mobility Anchor Points

The *Mobility Anchor Point (MAP)* is developed in ULOOP to be extended and adapted in the mobility anchor function of an existing mobility management solution. It interacts with resource management to get the resource information in the gateway and registers its context and sends keep alive message to the *Mobility Coordination Function (MCF)*. This sub-module takes care of coordinating the selection of MAPs based on resources; trust aspects; QoE. Based on the number of currently known active MAPs it is responsible to perform a MAP selection decision for the ULOOP node upon receiving MAP request from the MAG, which are then enforced on the data path.

4.4.2 Estimating Node Movement

ULOOP has addressed how to assist the network and the user in terms of mobility, by allowing devices to infer future roaming behavior, based on a selection optimization that simply relies on data available to devices, and which concerns visited networks [18]. Concrete examples of network parameters include, but are not limited to: number of visits performed over a specific period of time, e.g., 24 h; average duration of one visit; visited network attractiveness, e.g., trust level that a node has towards a specific gateway that is regularly visited; number of visits accepted/authorized; time elapsed since the last visit to a specific visited network.

For each visited network, nodes compute locally, seamlessly, and periodically a cost (a ranking parameter) based on a specific formula that relies on the collected network parameters. That ranking parameter is also stored in the listing of visited networks. As proof of concept, ULOOP has worked these concepts and integrated them into the end-user background application MTracker (Mobility Tracker), currently available to be applied in the majority of portable devices, Android included. Based on data available and passively collected, the MTracker application then tries to predict in how much time the node will change the network connection, and which will be the next network, passing this information to the server side, e.g., to the MCF, which then decides how to handle such information.

5 Summary and Conclusions

UCN envisions increasing the potential of future Internet architectures by devising communication and networking technologies, which support the creation of techno-social communities. For such purpose, UCN considers knowledge derived not only from networking paradigms, but also from social trust modeling as well as by taking advantage of an adequate estimation of potential communication opportunities (e.g., sharing of Internet access and relaying resources) even if devices belong to users that are not socially acquainted. The expectations concerning UCN are, from a business perspective, the opportunity to develop new community services and hence to derive new business models for Internet stakeholders. From a technical perspective, software defined ways to improve the network operation e.g., in regards to spectrum usage or energy-efficiency.

The paper provides notions concerning UCN as well as describes an implementation for the ULOOP software architecture, representing an instantiation of UCN which is currently available to the community.

Relevant research opportunities in the context of UCN relate with the application of trust as a potential parameter that stemming from social sciences can be applied to QoS as a way to create more robust Internet architectures. Another relevant field to be addressed is direct trading of resources on the network, as a way to develop new business models.

References

1. Aldini A (2013) Formal approach to design and automatic verification of cooperation-based networks. IARIA Int J Adv Internet Technol 6(1 and 2):42–56
2. Aldini A, Bogliolo A (2012) Trading performance and cooperation incentives in user-centric networks. In: International workshop on quantitative aspects in security assurance (QASA'12), Sept 2012
3. Aldini A, Bogliolo A (2012) Model checking of trust-based user-centric cooperative networks. In: 4th international conference on advances in future internet (AFIN'12), IARIA, pp 32–41, Aug 2012
4. Aldini A, Bogliolo A (2013) Modeling and verification of cooperation incentive mechanisms in user-centric wireless communications. In: Rawat DB, Bista B, Yan G (eds) Security, privacy, trust, and resource management in mobile and wireless communications. IGI Global. University of Urbino "Carlo Bo", Urbino. ISBN13: 9781466646919; EISBN13: 9781466646926
5. Ballester C, Seigneur J-M (2013) Dispositional trust self-adaptation in user-centric networks. In: Proceedings of AINA, Mar 2013
6. Ballester C, Seigneur J-M, di Francesco P, Moreno V, Sofia R, Bogliolo A, Martins N, Moreira W Jr (2013) A user-centric approach to trust management in Wi-Fi networks. In: IEEE INFOCOM—Demos Track
7. Bogliolo A, Delpriori S, Klopfenstein L, Aldini A, Seigneur J-M, Moreira W Jr (2012) Crediting aspects in ULOOP. In: ULOOP white paper 09
8. Eu, IST FP7 ULOOP—User-centric wireless local loop project (2010–2013) grant number 257418. Available at http://uloop.eu/
9. Haci H (2012) Novel scheduling for a mixture of real-time and non-real-time traffic. In: IEEE GLobecom 2012, Best paper award, Sept 2012
10. Haci H, Zhu H, Wang J (2012) Resource allocation in user-centric wireless networks. In: VTC-Spring
11. Jamal T, Mendes P (2014) Cooperative relaying in wireless user-centric networks. In: Aldini A, Bogliolo A (eds) User-centric networking—future perspectives. In: LNSN, Springer
12. Jamal T, Mendes P, Zuquete A (2011) Interference-aware opportunistic relay selection. In: Proceedings of ACM CoNext (student workshop), Tokyo, Japan, Dec 2011
13. Jamal T, Mendes P, Zuquete A (2011) Relay spot: a framework for opportunistic cooperative relaying. In: ACCESS conference, Luxembourg, pp 19–24
14. A. Matos (ed) (2013) D3.9: ULOOP software suite. ULOOP European project deliverable (gr. Nr. 257418), Oct 2013
15. P. Mendes (ed) (2013) D3.8: ULOOP framework specification and validation. ULOOP European project deliverable (gr. Nr. 257418), Oct 2013
16. Osman H, Zhu H, Haci H, Lopes L, Sofia R (2013) Method and apparatus for communication in a wireless network. In: (EP 13191667.8), Aug 2013
17. Ribeiro A, Sofia R, Zuquete A (2011) Improving mobile networks based on social mobility modeling. In: IEEE international conference on network protocols
18. Sofia R (2013) Method and apparatus for ranking visited networks. In: (EP 13186562.9), Aug 2013
19. Sofia R, Mendes P (2008) User-provided networks: consumer as provider. Feature topic on consumer communications and networking—gaming and entertainment. IEEE Commun Mag 46(12):86–91
20. Yildiz M (2012) Cooperation incentives based load balancing in UCN: a probabilistic approach. In: IEEE Globecom 2012, July 2012
21. Zhu H, Wang J (2009) Chunk-based resource allocation in OFDMA systems—part I: chunk allocation. IEEE Trans Commun 57(9):2734–2744

User-Centric Networking: Living-Examples and Challenges Ahead

Rute Sofia, Paulo Mendes and Waldir Moreira Jr.

Abstract This paper provides a description of operational examples of User-centric networking models, providing a specific characterization of their main features. The paper provides also a set of guidelines concerning the deployment of future user-centric networking models, having in mind to assist in the integration of these models into future Internet architectures.

Keywords User-centric networks · Internet architectures · Virtual operator

1 Introduction

Throughout the history of the Internet, several technologies and networking architectures have emerged, some of which have been widely deployed, and others have only made its reach to niche markets. Some generic cases that can be cited are multicast, IPv6, java, or even C++. Clearly, the adoption of such technologies relates not only to technical merit but also to a variety of parameters such as ease of deployment, or even interest of potential market stakeholders.

In what concerns the specific case of wireless networks which today abound as local access networks with limited mobility support, their utilization in more disruptive scenarios is yet to see a mass deployment. For instance, the *ad-hoc* wireless

R. Sofia (✉) · P. Mendes · W. Moreira Jr.
COPELABS, University Lusófona, Building U First Floor, Campo Grande 388, 1749-024 Lisboa, Portugal
e-mail: rute.sofia@ulusofona.pt

P. Mendes
e-mail: paulo.mendes@ulusofona.pt

W. Moreira Jr.
e-mail: waldir.junior@ulusofona.pt

A. Aldini and A. Bogliolo (eds.), *User-Centric Networking,*
Lecture Notes in Social Networks, DOI: 10.1007/978-3-319-05218-2_2,
© Springer International Publishing Switzerland 2014

architecture concept has emerged around 30 years ago and is yet to see a generalized deployment.

The lack of a mass deployment of ad-hoc networks may possibly be due to the fact that entities managing different networks (be it individuals or large organizations) lack clear incentives to participate in the creation of ad-hoc wireless networks. The result of this is actually the rise of simpler and more autonomic wireless architectures, *mesh networks*, in which all devices are static (placing lower technical challenges) and normally belong to the same administrative entity (thus reducing the adoption problem). Previous experience therefore shows that spontaneous deployments of infrastructure, mesh or ad-hoc wireless networks are changing the perceived applicability of new types of wireless emerging architectures, but there is still not a clear perception on how such adoption may progress, nor a clear perception on what are the incentives (from a social, economic and also technical viewpoint), both from an access and end-user perspective, to adhere to these networks. Last but not the least, there is also not a clear perception on the dynamics of these networks.

Despite the aforementioned aspects, a recent trend related to autonomic wireless architectures is giving rise to wireless community initiatives with the purpose to provide broader connectivity. In these type of architectures (*User-provided or User-centric networks, UCNs*) [24], **users and/or communities share subscribed access in exchange of specific incentives**.

UCNs disrupt Internet communication models in several ways. Firstly, any regular end-user device may behave as supplier of Internet connectivity and other services, and consequently, the user device becomes part of the network. In contrast, the architectural core of the Internet, the end-to-end principle [21], describes a clear splitting between network and end-user systems. Secondly, UCNs grow spontaneously based on the willingness of users to share subscribed Internet access. Thirdly, connectivity is expected to be intermittent given that UCNs are spontaneously deployed. These are some intuitive aspects which show that UCNs are a new paradigm at least in terms of Internet connectivity models. However, there is today still not a clear understanding of the extent and of the impact that this rising trend may have in the Internet design. As an initial step towards a better understanding of what such impact may be, this paper is focused on an analysis of existing examples of UCNs, their operation as well as advantages and disadvantages. Hence, the paper is organized as follows. Section 2 describes related work. Sections 3–5 go over the different examples of UCNs categorizing them according to their main properties. Then, Sect. 6 provides a list of derived assumptions and requirements based upon the four main UCN properties. Conclusions are provided in Sect. 7.

2 Related Work

The quickest distinguishable feature of UCNs is Wi-Fi sharing. Different terms found in the literature that point to the UCN concept are *social wireless networks, community Wi-Fi sharing*, and also *Peer-to-Peer Wi-Fi*.

The technical benefits of Wi-Fi sharing have been analyzed to some extent [11, 25] and measurement results show that such sharing is a cost-effective solution in particular for densely covered areas. These contributions relate only to specific examples of controlled deployment of public wireless networks.

Camponovo and Cerutti provide an analysis of regulation aspects concerning hotspot sharing [2]. The authors analyze hotspot sharing based on two regulatory (European) regimes, addressing different sharing models and explaining regulatory gaps.

Specifically concerning UCNs, user-centric wireless models [24] show that today the Internet end-user is already not only a consumer but also a provider of Internet services in the sense that he/she can share his/her subscribed Internet access based upon specific (community) sharing incentives.

Our work provides a reality-check about the current status of UCNs by analyzing real deployment cases and by contributing to a better understanding and characterization of this emerging type of network architecture.

3 The Global Concept of a User-Centric Network

This section provides a characterization of user-provided networks, including a set of assumption and requirements. To clarify main differences against other autonomic networks, the section also provides a comparison of connectivity features for user-provided networks against ad-hoc and other forms of multihop networking.

A UCN may be represented by a time-varying graph where *nodes* are wireless devices belonging to Internet users (individuals or communities), and where *edges* represent trust associations. The edge cost is a measure of the trust association strength. From a pure connectivity perspective, nodes have two roles: *regular* and *gateway*. Regular nodes use network resources provided by a gateway. Gateways provide networking services to a specific community of users, e.g. share bandwidth and provide mobility management solely based on their owner's (the *micro-provider, MP*) subscribed Internet access or in a coordinated way with one or several access providers. From a management perspective, a third entity integrates the UCN concept, the *Virtual Operator (VO)*. The VO is a role that provides some form of coordination (e.g. access point registration) to specific UCN communities without owning a specific infrastructure nor providing services. Therefore, a VO is simply a role assumed by an entity, by an *Internet Service Provider (ISP)*,[1] or by a *Service Provider (ASP)*.

User-centricity is a key aspect of UCNs: these architectures emerge based on user empowerment made possible by the ease deployment of new types of wireless architectures and by the user willingness to cooperate due to some form of communal or individual benefit (*incentive*). In addition to user-centricity, UCNs hold four other properties: *network resource sharing*, *cooperation*, *trust*, and *self-organization*.

---

[1] Internet access or connectivity is here defined as a networking service consisting of IP address space allocation, Domain Name Service and routing assurance.

Network resource sharing today reflects mostly Internet access or connectivity sharing. However, as these architectures evolve, we will likely observe sharing of additional network resources (e.g. energy) or of additional network services (e.g. mobility management).

Cooperation relates to the user's willingness to participate in UCNs, both sharing and profiting from available resources. Incentives to cooperate can be related to trust (e.g. social association), to some form of compensation (e.g. broader Internet access), or even to a more efficient network operation.

Trust management is today performed by having users signing up to a "community". However, to create UCN secure environments, user identification and traceability are issues that have to be addressed. Hence trust management relates to three main security concerns: *(i) assist users in terms of traceability; (ii) guarantee user privacy; (iii) provide data confidentiality when/if necessary.*

Self-organization relates to the capability to coordinate connectivity in scenarios where it can hardly be predicted given that it is based on the user's willingness to cooperate or adhere. In regards to self-organization, a key aspect is that UCNs rely on existing (private) deployments. For instance, in a campus it is likely that a UCN would rely on the existing Wi-Fi infrastructure mode. While within a village, a mesh solution could make a more adequate UCN.

3.1 UCN Taxonomy

The roles of the UCN stakeholders (users, MP, operators) and their relationships shape the Internet design, as these relationships impact the communication in ways that were not foreseen, e.g., by placing both the upstream (from user to network) and downstream (from network to user) flows at an equal level. To better understand what may be the impact of UCNs over the Internet, we analyze some of the most representative UCN examples as of today, and categorized them into five different sets: hotspot UCNs; mesh UCNs; Social networking UCNs; Mobile/Provider UCNs.

Hotspot UCNs represent most likely the majority of UCNs around us. In this category, UCNs are built upon existing Wi-Fi hotspots. Hence, hotspot UCNs are based on Wi-Fi infrastructure-mode, where an *Access Point* (AP) mediates all communication to and from a set of end-user devices. Users adhere to hotspot UCN models due to roaming incentives: a user shares his/her Internet access subscription in exchange of roaming across other hotspots shared within the same community.

Mesh UCNs are possibly the oldest UCN category. It should be noticed that UCNs based on mesh relate only to *user-centric mesh networks*, given that communities of users autonomously deploy the network. The wireless infrastructure is often deployed by the community to allow Internet expansion in relatively large areas. Incentives relate to low-cost Internet expansion (capillarity) and not so much with roaming.

Social networking UCNs rely on social networking to seamlessly distribute Wi-Fi credentials and to allow UCN expansion. The MP shares Internet access based upon credentials that he/she controls and in situations where the gateway can be simultaneously used by different users, as happens in a household.

Fig. 1 UCN categorization

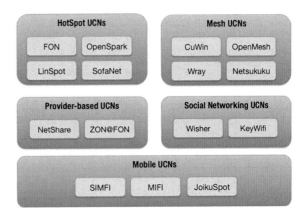

Mobile/provider UCNs are the most recent UCN category. The potential of this type of UCNs is still to be unveiled but what is clear is that the solutions here described will drive the deployment of UCNs. In this category, the network is formed anytime/anywhere by users according to their needs and following the subscription rules of their cellular providers. Mobile-based UCNs are often provided by operators due to offloading reasons, as a way to lower Capital EXpenditures (CAPEX).

4 UCN Living Examples

Finally, *Provider-based UCNs* are examples of concrete business applications of UCNs where the provider (access or service) is also a *virtual operator (VO)*.

This section provides an overview of today's living examples of UCNs. Such examples have been analyzed and grouped into different categories according to their operandis mode, and the proposed categories are illustrated in Fig. 1. As illustrated we consider five main models to categorize living examples of UCNs. The next sections will describe each of these categories, addressing operation for each of the examples, and summarizing the main aspects for each category.

Within each category we have grouped the known examples of UCNs. Then, within each category examples of operation as well as a description of the main features are debated. Such features contemplate resource management, cooperation incentives, mobility management, security aspects, and additional features which are relevant to cite as differentiators.

4.1 Hotspot UCN Model

Hotspot UCN models rely on regular hotspots owned by users (or available in public spaces) and give the opportunity for users belonging to the communities coordinated by VOs or ISPs to take share subscribed access and to take advantage of such sharing. In this category, the architecture relied upon corresponds to the regular infrastructure-mode of Wi-Fi, where an access point mediates all communication to and from stations. The access point (co-located with an access router) is normally provided by the VO. Moreover, the VO takes care of authentication of users within the community it coordinates. Today, most households represent a hotspot UCN.

The incentives for this category relate with roaming: a user shares an already subscribing access in exchange of roaming across other shared accesses. It should be remarked that a user is only entitled to roaming if (i) he/she acquired a gateway to a VO; (ii) such gateway becomes active and is registered (on the VO backend systems) as a shared hotspot.

In this category, there is no end-to-end data confidentiality nor privacy. Moreover, the MP is the responsible for any traffic to the access. In other words: access providers can only consider accountable the MP: any violation of the subscribed access is input ed to the MP.

FON

FON [5] is one of the most concrete and successful examples of a hotspot UCN model, and in. As of July 2013, FON claimed to have the largest Wi-Fi network with over eight million hotspots worldwide.

In order to be part of FON, all any user has to do is to register as part of the community, which implies obtaining credentials. In addition the user has to acquire a FON "social" router (FONERA). This is in fact an Access Point/router (AP) which is ready to allow Internet connectivity sharing and that will also be the basis for the free hotspots that are available to all FON users. Each FON social router is associated with specific user (username/password) credentials and as soon as it gets registered, it becomes a FON access point. The access to the FON social router is performed by the regular redirection to a Wi-Fi portal: each time a user is within the range of a social router and connects to the given signal, the browser is automatically redirected to the main portal page, which asks for the user's credentials.

Users that register in FON are expected to already hold an Internet access subscription. Hence, FON is not an operator nor a *Service Provider (SP)*. FON simply provides a box and a virtual registration to a community, and hence is here coined with the notion of *Virtual Operator (VO)*.

FON subscribers are then split into two categories, *Linus* or *Bills*. Linus are users that hold a broadband connection and a FON social router. By means of acquiring the social router, they are entitled to roam across the FON Wi-Fi hotspots for free. This is the regular case of a residential user that simply wants to have the possibility to roam freely. Here, the incentive to share is simply broader roaming.

Bills are users that want to do more than roaming, namely, they are looking for a financial incentive to use FON. Bills receive 50 % of the net revenue due to end-users (*Aliens*) that access FON by mean of previously acquired FON passes. Such end-users then access FON by means of the Bills' APs. In addition, FON gives Bills the means to personalize their access points and sign-in pages—Bills can then advertise their products and services within their neighborhoods. Overall, only Linus users have unlimited/free access to FON APs. Bills in contrast have to pay the access as normal users (but receive part of the income).

Today FON deploys their firmware on a large variety of low-cost APs, ranging from Linksys to Netgear. The firmware is based on the OpenWRT operating system. *Operation* Let us provide a concrete example for Bob, a user that wants to use FON to be able to profit from broader roaming. Bob goes to the FON site and registers freely as a new user providing the regular details such as e-mail address, postal address, etc. Upon registration, Bob becomes an Alien, i.e., Bob is registered but is not yet a FON subscriber. In order to become a Bill or Linus, Bob needs to acquire a FON social router (FONERA) and to activate sharing, i.e., to establish a *FON spot*.

Upon arrival of the new AP to his house, Bob connects it following FON's guidelines. Bob then enables sharing and registers its FON social router on the FON online account, thus becoming a Bill and obtaining access in any FONSpot worldwide.

When Bob roams and wants to access other FonSpots, he has simply to enter his credentials on the captive portal of FON. There is no security however, for the users profiting from shared connections.

In regards to resource management, FON provides very simple support, which only considers a simple priority scheme to take into consideration potential downstream and upstream restrictions. More relevant is the fact that a MP can limit (manually) the amount of bandwidth being shared.

Incentives to cooperate in FON are two-fold. Broader roaming is no doubt the main incentive to address. In addition and specifically for the case of Bill users, there is a financial incentive: FON allows Bills to keep half of the voucher generated revenue. Mobility management is not integrated in FON. The user can roam but there is no session continuity, nor any guarantees in terms of nomadism.

In terms of security, FON protects the wireless connection of the MP with regular means, e.g., WPA, WPA2, WEP. The shared access is, however, open.

Additional features supported by FON that are relevant to cite are that sharing is an optional feature. In order to roam a user has to acquire an AP from FON. However, if that AP is not active, the user can roam within the FON community. Moreover, the last APs incorporate a number of collaborative tools (e.g. Twitter, bitTorrent clients) and the user can download content from the Internet independently of his FONERA being or not being active.

In terms of policing and monitoring, FON equipment provides regular traffic management statistics (but does not differentiate between MP and shared connections.

OpenSpark

Being a predecessor of FON, OpenSpark [13] provides a similar type of architecture in Finland. OpenSpark is managed by MP-MasterPlanet Ltd, a Finnish ISP/WISP which manages the Wi-Fi network SparkNet. OpenSpark was born of SparkNet, and follows a model where the usage incentive is free roaming. Being a simple case of FON, OpenSpark provides less features than its peer.

Generic Operation Similarly to FON, in order to use OpenSpark Bob needs to have an Internet access subscription and to acquire an AP that has been personalized for use in OpenSpark. Bob then needs to activate the AP at home, to take care of its configuration, and to register itself (by means of the new AP and of the captive portal of OpenSpark) as an OpenSpark user.

OpenSpark does not provide any resource management, and the single type of incentive is broader roaming for users adhering to the community. Similarly to FON it does not contemplate any mobility management features. A key differentiator aspect to FON is that sharing is mandatory: the acquired APs must be up all times. Another differentiating aspect is the fact that OpenSpark allows expansion of hotspots by means of bridging between multiple APs (which will be seen as a single virtual AP).

LinSpot

LinSpot [14] is intended to be an easy off-the-shelf solution that allows a user to profit from an already existing subscribed Internet access, by turning the local AP into a paid and profitable hotspot. Compared to FON's 'Bill' model, LinSpot is cheaper to setup, as it is software-based and does not require the purchase of new hardware. On the other hand the MP must have an existing (compatible, open-source) AP/AR and a computer with an Ethernet interface. This computer will have the LinSpot software installed and is the manager of a LinSpot hotspot—the computer must be up all times. All the APs in the LAN segment therefore become LinSpot APs.

The business model of LinSpot is that LinSpot fixes the prices of Internet access in LinSpot-enabled APs. These prices are said to be half of those of commercial WISPs. The MP gets 85 % of all profit, being that the remaining 15 % goes to LinSpot. Payments are done via the PayPal system and profits are immediately transferred to the MP's account.

In its current version, LinSpot is available only to MacOS X operating system and does not support any form of wireless security.

Generic Operation Firstly, Bob checks that his computer is connected to his AP/AR through an Ethernet cable and that the computer is conFig.d to have a static IP configuration. In addition, in his router, DHCP, DNS, and security must be disabled (as LinSpot packs a DHCP server) .

Then Bob is ready to download the LinSpot software (currently only available to MacOS X) and follow the installation instructions. If not already registered in PayPal, Bob must create an account and register in LinSpot, in order to receive the payments. After this setup, anyone connecting to this access-point will be redirected to a web page with the billing information and, after paying the fees, allowed to use Bob's Internet access.

This means that every device in Bob's LAN will be redirected to this page. To avoid this, the LinSpot system is prepared with a free-access list with devices that are allowed to connect freely. Bob must add to this list every other computer he might have.

By installing LinSpot, Bob is opening a specific LAN (e.g. his home hotspot) to everyone around. Together with the fact that no encryption is being used, this means that Bob must be careful with the resources he is sharing (e.g. files and printers) because everyone will have access to them. Furthermore, Bob will have to leave his LinSpot-enabled computer turned on all the time, or his hotspot won't work.

The cooperation incentive for LinSpot is financial: an MP can get 85 % of revenue for users that rely on his/her hotspot. Moreover, roaming is also considered as an adhesion incentive, for users willing to pay monthly access fees. LinSpot provides no security at all. As additional features, LinSpot incorporates logging, as well as billing reports which are provided to MPs.

SofaNet

SofaNet [23] is one example of a UCN model in which a user needs to pay to have connectivity. To be able to use Internet connections made available by SofaNet members, one has to buy a flat rate voucher "Zeitflatrate" with SofaNet for 15 Euros. Vouchers can be used for 90 days, around the clock, but they come with a transmission volume limit.

The major condition to become a SofaNet member is to have an Internet subscription from T-Com or a reseller such as AOL, Freenet, or 1&1. SofaNet members can use their own DSL identification for their own Internet access. A new DSL identification, bought from SofaNet, will be used by guest users. Both DSL identifications can be used simultaneously over the same DSL connection. For that, two WLAN routers need to be conFig.d with the different DSL identifications. The DSL identification (e.g., PPPoE identification) identifies the user as well as the tariff (e.g. flat rate), which allows the SofaNet member to still use its own tariff plan, while guest users are allowed to use a different one. Any SofaNet member has to respect the SofaNet rules and must behave according to the conditions of the DSL operator. SofaNet is registered with the Federal Network Agency in Germany and provides the required storage of connection data. A SofaNet member is responsible must not give Internet access to anonymous users and should encrypt all the traffic.

For guest visitors, SofaNet relies on the pre-paid model. Pre-paid packages go from 1–2,5 Euros for 500 MByte transmission volumes for 24 h to a 15-37,50 Euros package with 6 Gbyte volume for 90 days.

The major difference to previous examples (such as FON or OpenSpark) is that normally UCN hotspot models are directed to communities of users willing to roam for free, and outsiders have to pay for Internet access. In contrast, SofaNet members always pay for Internet access, although it is only 17 cents a day. Moreover, while the core business of FON is to sell their hardware (Foneras), SofaNet sells DSL identifications. With the usage of two different DSL identifications, it is easier with the SofaNet model to identify who is using the shared Internet access (the MP or the guest users). It should be noticed however that their model of operation is tied to

German regulation, where there is still a clear splitting between the access line and the Internet services subscriptions.

Generic Operation To become a member of the SofaNet community, Bob first needs to buy a DSL connection from T-COM or a T-COM reseller. Hence, subscribers of other DSL operators in Germany, such as Arcor, Alice, Netcologne, and Versatel cannot become SofaNet members.

Bob must use its old DSL identification, while a new SofaNet DSL identification is to be used by guest users only. Once Bob gets its T-COM DSL connection and DSL identification, he has to buy two APs from SofaNet. One AP will be used by Bob only (protected with WEP/WPA) and the other is to be used for Bob's guests (open access). Both APs are expected to be connected to the same DSL service (e.g. by means of a switch). Therefore, there is a physical traffic separation up to the APs that Bob acquired. Nonetheless, the wireless link for guests is unprotected.

Resource management can be supported by SofaNet by deploying two VPNs to the different APs. In terms of cooperation, roaming is again the main incentive provided. In addition, there is a financial incentive which gives SofaNet members the possibility to provide their own pricing model to guest visitors. Security aspects in SofaNet can be provided end-to-end (by means of VPNs and of DSL identification). By default the access for guests is left open. An additional feature to state is that members are offered 1 Gbyte free traffic volume per month as incentive for roaming, as long as the user keeps the hotspots active more than 95 % of the time.

4.2 Social Networking UCN Models

Social networking UCN models are examples of UCNs which rely on social networking aspects to distribute Wi-Fi credentials and to allow connectivity models to expand. The user therefore shares connectivity for which credentials are known and also for situations where the AP can be used simultaneously, as happens for instance at a residential household.

In this UCN category, there is still a VO which is responsible for backend management, mostly to register shared APs and to take care of the basic authentication. The MP relies on his/her own equipment to allow the sharing to happen, but however, such sharing is completely independent from the existing connectivity model and also from the existing wireless architecture.

Wifi.com

Wifi.com[2] [28] is a brand for a UCN concept which goes a step further than FON in terms of empowering the end-user as connectivity provider, given that it bypasses the need to acquire hardware: Whisher's functionality is purely software-based and integrated into the end-user device. By relying upon a specific client software component

---

[2] Formerly known as Whisher, Wifi.com is a brand reselling for the plugin Whisher.

(plugin[3]), Whisher gives the means to establish shared connectivity among users that belong to the social sphere of an MP. The Whisher plugin basically gives an MP the possibility to exchange, in a secure way, Wi-Fi credentials of an AP the MP controls to users from his/her social network(s). Connectivity sharing therefore follows a social networking model.

Wifi.com has three flavors of access grant, namely *public*, *buddies-only* (currently unavailable), and *private*. As its name suggests, public access grant allows anyone to use the hotspot as long as he/she has the Wifi.com client software and a registered account. For the case that the owner is still not comfortable about fully sharing his connection, he can provide access to only his family and friends through the buddies-only option. This option is also useful in the case of a small network among neighbors. For that, the owner has to add these users to his buddy list, and only the users on this list will be allowed to access the hotspot. As for the private option, only users set as VIP will be able to access the hotspot. In order to add a user to the VIP list, the owner of the hotspot must select them from the local Wifi.com users' list. Such option is available for the case when the owner wants to have the most of the bandwidth available to him, but without letting other important users (i.e. the VIP ones) without access.

The owner of the hotspot is responsible for setting security measures (e.g. WPA/WPA-2, WEP) for his network. With that, the Wifi.com software will allow users access but without letting them know about the key.

Generic Operation Bob goes to the Wifi.com site and downloads the Wifi.com plugin for his machine, installing it. Once active, Bob is asked to register (obtain username and password) on Wifi.com. After registration, Bob provides information concerning the hotspot(s) he is willing to share, by issuing a specific identifier and a welcome message to other users belonging to his Wifi.com social network(s). The connection sharing is managed automatically according to the levels of trust specified by Bob, and the Whisher manager provides secure connections to all users that Bob is sharing connectivity, by means of WEP.

As mentioned, only the owner of the hotspot (Bob) can decide who is going to be allowed to access his hotspot. Another interesting feature is the ability to be connected in areas where there is no other users sharing connectivity. This is possible through the partners registered (e.g., Starbucks, Accor hotels, Best Western hotels) that offer connectivity at a very reasonable price. All the user needs is charge his account with premium Wi-Fi minutes (wifi out credit in whisher) which will be used in a per-minute based form according to the needs of the user. it should be noticed that Wifi.com does not allow simultaneous access, given that it inherits the provided access rules—there is no connectivity model change.

Wifi.com simply relates to the exchange of credentials and therefore there is no concern in terms of resource management. However, the MP can trace the usage of his/her shared hotspots and can block greedy users from accessing a hotspot. Moreover, the MP can also define the maximum amount of bandwidth he/she is willing to share with each category of user. Incentives to cooperate are again related

---

[3] Currently only available for MacOS and Windows XP.

with broader roaming only. In terms of security aspects, there is a clear bet on security, given that the credentials of shared hotspots are only known to the MPs and are exchanged securely.

Key WiFi. KeyWiFi [12] is a service provided by KeyWiFi LLC, New York, USA. KeyWifi has as mission to provide low cost broadband Internet access connections through a provisional patent-pending wifi sharing platform. In its essence the purpose is, as WiFi.com, to provide WiFi credentials to users roaming. KeyWiFi appeals to the regular end-user willing to become an MP and to get some profit out of such sharing. The argument of KeyWiFi relates to the unused spectrum available around, in particular in dense areas.

KeyWifi follows an ebay philosophy in terms of shared keys. Users willing to share their access (and to get revenue out of such sharing) register in KeyWifi as suppliers (MP). KeyWifi provides suppliers with a specific Keywifi key for each hotspot that the MP registers. The MP also pays KeyWifi $9.98 per month. Users that are willing to profit from KeyWifi shared hotspots will pay also $9.98 to KeyWifi, but two thirds of the fee will go to the user sharing a hotspot. KeyWifi therefore expects to be a catalyze for sharing of hotspots that today are simply private.

Generic Operation Bob has wireless Internet access both at home and at office by means of two different APs. Bob normally uses Internet access in his office during the morning, and during the afternoon and evening, he goes back home. Hence, both APs are in average unused 50 % of the time. Bob therefore decides to make some profit out of it and joins (registers) on KeyWifi, agreeing to pay 9.98 $ per month. He registers both his home and office hotspot in KeyWiFi, and users belonging to the KeyWifi community (also registered) can access his shared hotspots. Then, registered users of the KeyWifi community can access Bob hotspot. Two thirds of their Keywifi monthly fee then go to Bob.

Being a very recent model, there are still a few open issues. For instance, simultaneous use of several hotspots by a single user requires some management in terms of the revenue obtained by the MP.

4.3 Mesh-Based UCNs

Mesh-based UCN models are examples of UCNs based on mesh networks and hence are possibly the oldest way to deploy the UCN concept. The wireless architecture is mesh-based but the deployment itself is performed by communities of users. Hence, VOs in mesh-based UCNs are municipalities and also even specific communities of users. Also, the MP tends to be the VO for this category, as shall be explained.

In this category, the wireless infrastructure is often deployed by the community to allow Internet expansion of already existing places. The incentives relied upon in this category relate often to Internet expansion at low-cost and not so much with roaming.

Wray Village

The Wray village project [9, 29] was born from University of Lancaster Wray Broadband Project [18] in 2004. This project deployed a small number of mesh devices [9] having in mind to provide the village with a reliable network infrastructure at relatively low-cost. Wray covers an area of around two square kilometers and Internet access is provided by means of a radio link (5.8 GHz) which reaches Wray school main building, located on a hill. Then, wireless mesh nodes based on Locust Mesh firmware are located in strategic locations, chosen both due to geography and also due to expected location. Overall there are around 10 mesh nodes and it serves around 100 users, both from a residential and business coverage perspective. In order to assist in overcoming blind spots and also in providing better reliability, the Wray mesh runs the routing protocol *Ad-Hoc On demand Distance Vector (AODV)* protocol, which has been manually conFig.d to ensure better network operation. Most of the nodes are claimed to achieve 3Mbps, restricted by manual policing in order to ensure fairness. Moreover, experiments [9] done in Wray state that the performance achieved by Wray users vary depending on equipment and location. End-users who installed an exterior antenna achieve around 2 Mbps, while users relying on PCI or USB network cards achieve data rates as low as 0.5 Mbps. All maintenance in the mesh network is done by community users. This is therefore a particular case of a UCN, in the sense that users are the heart of the network operation.

Generic Operation Bob has recently moved to Wray and would like to have Internet access. All he has to do is to ask the Wray volunteer management council for an USB device which assists him with the configuration to the closest mesh node. Bob's data is then transferred across the mesh network with the assistance of AODV, up to the Internet access located in the Wray school. Policing rules are enforced to ensure data fairness.

Resource management is ensured manually and enforced in a way that is fair based upon local usage. Each individual user in the mesh network is provided with a maximum data rate of 3 Mbps.

Incentives in Wray relate to having Internet access in a remote area, at a low-cost. There is no specific need for mobility aspects. In Wray, there is a single MP—the community—which is also the VO. Hence, security is established by means of the regular Wi-Fi security ways, namely, WPA or WEP. Guests will have open access.

It should be noticed that in contrast with the previous examples of UCNs, there is not a global concept of bandwidth sharing in the sense that a group of users in a community is willing to share an existing access. In Wray, the MP is the community.

CuWIN

The *Champaign-Urbana Community Wireless Network (CuWin)* [4] is a volunteer coalition composed of networking researchers-developers and community volunteers that are "committed to provide low-cost, non-proprietary, do-it-yourself, community-controlled alternatives to contemporary broadband models". The CuWIN vision consists in improving networking technology in a way that community of non-expertise people can deploy a wireless mesh network by themselves in order to satisfy their connectivity necessities.

CuWIN provides any regular user worldwide with the possibility to voluntarily host CuWIN nodes. In other words, upon demand CuWIN may provide their firmware to any user. In addition, CuWIN also sells a complete box solution, which includes CuWin firmware and hardware from Metrix (Metrix Mark II).[4]

CuWIN firmware (CuWinWare) is compatible with a reasonable number of chipsets, but however, it requires that the chipset of the wireless card to be supported by NetBSD ad-hoc drivers. CuWIN recommends installing on nodes that are expected to be static, e.g. old PCs. Installation can be performed by PXE booting, by CD, or by online updating.

Nodes run on 802.11b, channel 11 (defined at compile time). Moreover, CuWIN relies on a particular type of routing protocol, the *Hazy Sighted Link State (HSLS)* [22] protocol. HSLS is a link-state protocol developed by BBT technologies which uses both reactive and proactive routing to minimize route updates.In contrast to other link-state solutions, it prevents flooding the network by making changes in "far away" links become less relevant than changes on links "nearby". By applying the fuzzy-logic principles, propagation of link state changes is done in a quicker way to nearby nodes, than to nodes far away. This gives room for CuWIN to claim to scale up to thousands of nodes.

The CuWIN architecture is a two-layer hierarchical network. It contains a back-haul layer (identified by cuwireless.net) which is solely composed of CuWIN routers. The second layer is used by regular users (end-user layer, cuwireless), and may just contain off-the-shelf APs, or even another wireless networking solution. Internet access is provided by users (and organizations) belonging to the CuWIN community which are willing to share their service. They therefore become CuWIN gateways, from a mesh network perspective.

Currently, the most prominent projects that are based on CuWIN are the Urbana project, Wireless Ghana, Mesa Grande Reservation.

Generic Operation To setup a new wireless CuWin mesh network in a given community Bob must first download CuWin software [3] or acquire the CuWIN kit. Bob is a "do-it-yourself man" and opts for the software installation on an old desktop PC equipped with a wireless card that holds an Atheros chipset. After booting, Bob's PC becomes a CuWIN node. Bob is then part of the CuWin backhaul in his town and opts to share his own Internet access with the community. He then convinces a few other users around to join the network and hence the network grows steadily.

Ana is a visitor in town and wants to read her e-mail. She simply opens her PDA and can see the *cuwireless* SSID available, so she connects to it and immediately can use the available network.

CuWin provides no resource management at all, and hence the quality of the shared infrastructure depends upon the simultaneous number of users, as well as the type of traffic being used. Incentives to adhere to CuWin are free roaming and the possibility

---

[4] the CuWIN Metrix kit costs 499.

to expand the network at a low-cost. An interesting additional incentive is the promise of services in CuWin, such as a VoIP platform. Security is left to the user sharing his/her access. CuWin is an interesting platform in particular for communities willing to expand existing Internet access in remote areas.

FreiFunk

Freifunk [6] is a non-commercial initiative which has as motto to provide free wireless access globally. The project is based upon mesh networking but adds the interesting feature of *picopeering-agreements (PPAs)* [20]. A PPA is an attempt to assist in allowing different wireless communities to interoperate, particularly having in mind free networks. Basic rules of a PPA are that an MP (owner of the PPA) must agree to provide free transit across the free network, and not to modify any data being transmitted; communication must be open; no warranties are provided. The PPA defines, in addition, terms of use which provide the MP with some flexibility in terms of formulating acceptance use policies. Freifunk provides users with a firmware (Freifunk Firmware, FFF) based on OpenWRT. The firmware originally included OLSR as the multihop routing protocol, a Web interface which provides AP configuration, as well as additional features such as traffic shaping and statistics, Internet gateway support. Today, FFF also provides support for Batman (in addition to OLSR).

Freifunk is therefore tailored for communities of users willing to share an Internet access, and there are several Freifunk communities, in particular in Germany. For instance, in 2004 the Berlin community consisted of 500 Freifunk APs, and Freifunk claimed having several thousand users accessing the Internet for free [7]. Freifunk claims 5945 users (shared APs) registered worldwide, being circa 2200 in Berlin alone.

Generic Operation

Bob decides to become part of the local Freifunk municipality. Therefore he downloads the FFF and uploads it to his local and compatible AP. Once the AP is active, a new firewall configuration is provided, splitting the local network from the exterior network. In case NAT is present, Internet access is possibly by means of neighboring stations that announce Internet access. Bob does not own an Internet access, but Ana, his neighbor, does. She is willing to share that access with neighbors (such as Bob). Therefore, all she does is to plug her Internet access router to the AP where she installed the FFF. The FFF AP automatically receives a default gateway via DHCP, becomes a gateway, and relies on OLSR HNA4 to provide an announcement of the new gateway to other nodes on the network. The connection to the default gateway is continually checked using "arping". If the connection disappears, then the HNA4 announcement is discontinued.

Freifunk is an interesting and one of the oldest concepts of UCNs based on mesh networking. However, its main incentive relates to low-cost deployment. There is no security in place, nor any type of resource management features.

Open-Mesh

Open-Mesh claims to be a "ever-growing group of people dedicated to community-owned WiFi, not owned or controlled by any one corporate entity". Open-Mesh products relate to the management platform required to assist an autonomous growth of community-owned mesh networks. Open-Mesh provides an AP/Access router based on different types of hardware and on the open source GPL *Routing OLSR and Batman INside (ROBIN)* [1] software platform. ROBIN is deployed on the operating system OpenWRT (Kamikaze version) and runs on any Atheros AP51 router. In terms of multihop routing, ROBIN gives the possibility to rely on *Better Approach to Mobile Ad-hoc Networking (BATMAN)* [8].

Generic Operation Bob decides to be a part of Open-Mesh. Therefore, he acquires an Open-Mesh AP or opts for just installing the Open-Mesh firmware in one existing compatible AP (e.g. Meraki, Ingenious, Ubiquiti). He then registers in Open-Mesh, to conFig. his mesh hotspot, by adjusting parameters such as SSID, location of owned APs, setting up WPA security, as well as the captive portal options. Moreover, Bob can specify MACs of the devices that can access a specific AP (SSID). Bob wants to create a mesh with 3 APs and hence registers the MAC of these APs, providing also a location for each. Then Bob conFig.s the common (public) SSID for the three APs, and a private SSID (for his use only). Both SSIDs will provide the basis for secure networks, based on WPA (personal) credentials. Bob can also conFig. channel and fix the Wi-Fi rate to 5 Mbps (or opt to have the regular auto adjustment).

Open-Mesh [19] does not have *resource management* in terms of sharing the infrastructure. Cooperation incentives relate to the possibility to deploy a low-cost wireless network in a user-friendly way. Security is provided by means of WPA and the MP can also block clients by means of MAC filtering. Moreover, Open-Mesh gives the MP the possibility (Web-based) to conFig. its shared network and to have a basic perspective on the usage of the different nodes that compose the network.

NetSuKuKu

The Netsuku project [15] envisions a completely autonomic network with no centralized control whatsoever, so that each node takes the same role on the network. It devises a set of protocols and architecture that should allow a network to scale to massive number of nodes and still require very little CPU and memory usage from each node.

Netsukuku relies on the end-user devices (personal computers) and is based on ad-hoc technology, meaning that, in order to be a part of the Netsukuku network, the user needs just to be at reach of another Netsukuku node and install the Netsukuku software. To allow the network to grow even more, in particular on its embryonic stage, it is possible for a node to be a part of the Netsukuku network through a VPN tunnel.

Netsukuku follows a fractal hierarchy in terms of topology, for the sake of both routing and naming efficiency. Nodes are grouped hierarchically into groups of 256 nodes, called a *gnode*. Then, gnodes are grouped in groups of 256 and so on, into n levels of hierarchy. The maximum level, n, depends on the number of addresses available, meaning that in IPv4 we have 2^{32} IPs, so $n = 4$. Similarly, using IPv6 we have 2^{128} IPs, meaning $n = 16$.

This hierarchical grouping of nodes was useful when creating both its own name resolution and routing protocol. The naming resolution scheme, named *A Netsukuku Domain Name Architecture (ANDNA)* [16] is designed as non-hierarchical and decentralized name resolution system and is a full replacement of the hierarchy *Domain Name System (DNS)* commonly used on the Internet. Moreover, Netsukuku relies on their own routing protocol, the *Quantum Shortest Path Netsukuku (QSPN)* [17], which was specifically designed for the hierarchical topology of Netsukuku and is designed to be decentralized and demand very little resources from each node. From the perspective of naming and routing, gnodes represent true nodes on the topology, and within each topology level, routing performs independently.

The network formed by all Netsukuku nodes can be connected to the Internet, through some gateway nodes, but may also co-exist in a completely independent way. This means that Netsukuku nodes can communicate directly, and use any IP application over it.

Generic Operation Bob wants to expand its Internet access, and so, Bob installs the open-source Netsukuku software in his desktop computer. Bob's computer has an external high-gain antenna connected to its network card, and as a result, Bob can instantly connect to his Netsukuku neighbors, at reach of his wireless signal. The network is self-conFig.d and there is nothing else that needs to be done in order for Bob to be able to communicate to every other Netsukuku node.

An interesting aspect for this UCN is that routing considers resource management based upon each node's available bandwidth. Incentives too cooperate in NetSukuku relate to the low-cost deployment in a completely plug&play way. In terms of security, Netsukuku follows the design principles of any mesh network: routers can sniff and misuse traffic. However, Netsukuku is developing a specific cryptographic layer (Carciofo) to provide end-to-end user anonymity and privacy to users, based on IP-in-IP tunneling. Interesting features of this particular case of a UCN are its decentralized naming scheme (ANDNA) and its particular routing protocol (QSPN). It should also be noticed that QSPN does not support mobile nodes and considers that networks are stable for a reasonable time (updates take several minutes), i.e., Netsukuku is not tailored for dynamic mesh networks, where nodes may join and/or leave frequently.

4.4 Mobile/Provider UCNs

This section provides a glimpse of the most recent category of UCNs. We described three products which assist the development of mobile-based UCNs. As shall be realized, the potential of this type of UCNs is still to be unveiled but what is clear is that the products here described will give a push to the deployment of UCNs.

MIFI

MIFI [26][5] is a product (smart AP) of Novatel currently being offered by the operators Verizon and Sprint to their 3G customers, as a service differentiator. The product is tailored for users on the go, to provide shared connectivity based on an existing 3G access. Verizon provides an access management platform (Web based) which can be accessed by Wi-Fi. The MP can then check usage, perform trust management for the shared devices, and specify concrete rules for sharing connectivity. Hence, this is a product tailored for users which temporarily need to provide Internet access to a few devices on the go.

In 2011 Verizon and Sprint charged $59.99/month for a 3G service with Mifi, up to 5 GB allowed [27]. Verizon also provides additional plans, e.g., daily vouchers. In terms of rates, Mifi 2200 [27] is constrained by the offered 3G rates and hence, users cannot profit from Wi-Fi data rates completely.

Generic Operation Bob is traveling with his family and spends the night at a hotel which has no Internet access. Bob has is Sprint 3G phone and can access the Internet. However, his wife Linda would also like to access the Internet. Hence, Bob activates his Mifi 2200, which provides a way to share his 3G phone Internet connection with Linda seamlessly.

MiFi does not integrate intelligentresource management. However, the MP can check the status of its subscription (data usage) and can manually perform admission control. Incentives to adopt Mifi simply relate to the sporadic need users may have to share Internet access based on 3G. This is not however a product tailored to allow global shared usage of existing private hotspots.

The security provided is simply based on MAC filtering, being the MP the one that states who can access his/her Mifi 2200. As additional features, Mifi 2200 provides an advanced policing and monitoring platform, which gives the means for the MP to prevent breaks in terms of subscription data rates limit.

JoikuSpot

JoikuSpot [10] is a software-based solution that allows users to automatically deploy a UCN on their mobile phone. The software is in 2013 available in most mobile devices, and is free. In its regular version Joiku allows only HTTP/S connections and

---

[5] Stands for "My Wi-Fi", read "maifai".

prevents to set up some hotspot configuration parameters (e.g. change ESSID name, turn on encryption), or in its full-featured "Premium" version that cost 9 euros. In addition, an optional module called JoikuBoost can be installed on top of JoikuSpot for allowing aggregation of multiple 3G connections.

Speakeasy Netshare

Speakeasy, a Seattle based Service provider was one of the first providers allowing wireless access sharing. In 2003, Speakeasy unveiled the Netshare WiFi plan which had as main purpose to allow a user to share his/her access with neighbors. The incentive to become an MP would be a revenue from 50% in terms of access costs. Security was left to the MP, even though at the time there was a recommendation for using 128-bit WEP.

In addition to Internet access, Speakeasy would provide each new user with e-mail and newsgroup access, as well as backup dial-up access. Moreover, the sharing was available not only via Wi-Fi, but also via Ethernet, Homeplug, etc.

Netshare is therefore a first example of a provider based UCN, where an ISP specifically states that the MP is responsible for any infringement of the existing Internet access subscription. The main Netshare incentive was revenue to the subscriber. The most relevant feature to cite is the fact that the provider would give each user (MP and regular users) e-mail and news access.

The ZON@FON Case

The ZON@FON is a concrete example on the application of FON by means of an access provider. ZON is an alternative Portuguese access provider which holds several services, ranging from digital TV to Internet access. In terms of Internet access, ZON provides relies on advanced cable technology (optical fiber and Eurodocsis 3.0) to provide residential customers with up to 1 Gbps. On the last hop to the Internet user, ZON provides different technologies, being Wi-Fi one of them. Recently, ZON partnered with FON to exclusively provide FON services in Portugal. Being a provider, ZON claims the deployment of around 100,000 ZON@FON hotspots in Portugal. The widespread deployment was achieved by providing each ZON subscriber with a specific AP/AR (based on the FON firmware, but updated to suit ZON's requirements), and also by having deployed a large number of APs in public locations, and specific neighborhoods in Portugal.

Even though technically ZON@FON follows the FON model, this is a concrete application of a UCN where the initiative is provided by the provider (access or service provider) and not by means of a user, or a community of users.

5 Comparative Analysis and Evolution Discussion

This section provides a comparative analysis of the five identified UCN categories. Such analysis is performed based on the main properties of UCNs, and on the different roles of users, MP and VO. The comparison is summarized in Table 1.

Let us start by explaining the differences in terms of who holds the VO role. In the hotspot category, the VO is a specific entity that manages credentials, initial authentication and AP registration. The same functionality is provided by the VO in the mesh-based model, but in this case the VO is normally a community of users. Similarly, in the social networking category, the VO is also a specific entity but its main responsibility is to ensure a secure exchange of credentials. In the case of the mobile-based and provider-based categories the role of VO is assigned to access operators. Therefore, the VO has a similar role across all categories. It has only a coordinating role and does not have any impact on the way traffic is transmitted in the communities or across the Internet. Moreover, the VO does not account for any end-to-end measures, such as data privacy or traceability. However, our understanding is that this role will evolve and the VO will, in the future, have responsibilities that go beyond initial setup and will become service differentiators, e.g. distributed mobility management across communities.

The set of MPs in the hotspot, social networking, and mobile/provider models fairly corresponds to the global user database, given that all users must share access to obtain a specific benefit. While in the mesh and mobile categories the MP set corresponds to a subset of the global community. Moreover, changes in the MP set are expected to be rare in most categories, being the exception to the rule the mobile-based category. A relevant aspect to mention is the impact that changes on the MP set may have on the overall network operation. In the hotspot model, and despite the fact that any user is an MP, the impact of changes to the MP set are not expected to affect significantly the network operation, given that such operation is tied to the Wi-Fi infrastructure mode, which splits the network operation into islands (hotspots). The same occurs in the social networking and provider-based categories. While in the mesh category, and despite the fact that MPs are a subset of the user universe, changes in the MP set are expected to bring high penalties to the network operation. This may possibly be counter-balanced due to the fact that most MPs are expected to exhibit a static behavior (roaming frequency may be low or scoped in nature).

In the mobile-based category changes to the MP set will introduce high variability into the network, given that users are expected to roam frequently—such variability is tied to the mobility pattern of all users.

From a global perspective, today, and due to the fact that the MP role is simply tied to the connectivity model, changes to the MP set are not significant from a global network perspective. However, the MP role is expected to evolve into a multi-user operation setting, where some forms of networking services, in particular in the control plane (e.g. AAA, mobility management) are to be sustained by the MP in cooperation with the access, as starts to happen when smart APs (such as Femtocells) are deployed.

Table 1 Comparison of the different UCN models

	VO	MP set	Resource management	Cooperation incentives	Mobility management	Security aspects
Hotspot	- Specific entity - Manages credentials and initial authentication - Manages AP registration	- All users - Changes in the MP set are not frequent - Changes in the MP set affect the network only partially	- N/A	- Broader roaming - Revenue for MPs	- N/A	- Mostly based upon private AP security (WEP) - Security for the MP; open access for guests
Mesh	- Community of users - Manages credentials and initial authentication - Manages AP registration	- A subset - Changes in the MP set are not frequent - Changes in the MO set affect all the network operation	- N/A	- Low-cost network deployment	N/A	- No integrated security
Social	- Specific entity - Manages credentials for access to private APs following social networking	- All users - Changes in the MP set are not frequent - Changes in the MP set affect the network partially	- N/A	- Broader roaming - Revenue for MPs	N/A	- Third-party support for Wi-Fi (WEP, WPA) key exchange
Provider	- Access or service provider - Provides any user with the possibility to share an existing 3G Internet access (limited sharing) - Differentiates an existing service by relying on UCNs	- All users - Changes in the MP set are expected to be frequent - Changes in the MP set may bring heavy penalties to the network operation	Based on subscription rules broader roaming	- On-the-go shared connectivity - Revenue	3G roaming	End-to-end if mobile access; otherwise requires VPN

Let us now provide some considerations in terms of adoption incentives. The promise of wider and free roaming is today the most common incentive. A second incentive observed is extra revenue or rebates for access cost. This incentive is more prominent in categories tied to the access stakeholders, e.g. provider-based categories, but it also appears in the social networking category. A third type of incentive is low-cost expansion of Internet access, and this is in fact the single incentive observed for examples that fall in the mesh category.

As these networks grow, incentives are also expected to evolve based upon new responsibilities that MPs may attain. For instance, incentives based on bandwidth tokens are feasible in a short-range time period, given that today such incentives are already present in a variety of collaborative tools, from an application layer perspective.

An interesting technical aspect is that most of the categories do not consider resource management even in simple forms. The categories that integrate some form of resource management are the ones where the VO is the provider—mobile-based and provider-based categories—which is somewhat obvious given that the control of the network is still centralized and hence, easier to perform: management of network resources in the mobile-based and the provider-based models is based on subscription rules. Resource management is, however, a field which is essential to assist UCN growth. Providing MPs with the automatic capability to share bandwidth in a clever way is also an incentive that will be deployed in future models. Another relevant aspect to consider is that intelligent and dynamic resource management is a must to assist in preventing access technical infringements (e.g. going over the average rate stipulated in the Internet access subscription).

In terms of security and data privacy, today UCNs rely on the available schemes, namely WEP for the MP and, most of the times, open access to regular users. Moreover, there are categories (such as mesh-based) which do not even integrate security, simply leaving the choice to the user and advocating the use of application layer privacy tools. We argue that for future models it is necessary to ensure three basic properties: confidentiality, non-repudiation and traceability. These are aspects that can be dealt with by adequate trust management models.

6 UCNs Follow-up: Assumptions and Requirements

This section aims to identify a set of regulatory, social, and technical assumptions as well as requirements that must to be taken into account for the development of UCN solutions. The analysis of relevant assumptions is provided based on the four major conceptual properties of UCNs: connectivity sharing and relaying, cooperation, trust,and self-organization. Such aspects are summarized in Table 2, and discussed in the next sub-sections.

Table 2 UCN Follow-up, assumptions and requirements

	Assumptions	Requirements
Connectivity sharing and relaying	- MPs hold an Internet access subscription that is not regulated by the VO	- Traceability across multihop sharing
	- The set of MPs is essentially static over time	- Security and non-repudiation
	- Sharing is performed on specific communities (coordinated by VOs)	- Trust management
	- Sharing today is based on a single-hop, but future cases will incorporate multihop sharing (relaying across a few hops)	- Communities: UCNs must be able to support autonomic growing of communities of micro-operators, which decide for a common identity based on a set of cooperation parameters
	- Equipment used in Internet access sharing can be fix (e.g. an AP) or mobile (devices carried by users)	- Software defined: UCNs deployed must be based upon low cost hardware of easy deployment, and software that is independent from radio technology, operating system, and devices
	- The access operator does not control UCNs	- Shared services: UCNs should provide a set of services, additionally to Internet connectivity, such as local DNS, Privacy, and data caching
		- Coverage: UCNs may provide resource sharing within a few hops away from an Internet access point based on routing or simpler relaying methods
		- Connectivity: UCNs should be able to keep active the offered services even in the presence of intermittent connectivity. Moreover sources and destinations may establish connectivity without the need to go to the access network
		- Traffic: UCNs must be able to provide detailed accountability and traceability (e.g. useful for traffic imputation)
Cooperation	- Users are willing to share subscribed access against a specific benefit (currently, it is wider roaming, revenue, or rebates)	- Technical incentives should be improved to assist in a win–win match for the sharers and the ones profiting from such sharing. Such incentives have to be considered from the end-user and from the access perspective
	- Network topology makes it possible to share subscribed access - networks are dense	- UCNs must integrate mechanisms that reward adequately users that cooperate the most
	- Users move across different locations (different APs)	- Scalability must be considered also from an adoption (spreading perspective). UCNs should integrate mechanisms to assist adhesion/adoption

(continued)

Table 2 Continued

Trust management	- UCNs only provide partial data confidentiality	- UCNs must allow sharing of Internet services based on well defined trust levels derived from social networking tool as well as from nodes networking behavior, and user's social behavior and interests
	- Current models disregard user anonymity aspects	- UCNs should implement trust models that not only consider community beliefs, but are actually dependent upon surroundings and the level of confidentiality that the user expects on a specific moment and for specific application. For instance, trust models may be influenced by local conditions such as the degree of current connectivity and the current reputation level, as well as by external conditions such as the overhearing probability around a specific node
	- The MP is the accountable entity from an operator perspective	- UCNs must implement *grassroots* trust mechanisms, aiming to increase the scalability and robustness of the system, without being confined to a specific community
		- UCNs should implement reputation mechanisms (e.g., similar to the E-Bay model) to identify the networking behavior of nodes
		- UCNs must ensure data confidentiality end-to-end and not only on the wireless link to the MPs. Confidentiality should be provided by mechanisms that do not reduce the user privacy
		- Reputation: UCNs must allow micro-operators to claim not having done a specific action (*repudiation*), otherwise he/she may see his Internet connection blocked or event have to deal with legal actions
Self-organization	- UCNs form autonomously based on community behavior and driven by community needs	- UCNs must be able to self-organize recurring to incentive schemes to ensure a smooth operation
	- Users are not fairly balanced across APs of MPs	- UCNs must be able to provide a better usage of available resources, such as by taking advantage of aggregated backhaul capacity or load-balanced across the micro-providers in order to respect the traffic limits described in Internet subscription agreements
	- The MP set and user set varies frequently with time	- UCNs should be able to estimate its dynamic morphology, namely the number of users at specific periods of time, the node degree, as well as the average bandwidth that nodes can take advantage from

6.1 Connectivity Sharing and Relaying

Current examples of UCNs rely on one-hop wireless connectivity sharing by individual users, communities or organizations. In other words, an MP shares subscribed Internet access with specific communities. Such sharing can be based on equipment that is fixed in a specific place (e.g. residential household) or even just based on equipment carried by humans. Hence, MP elements may move while sharing Internet access. Moreover, it is assumed that the MP already subscribes Internet access e.g. based on mobile, wireless, or fixed access technologies.

Another relevant aspect to consider is that the set of MPs today remains essentially static over time. In other words, the universe of users profiting from the sharing tends to be the same universe of users sharing, as we shall see with the examples provided. Sharing today is basically related to the offer of roaming services. By sharing access within a specific community, the user can profit from wider roaming.

6.2 Cooperation

Today's UCN cases are supported by the willingness of the end-user to become part of existing communities. Motivation to do so relates to the roaming incentive, and to the end-user's belief that the benefit of using UCNs is higher than the risk incurred. However, this is a fake sense of security, and with time, UCNs will evolve and users will become more intransigent in terms of incentives to cooperate.

6.3 Trust Management

Data privacy in UCNs is normally partial, given that it is only ensured on the wireless link and to the MPs. A user can of course cope with this gap by relying on specific privacy mechanisms, e.g., using some specific application or establishing a tunnel to a specific, trusted entity (e.g. a VPN to an enterprise). The flip-side of this is the related overhead both in terms of configuration/processing time, and in terms of data. It should also be noticed that despite the fact that the MP communication is protected, malicious user traffic may pass by the MP device (e.g. access point) and thus may result in serious violations. Another relevant aspect is traceability and non-repudiation, which becomes even more serious if one considers future multi-hop UCN scenarios. Trust management can assist in lowering the barriers of these concerns without the need to consider complex third-party certifying entities or revocators. Currently, trust within a specific UCN is confined to a specific community which is normally managed by a VO. In the future, trust models should not only consider community beliefs but actually be dependent upon surroundings, level of confidentiality that the user expects on a specific moment and for specific applications. Hence, the most adequate trust management models to consider in terms of

UCNs are decentralized ones, where each peer holds specific trust values and metrics that other peers can access. In addition, ways to fight back selfishness of peers (fight back the *tragedy of the commons*) have to be considered, given that UCNs have a highly dynamic character in terms of who composes communities.

6.4 Self-Organization

Self-organization is a key aspect in UCNs in particular due to the fact that the network operation is being placed closer to the user, and driven by community needs, and community network usage. Today, self-organization is applied in UCNs mostly for improving resource management aspects e.g. in case of offloading. Cooperation models require, however, self-organization to be addressed both from a network as well as from a user perspective, thus providing more robustness to UCNs, where the control functionality may be split across different physical devices which are not necessarily owned by a single operator, or even by an operator.

7 Summary and Conclusions

This paper provides an overview and categorization of UCN living-examples as a way to assist the robust development of these novel architectures and as a consequence to assist in a shift concerning Internet architectural design.

A first conclusion to draw is that there is a clear paradigm shift due to cooperation amongst Internet communities which is changing the network operation and giving rise to new networking opportunities. A second conclusion is that there are clear technical and economic limitations to today's UCNs. Technical aspects to improve relate to adequate resource management, mobility and security. From an access perspective there are technical advantages that must also be considered, such as solutions to keep traffic "local" (confined to specific communities) or even methods to make networks more robust by exploring cooperative networking. Economic limitations of these architectures are today tightly related to the lack of understanding in terms of applicable business models. Therefore, a key aspect to analyze are ways to model incentives in a way that becomes profitable to both the individual user and the community, as well as to the access.

Research work in this field should consider how to optimize available resources (bandwidth and energy);how to devise trust models to lower operational complexity due to a need to reinforce security (traceability and liability); how to define incentive plans to adhere to these novel architectures.

Acknowledgments We thank all of the authors that contributed to the white paper on which this paper was based, report number SITI-TR-11-03 (2011): Rute C. Sofia, Paulo Mendes, Waldir Moreira Junior, Andréa Ribeiro, Saulo Queiroz, Antonio Oliveira Junior, Tauseef Jamal, Namusale Chama and Luís Carvalho.

References

1. Anselmi A Robin—open source wireless mesh networking. http://robin-mesh.wik.is/. Accessed Nov 2009
2. Camponovo G, Cerutti D (2005) Wlan communities and internet access sharing: a regulatory overview. In: International conference on mobile business ICMB, pp 281–287
3. CuWin (2009a) Cuwin ware kit. http://sourceforge.net/projects/wireless/files/wireless/
4. CuWin (2009b) Community wireless solutions. http://cuwireless.net/
5. FON Wireless Ltd (2009) Fon. http://www.fon.com
6. Freifunk (2006a) Freifunk community. http://start.freifunk.net/
7. Freifunk (2006b) Freifunk.net—a successful do-it-yourself approach for building wireless community networks in germany. http://start.freifunk.net/. Accessed Oct 2006
8. Freifunk (2008) Better approach to mobile ad-hoc networking (b.a.t.m.a.n.). IETF draft (Expired)
9. Ishmael J, Bury S, Pezaros D, Race N (2008) Deploying rural community wireless mesh networks. IEEE Internet Comput 12:22–29
10. Joikusoft (2007) Joikuspot. http://joikusoft.com/
11. Kawade S, Van Bloem JW, Abhayawardhana VS, Wisely D (2006) Sharing your urban residential wifi (ur-wifi). IEEE 63rd vehicular technology conference. VTC 2006-Spring, vol 1, pp 162–166
12. KeyWifi-LLC. Keywifi: unlocking hotspots near you. http://keywifi.com/
13. Kiviö M (2009) (MP-MasterPlanet) Openspark. http://www.openspark.fi
14. Linspot (2009) Linspot. http://www.linspot.com/
15. Netsukuku (2009) Netsukuku - close the world, txen eht nepo -. http://netsukuku.freaknet.org/
16. Netsukuku Project. Andna—a netsukuku domain name architecture, manual. http://netsukuku.freaknet.org/doc/manuals/andna
17. Netsukuku Project (2009) The qspn. Technical report, Sept 2009
18. Network Research and Special Projects Units (2006) University of Lancaster. Wray broadband project
19. Open-Mesh. Open-mesh, ultra low cost wifi. http://www.open-mesh.com/store/
20. Pico. Pico peering agreement v1.0. http://www.picopeer.net/PPA-en.html
21. Saltzer JH, Reed DP, Clark DD (1984) End-to-end arguments in system design. ACM Trans Comput Syst 2(4):277–288. doi:10.1145/357401.357402
22. Santivanez C, Ramanathan R (2001) Hazy sighted link state (hsls) routing: a scalable link state algorithm. BBN Technical Memo BBN-TM-1301. BBN Technologies, Cambridge, Aug 2001
23. SofaNet. Sofanet, sicherer verbinden. http://www.sofanet.de
24. Sofia R, Mendes P (2008) User-provided networks: consumer as provider. IEEE Commun Mag (Feature Topic on Consumer Communications and Networking - Gaming and Entertainment) 46:86–91
25. Solarski M, Vidales P, Schneider O, Zerfos P, Singh JP (2006) An experimental evaluation of urban networking using ieee 802.11 technology. In: 1st workshop on operator-assisted (wireless mesh) community, Networks, pp 1–10
26. Verizon. Mifi 2200, intelligent mobile hotspot. http://www.verizonwireless.com/b2c/mobilebroadband/?page=products_mifi
27. WiFi Planet (2006) Novatel wireless mifi 2200 for cdma 1x ev-do networks. http://www.wifiplanet.com/reviews/article.php/3824351. Accessed June 2009
28. Wifi.com (2009) Whisher. http://www.whisher.com/
29. Wray Village. Wray community communications. http://www.wrayvillage.co.uk/wraycomcomhome.htm

User-Centric Networking: Routing Aspects

**Namusale Chama, Antonio Jr., Waldir Moreira Jr., Paulo Mendes
and Rute Sofia**

Keywords Multihop routing · User-centric networks · Mobility · Energy-awareness

1 Introduction

Wireless revolutionizes local area communications permitting the general public to provide communications services as well as to become micro-providers in *User-centric Networks (UCNs)*. This emerging networking paradigm relies in the user's willingness to share connectivity and resources. In comparison to traditional Internet routing scenarios (be it based on wireless or fix line technologies), UCNs bring in forwarding challenges, due to their underlying assumptions, namely: (i) end-user device nodes may behave as networking nodes, (ii) nodes have a highly nomadic behavior, (iii) data is exchanged based on individual user interests and expectations.

Furthermore, emerging trends such as UCNs adding to the development of faster, more reliable wireless standards, miniaturization of devices, and reduced costs of hardware and services, is leading to a fast evolution of technological as well as

N. Chama · Antonio Jr. · W. Moreira Jr. · P. Mendes · R. Sofia (✉)
COPELABS, University Lusófona, Building U First Floor, Campo Grande 388, 1749-024
Lisbon, Portugal
e-mail: namusale.chama@ulusofona.pt

Antonio Jr.
e-mail: antoniocojr@gmail.com

W. Moreira Jr.
e-mail: waldir.junior@ulusofona.pt

P. Mendes
e-mail: paulo.mendes@ulusofona.pt

R. Sofia
e-mail: rute.sofia@ulusofona.pt

A. Aldini and A. Bogliolo (eds.), *User-Centric Networking*,
Lecture Notes in Social Networks, DOI: 10.1007/978-3-319-05218-2_3,
© Springer International Publishing Switzerland 2014

societal aspects in the way that people communicate. For instance, people expect to be able to send and retrieve information whenever and wherever they want. Yet, there are technological limitations which may affect this anytime-anywhere communication paradigm, e.g. gray areas (i.e., areas where the wireless signal strength is not enough to sustain connectivity); physical obstructions; limited battery devices; environmental aspects; limited resources and security issues. Related literature has been addressing aspects to mitigate wireless interference and to take advantage of cooperative diversity which may mitigate some of the problems posed by physical obstructions and coverage problems due to node mobility. However, it is imperative to say that, since information is relayed among nodes and these nodes can be highly dynamic, communication may experience delay, varying from short to long periods, as isolated areas (e.g., intermittent connected networks) may form in the case of node failure (e.g., damaged AP) or mobility (e.g., user changes position). Thus, to increase the performance of multi-hop communication, several improvements can be made, by taking advantage of transmission opportunities provided by moving nodes and accessible APs, for instance. This may occur, for instance, when a user is at a public location without Internet access. If other users are in the vicinity, and such users are part of a UCN, then some of them may share Internet access and data can be relayed until it reaches the closest Internet gateway. Another situation may occur when information is simply carried by users that happen to be moving towards the place where the destination is located. Nowadays, this is possible thanks to the size of devices which are making them easier to carry around, and also to the resource capabilities they have. For instance, the haggle EU project [36] exploits store-carry-and-forward capabilities (i.e., devices' powerful features, user willingness, trust among users, opportunistic contacts) aiming to provide communication in scenarios with intermittent connectivity. Haggle considered human mobility and the power of users' devices to perform forwarding of information independently of the network layer. So it is easy to see that the way people communicate is arriving at a point where such communication must happen independently of the infrastructure available, and depending on the capabilities of intermediary devices as well as their mobility pattern, interests and social ties.

In what concerns the network layers, this new communication paradigm demands more reliable and efficient protocols, as today we have areas where spectrum abounds and creates interference - dense networks, e.g. residential households, shopping malls) as well as areas where communication is only possible through the formation of clusters of users (e.g., intermittently connected networks). Even in a metropolitan area, intermittent connected networks exist due to wireless environments, unexpected disruptions, and areas where the networking infrastructure is sparse (e.g., city parks).

Understanding human mobility and social interaction is important for the adequate operation of data transmission in the context of UCNs. Different types of models have been proposed in the last decade, from those founded on purely synthetic movements, such as the ones based on purely random movements of the nodes, to the ones aiming at reproducing the mobility patterns inside specific places. New insights and more refined and realistic models are still needed, bringing together real world measurements and mathematical characterization of node mobility. So now, besides considering the characteristics of the wireless medium and aspects that influence its functionality, other relevant characteristics as user' s interests and their social interaction (resulting from the way people move) have a key role when it comes to establishing communication and managing it considering resources available and trust among users. The former has shown itself rather useful when it comes to the dissemination and retrieval of information, while levels of social interaction with family, friends, acquaintances, and total strangers dictates when and how such dissemination and retrieval is going to happen. Last but not least, a routing solution for user-centric networks must consider two basic aspects: robustness from an end-to-end perspective, and intermittent connectivity support. This last aspect is a major requirement to operate a future Internet, which means that routing systems must be able to keep graphs connected.

2 Related Work

With the advent of Web2.0 and the rise, both in variety and in coverage, of wireless technologies and user-friendly devices, there is a change in terms of Internet user behavior: the user is becoming more than simple consumer of services, to have roles where he/she shares or even provides networking services. Sofia and Mendes [33] describe new user-centric communication models, in which the user is not only a consumer but also a provider of communication opportunities (user empowerment) and alerts to the need to consider user-centric communications as part of an Internet of the future. Our work builds upon the models the authors describe with a specific focus on user-centric routing. Since most of the users are currently connected by means of wireless links, it is important to investigate algorithms and metrics to increase reliability and performance over multiple paths dynamic wireless networks: (i) multi-path due to wireless diversity; (ii) dynamic due to users' behavior.

In what concerns multihop routing, the most popular approaches such as AODV and OLSR have been engineered to sustain better *Quality of Service (QoS)* but not dynamic node movement. A line of work has addressed this need based on the definition of metrics that make a network more robust by taking into consideration link duration [3]. Chama et al. [6] has provided an extensive analysis on parameters capable of tracking a few aspects of mobility in routing protocols as a way to derive metrics that can be applied to multihop routing approaches, to make them more sensitive to node movement and hence, reduce the need for path recomputation.

A relevant aspect addressed in related literature concerns the capability to allow the network to expand based on heterogeneous and portable devices. These are often carried and transported by humans. In regards to reducing the energy consumption in mobile devices, there have been efforts in physical and data link layers as well as in the network layer related to the routing protocol as has been detailed by Oliveira Junior et al. [25]. Most of the related proposals consider energy-efficiency from an engineering perspective, i.e. extensions of the existing on-demand and link-state routing make modifications in the protocols to devise energy-awareness.

One of the major assumptions in UCN and for which routing has to be prepared is the intermittent characteristic of wireless connectivity. This intermittent behavior may occur in sparse networks such as in small villages, rural areas, or disaster areas, as well as in dense urban networks. In the latter case, intermittent connectivity may be due to wireless interference. Ad-hoc routing protocols such as AODV [32] and OLSR [10] assume that a complete path always exist between a source and a destination, and try to discover minimum cost paths. This means that such protocols are useful for networks with low to medium dynamics. While in UCNs the movement of nodes follows their owners behavior - human movement patterns -, behavior that may lead to situations where some end-to-end paths may not be temporarily suitable for communication. A family of algorithms (e.g. epidemic, gossip, greedy) [5, 14, 16, 18, 35, 37] tries to make use of user mobility to route information in loose connected graphs, as the ones provided by end-users. The primary focus of this family of algorithms is to increase the likelihood of finding paths, using only information about spontaneous local connectivity. However, such algorithms are agnostic to the status of the network in terms of connectivity (potential contacts), storage and queuing capability of nodes and bandwidth capability of links. Their final goal is only to increase the probability that a message is really delivered to its destination. A more realistic scenario for user-provided networks is the one in which: (i) most of the nodes have resource constraints, and (ii) local connectivity may also be predictable or scheduled (e.g. connectivity provided every day while driving to work). However, routing solutions for these UCN communication scenarios have received little attention to date. The needed investigation should analyze the trade-off between delivery probability and resource usage: for instance, distributing messages to a few or large number of nodes will increase the probability of delivering a message to its destination, but in return, more resources will be used.

Balasubramanian et al. [2] propose an algorithm to replicate packets optimizing a specified routing metric in scenarios where nodes have limited resources. However, the authors do not consider node dynamics as well user behavior, such as willingness to cooperate in message forwarding. The latter limits the impact of the proposed algorithm, since nodes are autonomic in the sense that they can decide on their own whether to implement or not the rules of a routing algorithm [31]. In terms of applications support, most of the previous proposals consider only applications tolerant to changing network conditions, such as delay and losses. However, although some applications are tolerant to quality oscillations this does not mean that they would not take advantage from low delay and minimum number of expired messages. For instance minimizing delays reduces the time messages spend in the network, reducing the contention for resources.

Regarding evaluation frameworks, we highlight the most recent opportunistic routing proposals are based on *Evolving Graph (EG)* theory to design/evaluate least cost routing protocols. EG provides a formal abstraction for dynamic networks and reflects the different connectivity graphs in the time domain by considering node mobility. The result is that connectivity of links are transcribed into subgraphs for different instant in time. Thus, Newman et al. [24] take into consideration one of the formalized EG criteria (i.e., foremost) to determine journeys (i.e., future temporary connections between nodes that can form a path over time) in which data can quickly reach its destination. This evaluation framework provides designers with an algorithm that is able to reach good performance in scenarios where connectivity patterns are known beforehand. Additionally, the algorithm can be used as lower bound reference to compare opportunistic routing solutions. The work of Spyropoulos et al. [35] provides principles to help developers in designing routing solutions based on their classification of opportunistic routing, identified utility functions. The authors show that by knowing the application characteristics and requirements, the choice/design of routing solutions is eased. Still, both works lack a guideline of how performance metrics and experimental setups can be used.

3 Use-Cases, Assumptions and Requirements

This section provides a characterization of a UCN scenario, including a set of assumption and requirements, to assist in the debate concerning user-centric routing. To clarify main differences against other autonomic networks, the section also provides a comparison of connectivity features for UCN against ad-hoc and other forms of multihop networking.

Figure 1 provides a high-level perspective on potential cases of UCNs, to assist in the description of two potential applicability cases which are described next, namely, a scenario based on dense networks, and the other based on challenged communications. The intent is to assist in explaining routing assumptions and requirements that are today already present for UCN environments.

3.1 The CityRoam Scenario

CityRoam stands for an example of an applicability scenario based on community services. The citizens registered in the system have the right incentive [4, 39, 40] to forward data to other registered users via *Wireless Local Area Networks (WLANs)*. The users may also agree to share their subscribed Internet connection (mobile, fixed, or wireless) [34]. Here, UCNs exhibit the usual spontaneous growth based on the idea that the dissemination of information is expected to improve the citizen's daily life. For instance, by means of such spontaneous settings, citizens can get local information concerning e.g. traffic or up-to-date situations. They can also exchange messages independently of their location and terminal. This is therefore a context-based applicability scenario where there is a backbone provided by the city eventually in agreement with several operators. Then, User Equipment (UE) shares services based on specific policies and local trust/security management, thus giving users the opportunity to connect even in gray areas. Consequently, CityRoam is partially based upon end-users willingness to forward data and share their Internet access and results in increased coverage (capillarity) from the provider perspective.

Metropolitan Area: City + surrounding villages

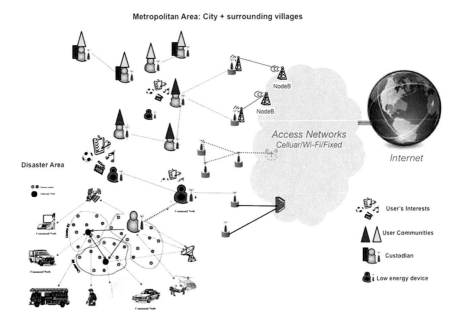

Fig. 1 High-level perspective on potential UCN infrastructures.

Description Maria is a professional in a highly demanding job. As her day is packed with meetings starting early in the morning, she would like to make sure she has all the necessary items with her before leaving her house. This includes her mobile, keys, glasses, wallet, and electronic company identification card. These devices exchange information, creating an opportunistic network. When leaving her apartment door, the door key initiates a check to see that she has everything. She has left her ID card on the counter, which causes an alarm, sending a notification to her mobile. Maria leaves home very early to go to the airport for a business trip. While walking to the bus stop the street lights shift on in a coordinated manner when she is passing by, due to a system of moving detection sensors, allowing her to walk safely to the bus stop while saving energy. At the bus station, Maria has access to the first news of the day by a city service that distribute news to bus stops. News are mostly about local activities in Maria' s neighbor (provided by citizens). Maria' s device interacts with the bus stop allowing the upload of news of interest to Maria. When entering the bus, Maria keeps reading the news previously downloaded, while some of them are forwarded to the device of another passenger with similar interests. As soon as Maria returns from her trip her mobile device offloads to the residential WLAN. Her phone attempts to locate her suitcase, also UCN enabled, but alas the suitcase last reported position was being loaded onto a flight to another city. Since the phone is aware of the description of the suitcase, it automatically notifies the lost luggage office of the airline with the description and a photo. The lost luggage office acknowledges receipt of the report, and thus avoiding a fruitless wait for her luggage to appear on the belt, Maria leaves to find the taxi from the company's that she always use.

Not having had a chance to buy groceries, her refrigerator, which is aware of her schedule and hence her return, knowing her meal preferences, orders the necessary grocery items and schedules the delivery for that evening when she will be home. A notification of this transaction is also sent to her mobile with an option of modifying the delivery time. Meanwhile, her luggage has arrived and based on the same schedule, her mobile arranges for the delivery of that the same evening. Following a busy day of meetings, she returns home in good time for the delivery of her groceries, and luggage. After dinner, while relaxing watching TV, her device, which is registered to the local UCN, receives a notification that some friends have decided to go to a movie in the local cinema. The device notifies the TV, which displays the information. By clicking OK in her remote, she notifies the others that she will join them.

3.2 Emergency Network

A second applicability scenario concerning UCNs and where routing is required relates with critical networking infrastructures. In such situations it is assumed that most of the available communication infrastructure in damage, which makes is difficult to setup communication in a reasonable and useful time-frame. UCN can, in this situation, assist devices in self-organizing to quickly establish a local infrastructure across UEs; wireless APs; vehicles equipped with Internet access (e.g. rescue teams). In such scenario, the plain application of current multihop routing approaches may not be enough to route data with some certainty due to the unpredictable availability of communicating devices. For instance, devices available may not be enough to sustain communication. Hence, in such situation, routing solutions need to take advantage of any transmission opportunity (*opportunistic communication*).

Description John is on a business trip to Lisbon, carrying a UCN enabled device which integrates a routing solution that allows the establishment of opportunistic communications, to route information on the fly. While driving from Lisbon to Sintra, a minor earthquake occurs near the coast and isolated fire situations spread out of control turning into a major wildfire and a major communication infrastructure is damaged. Devices of people running out of the affected areas collect sensory information (e.g. temperature) and sonic information (e.g. number and amplitude of human voices and wireless communications). This information is sent outside the affected area by any means available (e.g. undamaged cellular links, moving vehicles). Analysis of mobility patterns and available transmission opportunities gives indication about the best devices to consider in the communication process. The reception of spread information impacts the moving direction of rescue teams. While moving, these teams send rescue information towards each affected area, in order to allow people on the field to start providing first help. This dissemination is based on nodes that are moving as fast as possible towards the affected areas, and have good battery conditions. Social data provides information about spontaneously created helping groups, which can provide a more efficient dissemination of information.

3.3 User-Centric Routing: Assumptions and Requirements

Based on the two examples provided in Sect. 2, this section provides a set of assumptions and requirements for routing solutions to consider in UCN.

The primary focus of routing in the context of UCNs—*User-centric Routing (UCR)*—is to be capable of transmitting information in environments where the communication upstream (from user to the backbone network) becomes as relevant, if not more relevant, than the communication downstream (from the network to the user). UCN equipment is based on current networking technology (routers and switches) as well as on UE (relayers). Moreover, UCN equipment operation is highly dependent on social aspects, as the citizen controls part of that equipment. The operation relates with the citizen environment (e.g. urban or rural landscapes); social communities; roaming habits of the citizen.

Routing in UCNs requires networked devices to have embedded functionality that can make use of several environmental and communication interfaces, sustaining communications among an unlimited number of devices that are able to collect and process information without a constant human intervention. This is a reasonable assumption, as today any end-user device, mobile or not, integrates such features.

Assumptions In terms of communication capabilities, many of today's networks able to transmit environmental information (wireless sensor networks) are evolving toward a protocol-translation gateway model, similar to what happened before with computer networks. However, protocol gateways are inherently complex to design, manage, and deploy. The network fragmentation leads to non-efficient networks because of inconsistent routing, for instance. Moreover, the Internet today is more and more IP-based, being TCP/IP widely accepted as a flexible alternative to design scalable and efficient networks involving large numbers of communicating devices, as advocated by the *IP for Smart Objects (IPSO) alliance*, and as suggested by the action Plan for the deployment of the *Internet Protocol version 6 (IPv6)* in Europe [12].

A key assumption for routing in UCNs is that wireless devices are IP enabled, independently of their size and battery capacity. This is a key starting point although there is not a strong requirement for UCN technology to be based on IP.

We summarize the major assumptions to be observed when devising or adjusting routing solutions to UCNs as:

- UCN nodes must incorporate solutions (software or hardware based) that allow then to collect environmental and/or contextual data.
- Some devices are UCN gateways, i.e., they are capable of providing Internet access and/or routing data to controllers that can provide Internet access.
- UCN nodes may or may not have intermittent wireless connectivity.

- UCN nodes may or may not be limited in terms of battery, storage, as well as bandwidth capacity.
- Users involved in UCNs are willing to forward data to other users, passively, or actively.
- Some UCN nodes can relay data to others.

Requirements Routing in UCN aims to extend communication across the Internet assuming a humongous number of mobile devices as well as assuming that there is a relevant portion of data generated by and focused on the individual citizen.

A continuous exchange of information across such infrastructure, that has a high topological variability due to the self-organizing properties of UCNs, requires a paradigm shift in the way that routing is devised. Relevant aspects to consider relate with the need to assist data to be forwarded based on the interests expressed by objects and not instead by the reachability of a specific object, as occurs today.

Destination reachability based solutions, based on globally routable identifiers, as used today in the Internet, limits any effort to devise UCNs. Hence, one of the technological roles of routing when applied to UCNs should be to devise an information-centric routing framework that considers information based on: the user's roaming patterns (*mobility awareness*); the limited battery capacity of nodes (*energy awareness*); some aspects of social behavior, such as shared interests and the opportunities to disseminate such interests (*opportunistic data transmission awareness*).

These aspects can be summarized as the following requirements:

- Routing must be able to distribute information based on interests manifested by nodes and taking into consideration context-awareness.
- Routing must interface with or encompass some system to identify/track information blocks rather than objects.
- Routing must be able to forward information even in the presence of intermittent connectivity.
- Routing should support secure and private communications, whenever required.
- Routing performance must be ensured, independently of the number of nodes.
- Routing must be based on connectivity among peer objects, avoiding dependencies upon network devices with specific roles, in particular, dependencies on centralized gatekeepers.
- Routing must be aware of the intrinsic characteristics of nodes, such as battery, storage capacity, environmental capture capacities, processing power, and connectivity degree, in order to achieve an efficient control of available communication resources.

4 User-Centric Routing

Multihop routing approaches are usually considered in any wireless scenarios. In UCNs, as explained these approaches fall short due to the variable topological UCN behavior, where formed routes will be subject to more frequent breaking due to the fact that nodes in the network are now part of the *Customer Premises (CP)* as well as limited in terms of energy capacity. A routing protocol or framework will provide or compute a route when there is need to transmit data for nodes that are not in each other transmission range. Also, when a route incurs a break on one of the links on the route, the routing protocol will recompute an alternative path for data transmission to commence. It is imperative that a discovered route be a stable as frequent path re-computations weigh down the performance of routing in terms of control overhead, increased delay, lowered throughput and in some cases packet loss will also occur.

This section gives insight into how to make current multihop approaches more adaptable to UCNs, by going over the three dimensions mentioned in Sect. 3, namely: mobility awareness; energy awareness; opportunistic transmission awareness.

4.1 Mobility Awareness

Node movement and its impact on the network operation is often left to be taken care of by mobility management solutions (control plane). While such solutions assist in handing over data sessions, on the network layer the routing process will always experience link breaks independently of being temporary or permanent. In other words, current mobility management solutions assist in making applications agnostic to node movement up to some extent. However, the underlying layers experience such impact which will then have repercussion in the network performance.

Concerning routing, a potential way to overcome such impact is to investigate mobility metrics that assist routing in becoming more sensitive (more adaptive) to node movement patterns.

Prior work [6] has addressed potential mobility tracking parameters that can be used to derive adaptive routing metrics. Some of such mobility tracking parameters are *pause time, link duration* and *average number of link breaks.* From the mobility parameters that were reviewed (e.g. node degree, number of link breaks, link duration), link duration is a parameter that possesses some ability to capture properties that may assist in distinguishing between permanent and temporal link breaks. Another category of parameters that we have researched demonstrating some relevancy in terms of mobility awareness relates with the definition of the movement relation of a node towards its neighborhood, aspect which we identify as the *node's neighborhood mobility correlation* [7, 8].

Link Time Stability *Link duration (LD)* is a parameter that is tightly related to the movement of nodes; it is also, as of today, one of the parameters that is most popular in terms of tracking node mobility. By definition, *link duration is associated to the period of time where two nodes are within the transmission range of each other.* In other words, it is the time period that starts when two nodes move to the transmission range of each other, and that ends when the signal strength perceived by the receiver node goes under a specific threshold. Some authors then provide a variation of this definition by working the threshold value.

In order to assist in developing a cost associated to link stability, we have considered two different metrics associated to the notion of link break and duration, and to the relation of these two elements.

The first embodiment of link stability for link l, $s1_l$, comprises the ratio between the time a link is down (link break duration), lb, and the link lifetime lf for the duration that elapses between two consecutive breaks, as expressed in equation 1:

$$s1_l = \frac{lb}{lb + lf} \qquad (1)$$

Such ratio gives a measure of stability in the sense that the more prone is a link to break, the lesser is its stability. It is a simple metric which should assist in prioritizing links over time, and in choosing the ones that have a lower $s1_l$. The ratio will avoid short-lived links, since the duration in which the link is in broken state (lb) will be large. As nodes move, new links are formed and others are broken, meaning that link stability can change with time. A good link metric is one that captures the change in link stability. A link break means that there is a change in link stability. In our metric, link cost depends on the time the link has been down: links that incur long breaks will not participate in routing in the presence of links that are stable. Link stability depends on the time the link has been down and up. Implicitly, the metric captures nodes that are in group mobility. It can differentiate links that are formed between two mobile groups whose propagation path differ. It can also capture stable nodes that are static.

In a second embodiment of link stability based on LD, we introduce an additional parameter: the *number of link breaks*, nb_l. We refer to this embodiment as $s2_l$, provided by Eq. 2:

$$s2_l = \frac{lb * nb_l}{lb + lf} \qquad (2)$$

where $s2_l$ takes into consideration the time period that a link is active, and also the number of breaks incurred with respect to a specific time-window. In comparison to $s1_l$, $s2_l$ not only considers the percentage of time a link is active, but also the frequency of breaks during that period.

To provide a concrete example, let us consider two links i and j, with the same duration: $lf_i = lf_j = 10$ seconds and also with the same link break duration, $lb_i = lb_j = 2$ seconds. However, while in i such inactive time is derived from one single, longer break, for link j that has been the product of 2 link breaks.

The routing metrics were implemented in AODV where it was found out that that our metrics, and in particular the s2 metric assists AODV in a better path selection - paths that are more steady, thus reducing the need to have path recomputation. Observed was that mobility-aware routing metrics actually improved routing performance, in both protocols, although higher improvements were noted in AODV compared to OLSR.

Node to Neighborhood Mobility Correlation A node to neighborhood mobility correlation relates with the stability that a node can provide in a path and which is dependent not only of the node's individual mobility pattern, as well as of the neighbors individual patterns towards the node—the group mobility pattern. A routing solution must therefore take into consideration not only the own perspective of a node concerning movement, but also the group relation. To capture this neighborhood movement variability, we have address the notion of node degree stability as well as the notion of link breaks in the context of group movement. The relation between the node degree at an instant in time as well as the integration of new nodes in the neighborhood are parameters, that when combined, assist in understanding the mobility variability surrounding a node.

4.2 Energy-Efficiency Awareness

Multi-hop routing has been extensively analyzed and optimized in terms of resource management, but in terms of energy efficiency there is a lack of a thorough analysis in wireless environments. On the other hand, there is considerable related work in the fields of energy efficiency and energy awareness for sensor networks. Even though it is relevant to consider the results achieved in such networks, there are specific requirements of UCNs which make energy awareness and efficiency problems that are not trivial to be solved. Firstly, UCN nodes are heterogeneous in terms of resources such as battery capacity. Secondly, such nodes exhibit frequent movement and are also expected to frequently join and leave a network.

Routing in UCNs can be adjusted to be energy-aware. In our work we have considered organic ways to make multi-hop routing more flexible, namely, the inclusion of energy-aware routing metrics in current multihop solutions.

Energy Awareness Routing Notions A UCN node is expected to be interconnected via one or more networking interfaces. In the context of energy-awareness one can consider the edges that interconnect nodes to have an *energy-efficiency cost* which is a measure of energy expenditure of the nodes involved in the connection. Such cost can be computed based on the perspective of the source node—the node transmitting—, or based on the perspective of both nodes involved in a potential transmission—the *source/father* node, and the potential *successor* node.

Concerning the source node perspective, there are three main modes of energy expenditure. A node is in *Transmit mode* when transmitting information. Hence, *Transmit Power (Tx Power)* for a node corresponds to the amount of energy (in Watts) spent when the node transmits a unit (bit) of information. A node is in *Receive mode* if it is receiving data. Hence, *Reception Power (Rx Power)* for a node corresponds to the amount of energy (in Watts) spent when the node receives a unit (bit) of information. Particularly for the case of 802.11, there are two additional states a node may be at. When not receiving or transmitting, the node is still listening to the shared medium (overhearing) and is said to be in *Idle mode*. When the node is not overhearing, then it is said to be in *Sleep mode*. In this mode, no communication is possible but there is still a low-power consumption.

The way a node spends energy is based on an energy consumption model, which dictates how much energy (how many units) are spent for each mode per unit of data (transmitted, received, overheard). Then, different node metrics can capture such energy spending or savings, and thus can make a node energy aware up to some point. Feeney et al., for instance, provide a general model [13] for packet-based energy consumption, i.e., energy spent by a node when it sends, receives, or discards a packet.

Energy-aware routing metrics are normally associated to the perspective of a node and hence are known as *energy-aware node metrics*. The main energy-aware node metrics are (i) Transmission Power [9], (ii) *Residual Energy (RE)* [38], and (iii) *Drain Rate (DR)* [17]. These metrics are normally used to the problem of maximum lifetime routing, i.e., increasing the network lifetime. The transmission power metric aims at maximizing the network lifetime by minimizing the total energy consumption per packet. The residual energy metric goal is to extend the network lifetime by extending node lifetime and balancing the energy consumption per node. The drain rate metric aims at maximizing the network lifetime by predicting the node lifetime. The transmission power is commonly applied as a link cost (even though it is a metric that provides only a node perspective) in shortest-path computation. The residual energy and drain rate metrics are normally considered to be applied in min-max algorithms, which explicitly avoids the minimum energy problem by selecting the route that maximizes the minimum residual energy of any node on the route. Routes selected using min-max algorithms may be longer or have greater total energy consumption than the minimum energy route. This increases per packet energy consumption, but it generally performs better than minimum energy routing.

Out of the metrics mentioned, the most relevant to consider in the context of routing applied to UCNs are the drain rate and residual energy metrics, as explained in previous work [25, 26]. Still, in the context of UCN such metrics cannot be applied in isolation to provide energy awareness, in the context of multihop proposals. We would also like to emphasize that the *Internet Engineering Task Force (IETF)* is currently discussing energy-aware multihop metrics tailored for energy efficiency for routing protocols in the context of the working group ROLL.

Novel Metrics for Energy-Awareness Based on the notion that in UCNs nodes are heterogeneous in terms of energy capacity we have discussed and validated several metrics, summarizing the most relevant ones in this section. Initially, we have proposed heuristics which consider an energy-awareness ranking of node based on idle times, which a node provides a ranking in terms of the node robustness to optimize the node lifetime as well as the global network lifetime. Then, a second heuristic we have considered the impact of node degree history for ranking the node to extend the lifetime.

The *Energy-awareness Node Ranking (ENR)* metric [26–28] explores the fact that nodes that have been in idle mode for the majority of their lifetime, and that still exhibit a good estimate for their future energy level are the most adequate candidates to constitute a shortest-path. In ENR we estimate how much of its lifetime has node

i been in idle mode, to then provide an estimate towards the node's future energy expenditure, as this will for sure impact the node's lifetime. Such periods are the ones that are expensive to i in terms of energy. Hence, we consider the total period in idle time, t_{idle} over the full lifetime expected for a specific node, which is given by the sum of the elapsed time period T with the estimated lifetime of the node, as provided in Eq. 3. The estimated lifetime $C(i)$ was first provided by Garcia-Luna-Aceves et. al [17].

$$ENR(i) = \frac{T - t_{idle}}{T \times C(i)} \tag{3}$$

ENR is therefore a node weight which provides a ranking in terms of the node robustness, from an energy perspective, and having as goal to optimize the network lifetime. The smaller $ENR(i)$ is, the more likelihood a node has to be part of a path.

Based on ENR, we have developed the *Energy-awareness Father-Son (EFS)* metric, which considers a composition of the ENRs of both a father and successor nodes as specified in Eq. 4 [30].

$$EFS(i, j) = ENR(i) \times ENR(j) \tag{4}$$

EFS provides a ranking which we believe is useful to assist the routing algorithm to converge quickly in particular in multi-path environments, as the selection on which successor to consider shall be made up from, by the father node. The goal is, similarly to ENR, to improve the network lifetime without disrupting the overall network operation. Hence, the smaller $EFS(i, j)$ is, the more likelihood a link has to become part of a path.

These two sets of metrics and variations have been validated via discrete event simulations in the context of both AODV and OLSR. Operationally, these two protocols have a very different behavior, and applying global metrics to them independently of the protocol behavior is not trivial. However, from an energy-aware perspective, it is possible to do so, by considering that both families rely on shortest-path computation.

Hence, the line of thought considered in the development of our energy-aware metrics is that the principle of shortest-path computation must be kept. Instead of hop-count, a metric that can provide an energy expenditure cost to a node is considered. The main caveat related with this change is that in order to keep accuracy, one must ensure that the protocol synchronizes path status adequately. This implies considering either a time-window mechanism, or updates to a node's cost each time a change occurs. These are regular techniques, where it is essential to find an adequate commitment between accuracy and low overhead due to the required signaling.

In terms of the behavior of EFS vs. ENR there seems to be an improvement in particular when scenarios have larger distances, and when the network load is higher [30]. This implies that EFS seems to provide more robustness when scenarios have more variability (e.g. more nodes moving, and several successors at disposal). In terms of network lifetime and for the scenarios evaluated, ENR results in a small improvement. The greater advantage of applying EFS instead of ENR seems to relate with an improvement in throughput and a significant improvement concerning packet loss. Our belief for this gain relates to the fact that EFS allows nodes to react quicker to energy changes on a path—resulting paths will be more robust earlier in time, assuming that nodes have a reasonable out-degree (several successors available).

Thinking on real implementation, we have discussing the impact of energy awareness and operational aspects of the link-state and distance-vector routing families. Then, we have described and discussed the routing architecture specification for energy awareness submitted to the IETF working group ROLL [29]. The specification can be applied in any available multihop routing protocols, such as AODV, OLSR and RPL.

4.3 Opportunistic Communication Awareness

Related work concerning opportunistic routing aim to investigate the use of node contact metrics (e.g., frequency of encounters), resulting from node mobility, to reach a good trade-off between cost and rate of message delivery. These proposals started by investigating schemes based on which nodes send unique copies of messages (replicate once) until destination is found (e.g., single copy forwarding) aiming to reduce transmission costs (i.e., number of message replicas in the network) and schemes based on which nodes keep replicating messages to any encountered node (e.g., Epidemic) aiming to increase delivery probability and to reduce delay.

In a second stage, several proposals started to investigate methods to mitigate the cost of replication mechanisms, aiming to achieve the delivery probability and delay of epidemic approaches with the low network cost of single-copy ones. These replication-based approaches tried to exploit more elaborated networking aspects such as node encounter, resource usage, and social similarity, which is the latest trend identified in the last couple of years.

Then, we have further investigate the different opportunistic routing solutions, with particular emphasis on the social-aware approaches [20]. Another important aspect that we could observe while covering the state-of-the-art in opportunistic routing is related to the way proposals are evaluated: there no guidelines with respect how such proposals should be compared in order to provide a fair performance assessment. This led us to come up with a Universal Evaluation Framework [21, 22], which aids networks to evaluate new opportunistic routing solutions to already existing ones in way to have their assessment done respecting the limitation of each of the proposals involved.

In what concerns solutions based on social similarities, it is important to achieve a correct mapping between real node interaction and the social graph that aids routing. Hossmann et al. [15] show that the key for successful forwarding is related to the ability of mapping social interaction (resulting from the mobility process) into a clean social representation (i.e., that best reflects the mobility structure), which should capture the daily life routine of nodes. Eagle and Pentland [11] show that people have routines that can be used to identify future behavior as well as interaction with people with whom they share similar behavior and potentially the same community. In what concerns the latest argument, the identification of social structures encompasses the challenge of detecting and adjusting communities on-the-fly in a useful time frame. Current research efforts show the difficulty of constructing and adjusting social structures in short periods of time.

In UCNs, the user daily routine is an essential aspect and hence, our choice concerning opportunistic communication awareness falls the social approaches.

In this context, in previous work we have proposed two utility functions [19] that take into consideration the daily routines of users and the intensity of their social interactions to take forwarding decisions: the *Time-Evolving Contact Duration (TECD)* that weights social interactions among nodes considering the duration of contacts; and *TECD Importance (TECDi)* which estimates the importance of nodes according to the weight of the links to its neighbors and their importance. Experiments carried out to evaluate the two utility functions showed that routine has a positive effect on opportunistic routing, when compared against contact and social-based benchmarks.

This work evolved to become a new routing algorithm, dLife [23] which captures the dynamics of the network represented by time-evolving social ties between pair of nodes.

Another part of this work refers to point-to-multipoint communication, which is a desirable feature in opportunistic routing since it increases reachability of nodes interested in the content of the messages. Such feature has been shown to lead opportunistic routing to have better performance and wise use of resources [1].

dLife has been implemented in the context of the ONE simulator, as well as in the context of a realistic Delay Tolerant Networking testbed set in the Amazon region[1] holding 10 devices (3 personal computers with Ubuntu 10.10 Maverick, 3 smartphones Android 2.3.6 Gingerbread, 4 wireless routers with OpenWrt 10.03.1) with the purpose to exploit physical proximity, a key aspect of dLife to determine different levels of social interaction among devices.

5 Summary and Future Work

This paper addresses assumptions and requirements of routing in the context of UCNs, alerting to the need to revisit multihop approaches making them aware to node movement, to the limited battery capacity of the nodes, as well as to the potential value of opportunistic communications.

The line of thought followed is that in order to allow end-to-end routing on the Internet to adequate work in future Internet architectures, it is required to

---

[1] A DTN testbed was created in collaboration between COPELABS and the University Federal of Pará (UFPA) in order to test the performance of dLife in the extreme conditions of the Amazon region, in the context of the joint project UCR: User-centric Routing, 2010–2013, project funded by Fundação para a Ciência e Tecnologia (FCT).

consider new metrics that incorporate properties that allow nodes to choose other-than-shortest-path solutions, while at the same time ensuring backward compatibility with the current solutions.

The work described has addressed ways to integrate metrics capable of providing such sensitivity to current shortest-path based approaches showing that it is feasible to improve the network lifetime as well as to reduce path recomputation, by integrating simple metrics into existing approaches, derived from parameters that are, in their majority, passively captured by nodes. In other words, these metrics do not require active probing of the network.

A conclusion to draw is that the metrics proposed have shown good improvements of different network properties (e.g. lifetime, path recomputation). Such improvements were possible without adding any significant overhead be it to a node or network operation, as described in the different IETF specifications proposed.

Research work in this field should address ways to combine the three dimensions mentioned, as well as to validate the metrics proposed in the context of other multihop solutions.

Acknowledgments We thank all of the elements involved in the following projects: UCR, User-centric Routing (http://copelabs.ulusofona.pt/~ucr, 2010-2013), which involved COPELABS and University of Coimbra, Portugal; DTN-Amazon (http://copelabs.ulusofona.pt/~dtnamazon, 2012-2013), which involved COPELABS and Universidade Federal do Pará, Brazil.

References

1. Ammar MW, Zegura E (2005) Multicasting in delay tolerant networks: semantic models and routing algorithms. In: Proceedings of the 2005 ACM SIGCOMM workshop on delay-tolerant networking, WDTN ' 05, ACM, New York, pp 268–275
2. Balasubramanian A, Levine BN , Venkataramani A (2007) DTN routing as a resource allocation problem. In: Proceedings of ACM SIGCOMM 2007
3. Biswas S, Morris R (2005) ExOR: opportunistic multi-hop routing for wireless networks. In: Proceedings of ACM SIGCOMM 2005
4. Buchegger S, Boudec J (2002) Performance analysis of the confidant protocol(cooperation of nodes: fairness in dynamic ad-hoc networks). In: Proceedings of MobiHoc 2002
5. Burgess J, Gallagher B , Jensen D , Levine BN (2006) MaxProp: routing for vehicle-based disruption-tolerant networks. In: Proceedings of IEEE Infocom 2006
6. Chama N, Sofia R (2011) A discussion on developing multihop routing metrics sensitive to node mobility. J Commun 6(1):56–57
7. Chama N, Sofia R (2010) Redifining link duration: making routing sensitive to mobility. In: IEEE GLOBECOM SWN 2010 workshop
8. Chama N, Sofia R, Sargento S (2011) Impact of mobility on user-centric Routing. In: Proceedings of IEEE ICNP, PhD Forum
9. Chang J-H , Tassiulas L (2000) Energy conserving routing in wireless ad-hoc networks. In: Proceedings of INFOCOM 2000, 19th annual joint conference of the IEEE computer and communications societies, IEEE, vol 1, pp 22–31
10. Clausen T, Jacquet P (2003) Optimized link state routing protocol (OLSR). In: IETF RFC 3626
11. Eagle N, Pentland A (2009) Eigenbehaviors: identifying structure in routine. Behav Ecol Sociobiol, vol 63, pp 1057–1066

12. European Commission (2008) Action plan for the deployment of Internet Protocol version 6 (IPv6) in Europe, Brussels, May 2008
13. Feeney L, Nilsson M (2001) Investigating the energy consumption of a wireless network interface in an ad hoc networking environment. In Proceedings of INFOCOM 2001, annual joint conference of the IEEE computer and communications societies, IEEE, vol 3, pp 1548–1557
14. Ferriere HD, Grossglauser M, Vetterli M (2003) Age matters: efficient route discovery in mobile ad hoc networks using encounter ages.In: ACM MobiHoc
15. Hossmann T, Spyropoulos T, Legendre F(2010) Know thy neighbor: towards optimal mapping of contacts to social graphs for dtn routing. In: Proceedings of IEEE INFOCOM, San Diego, March 2010
16. Karp B, Kung HT (2000) GPSR: greedy perimeter stateless routing for wireless networks. In: Proceeding of ACM MobiCom, pp 243–254, August 2000
17. Kim D, Garcia-Luna-Aceves JJ, Obraczka K, Cano J-C, Manzoni P (2003) Routing mechanisms for mobile ad hoc networks based on the energy drain rate. IEEE Trans Mob Comput 2(2):161–173
18. Lindgren A, Doria A, Schelen O (2004) Probabilistic routing in intermittently connected networks. In: LNCS, vol 3126. Springer, Heidelberg, pp. 239–254
19. Moreira W, de Souza M, Mendes P, Sargento S (2012) Study on the effect of network dynamics on opportunistic routing. In: Proceedings of ADHOC-NOW, Belgrade
20. Moreira W, Mendes P (2013) Social-aware opportunistic routing: the new trend. In: Woungang I, Dhurandher S, Anpalagan A, Vasilakos AV (eds) Routing in opportunistic networks, Springer, Heidelberg
21. Moreira W, Mendes P, Sargento S (2011) Assessment model for opportunistic routing. In: Proceedings of IEEE LatinCom, Belém, October 2011
22. Moreira W, Mendes P, Sargento S (2012) Assessment model for opportunistic Routing. IEEE Lat Am Trans 10(3):1785–1790
23. Moreira W, Mendes P, Ferreira R, Cirqueira D, Cerqueira E (2013) Opportunistic routing based on users daily life routine. In: IETF draft draft-moreira-dlife-02, working in progress, April 2013
24. Newman MEJ (March 2003) The structure and function of complex networks. SIAM Review 45:167–256
25. Oliveira A Jr, Sofia R (2010) Energy-efficient routing in user-centric environments. In: IEEE/ACM international conference on green computing and communications
26. Oliveira A Jr, Sofia R (2011) Costa A Energy-efficient routing. In: Proceedings of IEEE ICNP
27. Oliveira A Jr, Sofia R, Costa A (2012a) Energy-efficient heuristics for multihop routing in user-centric environments. In: 12th international conference on next generation wired/wireless advanced networking (NEW2AN 2012), St. Petersburg
28. Oliveira A Jr, Sofia R, Costa A (2012b) Energy-awareness metrics for multihop wireless user-centric routing. In: Proceedings of ICWN, USA, July 2012
29. Oliveira A Jr, Sofia R, Costa A (2012c) Energy-awareness metrics global applicability guidelines. In: draft-ajunior-energy-awareness-00, IETF individual draft (working group proposal ROLL)
30. Oliveira Jr A, Sofia R, Costa A (2012d) Energy-awareness in Multihop Routing. In: 2012 IFIP wireless days conference (WD'12), Dublin
31. Panagakis A, Vaios A, Stavrakakis I (2007) On the effects of cooperation in DTNs. In: ICST Comsware 2007
32. Perkins CE, Belding-Royer EM, Das S (2003) Ad hoc On demand distance vector (AODV) routing. In: IETF RFC 356
33. Sofia R, Mendes P (2008) User-provided networks: consumer as provider. In: IEEE communication magazines feature topic on consumer communications and networking—gaming and entertainment
34. Sofia R, Mendes P, Panwar S, Kennedy D, Vallejo J, Trossen D (2009) User-provided networking: challenges and opportunities. In: Proceedings of 1st ACM Workshop on user-provided networking: challenges and opportunities

35. Spyropoulos T, Psounis K, Raghavendra CS (2005) Spray and wait: an efficient routing scheme for intermittently connected mobile networks. In: Proceedings of the ACM SIGCOMM workshop on delay-tolerant networking (WDTN" 05), pp 252–259, August 2005
36. The Haggle Project. Code available at http://code.google.com/p/haggle/
37. Vahdat A, Becker D (2000) Epidemic routing for partially-connected ad hoc networks. In: Technical reports CS-2000-06, Duke University, July 2000
38. Xie Q, Lea C-T, Golin M, Fleischer R (2003) Maximum residual energy routing with reverse energy cost. In: Global telecommunications conference 2003, GLOBECOM" 03, IEEE, vol 1, pp 564-569
39. Zhong S, Yang Y, Chen J (2003) Sprite: a simple, cheat-proof, credit-based system for mobile ad-hoc networks. In: INFOCOM 2003
40. Zhu H, Lin X, Lu R, Shen X (2008) A secure incentive scheme for delay tolerant networks. In: Chinacom 2008

Part II
Trust Management and Cooperation Incentives

Trust Management Support for Context-Aware Service Platforms

Ricardo Neisse, Maarten Wegdam and Marten van Sinderen

Abstract High quality context information retrieved from trustworthy context providers allows a more reliable context-aware service adaption but also implies a higher risk for the service users in case of privacy violations. In this chapter we present a trust management model that support users and providers of context-aware services in managing the trade-off between privacy protection and context-based service adaptation. We applied our trust management model in two trust-based selection mechanisms. The first trust-based selection mechanism support users of context-aware services in selecting trustworthy service providers to interact with. This mechanism supports the users in the selection process, taking into account the users' goals, trust beliefs, and the trust dependencies between the service users and the entities that collaborate in the context-aware service provisioning. The second trust-based selection mechanism supports context-aware service providers in selecting trustworthy context providers taking into account the trustworthiness of the context providers to provide context information about a specific user and quality level. To conclude this chapter we present the evaluation of the technical feasibility of our trust management model in trust-based selection mechanisms through a prototype implementation.

Keywords Trust management · Context-aware · Subjective logic

R. Neisse (✉)
European Commission Joint Research Center (JRC), Ispra, Italy
e-mail: ricardo.neisse@jrc.ec.europa.eu

M. Wegdam
InnoValor, Enschede, The Netherlands
e-mail: maarten.wegdam@innovalor.nl

M. van Sinderen
Centre for Telematics and Information Technology (CTIT), University of Twente,
Enschede, The Netherlands
e-mail: m.j.vansinderen@utwente.nl

A. Aldini and A. Bogliolo (eds.), *User-Centric Networking*,
Lecture Notes in Social Networks, DOI: 10.1007/978-3-319-05218-2_4,
© Springer International Publishing Switzerland 2014

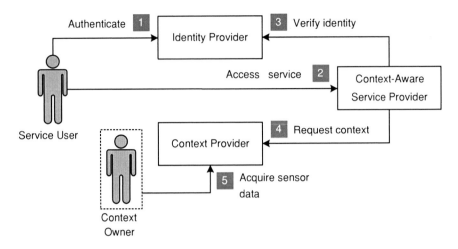

Fig. 1 Roles in a context-aware service platform

1 Trust Relationships in a Context-Aware Service Platform

Context awareness is the *ability of applications to utilize information about the user's environment (context) in order to tailor services to the user's current situation and needs* [5]. Figure 1 illustrates the five roles we distinguish in a context-aware service platform, namely *Service User, Context Owner, Identity Provider, Context Provider,* and *Service Provider.*

The arrows in Fig. 1 indicate the basic interactions between the roles when a user accesses a service provider. The box with a dotted line that surrounds the *Context Owner* represents sensors in the environment that collect context information about this entity. First, the *Service User* authenticates with the *Identity Provider* and receives an identity token (1). After the authentication is performed, the *Service User* requests access to a service provided by the *Service Provider* (2), which will verify the identity token of the user in order to grant access to the service (3). To be able to adapt the service to the relevant context, the *Service Provider* requests context information about the *Context Owner* from the *Context Provider* (4). This context information is retrieved by the *Context Provider* from sensors in the physical environment of the *Context Owner* (5) and might be, for instance, the current activity or location of the *Context Owner.*

Trust management is necessary in context-aware service platforms because users and service providers, which are expected to be pervasive and numerous, need to judge whether new, existing, and previously unknown entities are (un)trustworthy to interact with. In this chapter we focus on trust aspects related to identity provisioning, privacy enforcement, context information provisioning, and context-aware service provisioning. Figure 2 presents a summary of the trust relationships between the roles in a context-aware service platform that we want to address.

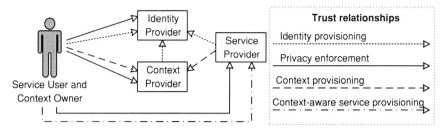

Fig. 2 Trust relationships in a context-aware service platform

In Fig. 2 the *Service User* and *Context Owner* roles are assigned to the same entity meaning that the service adapts its behavior to the context of the service user only. The *Service User* should trust the *Service Provider* to reliably provide a specific context-aware service. The *Context Owner* should trust the *Context Provider* and the *Service Provider* to handle his/her context information. The *Context Owner* expects that his/her context information is released only when his/her privacy preferences authorize the access, and (s)he can only accept his/her context information to be communicated if (s)he trusts that both the *Context Provider* and the *Service Provider* are able and willing to adhere to his/her privacy preferences.

The *Service User* and the *Service Provider* should trust the *Context Provider* to provide context information. This is important in order to assess if the context information is provided with the required quality characteristics and consequently can be used in the expected context-aware service adaptation. Trust in the *Context Provider* from the *Service Provider*'s point of view is also required in case dynamic context-based security solutions requiring trustworthy context information are in place. One example is a service provider that only authorizes access to a service if the service user is at a specific location, for example, inside an office building.

All the entities should trust the *Identity Provider* because it is responsible for the authentication and verification of credentials to allow the entities to identify themselves to other entities in the service platform.

The trust management model we describe allows the specification of trust relationships targeting the trust aspects depicted in Fig. 2. Each trust relationships focuses on a specific trust aspect depending on the functionality provided by the role. The set of trust relationships we address in Fig. 2 is by no means exhaustive. Other trust relationships targeting different aspects may be required in other service scenarios depending on the functionality provided by the roles. Our trust management model proposes a basic set of trust aspects, based on our reference context-aware service platform, and motivates the definition of trust relationships focusing on these trust aspects. Our trust management model also includes trust assessment support considering the dependencies between these trust aspects.

For each of the trust relationships presented in Fig. 2 it is possible to establish a trust value according to a certain aspect-specific metric. The following sections present a detailed analysis of the aspect-specific trust relationships we identify and shows example trust metrics from the literature for obtaining trust values related to the set of trust aspects we consider. We also refine further the trust relationships, giving more details about the trust relationships specified in Fig. 2. Our objective with this discussion is to motivate our trust management model and trust-based selection mechanisms described further in this chapter. We present an extension of the trust management solution proposed in [21] with an enhanced metamodel for the specification of trust relationships and a prototype implementation using the Eclipse Modeling Framework (EMF).

1.1 Trustworthiness of Context Providers

Figure 3 shows the interaction and trust relationships between the *Context Owner* and *Context Provider* roles. The *Context Owner* registers his/her identity with the *Context Provider* and informs the *Context Provider* of his/her privacy preferences. The *Context Provider* stores the *Context Owner*'s privacy preferences and is expected to enforce the privacy preferences when the context information about the *Context Owner* is released to third parties (authorization enforcement), or when context information is deleted, quality reduced, or anonymized [18] after a certain period of time (obligation enforcement).

We identify two types of trust relationships between the *Context Owner* and the *Context Provider*. The *Provide context at specified QoC level* trust relationship is related to the reliability and competence of the *Context Provider* to provide context information according to a Quality of Context (QoC) level. The *Enforce privacy preferences* trust relationship is related to the competence and honesty of the *Context Provider* to enforce the *Context Owner*'s privacy preferences. These two relationships are respectively a refinement of the *Context provisioning* and *Privacy enforcement* trust relationships in Fig. 3.

One existing approach [17] to evaluate the trustworthiness of context information providers takes into account the cryptographic trustworthiness of the context provider's identity. This approach is not adequate because the fact that the identity of a context provider uses trustworthy cryptography has no relation to the capabilities of this context provider with respect to the provisioning of context information. Other approaches to evaluating the trustworthiness of a context provider propose using the following metrics and mechanisms: reputation of the context provider established by a community, statistical analysis of the context information [8], and aggregation of context information from redundant context providers in order to increase the trustworthiness [9].

In order to simplify the modeling of QoC attributes we propose a simplified and concise QoC model [24] based on the existing literature that uses as a reference an existing ISO standard metrology vocabulary. Our QoC model clearly distinguishes the important quality attributes and clarifies the terminology of existing QoC models, providing a more accurate and simplified QoC vocabulary. This model directly benefits developers of context-aware service platforms because it improves the

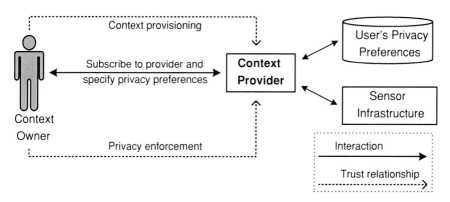

Fig. 3 Trust relationships between user and context provider

understanding of the QoC attributes. We applied our QoC model in the trust management mechanism we propose to support context-aware service providers in selecting trustworthy context providers.

In some of the existing approaches for context information management like the Context Handling Platform (CHP) [5], context-awareness is also realized using the concept of context situations and situation events. Context situations values are a composite of context information instances and the same trustworthiness evaluation approaches applicable to context information can be used. To evaluate the trustworthiness of situation detection components (a.k.a. Context Managers), trust values associated to the honesty and competence to observe and report situation events can be defined. Examples of these trust aspects are:

- Competence to observe situation events: a context manager component is capable of detecting all situation events meaning that all changes in the situation conditions are observed and there is no relevant context change that is not observed;
- Honesty to report situation events: a context manager component reports all observed situation events, is not omitting the reporting of observed events, and is also not deliberately reporting non-observed (fake) events;
- Competence to report events: the quality of context information values reported in situation events conforms to a specification.

Trust mechanisms for evaluating privacy enforcement trustworthiness take into account the existence of information handling privacy policies defined by the context provider (e.g. P3P policies) [28]. The assumption made by this approach is that the existence of these privacy policies alone already contributes positively to the trustworthiness of the context provider. Furthermore, if the privacy policies defined by the context provider match the privacy requirements of the context owner, this is not a guarantee that the privacy requirements will be met. We believe that such a guarantee can only be given if tamper proof auditing mechanisms (e.g. using TPM devices [23]) are in place to verify the enforcement of the privacy policies. Trust with respect to privacy enforcement can be also increased if manual Electronic Data

Processing (EDP) audits are conducted (contrary to technology audits) and if the stated privacy policies are bound to legal liability of the context or service providers in case privacy violations are observed.

The following metrics are proposed by Daskapan et al. [4] and Kolari et al. [28] to calculate trust values regarding privacy enforcement aspects: user interest in sharing the information, confidentiality level of the information, number of positive previous experiences with the information consumer, number of hops between a provider and a consumer, a priori probability of distrusting, and service popularity in search engines. The number of hops is related to identity certification and the chain of certificate authorities between the information source (provider) and the target of the information (consumer).

Indirect privacy enforcement trust values can also be obtained through trust recommendations received from trusted third parties specialized in privacy protection issues. Privacy protection organizations take care of privacy policies certification in the same way identities are certified today by certification authorities [27]. Privacy recommendations might be provided also by informal organizations such as virtual users' communities and consumer protection organizations.

1.2 Trustworthiness of Identity Providers

Figure 4 shows the interaction and trust relationships between the *Service User* and the *Identity Provider* roles. The *Service User* subscribes and registers his/her identity profile information with the *Identity Provider* and provides his/her privacy preferences. The *Identity Provider* delivers as a result a digital identity, which can later be verified cryptographically by anyone. The *Identity Provider* is supposed to enforce the privacy preferences when someone requests access to the identity profile information, and should release only the allowed information if any information at all.[1]

We identify two types of trust relationships between an identity provider and an identity holder. The first trust relationship is related to the provisioning of identities that are reliable in the sense that they correctly identify an entity and the entity's profile attributes. The second trust relationship is related to the competence and honesty of the identity provider in enforcing the identity holder's privacy preferences for the identity attributes.

One metric that influences trust in the identity is the authentication method used. Identity providers that use very strong authentication, e.g. using Smart Card technology, can be relied on more to securely authenticate someone than identity providers that use only username/password authentication. The user registration policy also influences the identity provisioning trust. Identity providers that allow users to freely register without verifying the identity attributes of the user (e.g. Google and Yahoo) might not be trusted as much as identity providers that do not allow registration without doing some form of identity proofing (e.g. an university or a bank). With respect

---

[1] One privacy preference might state that the identity should be completely anonymized.

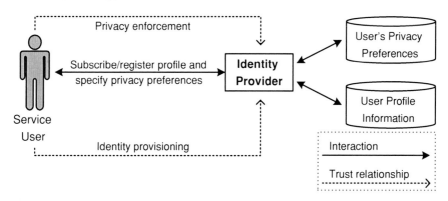

Fig. 4 Trust in identity provider

to privacy enforcement of identity attributes, similar techniques to those described for the context providers can be used.

The trustworthiness level of an identity provider is related to the levels of assurance concept proposed by the Kantara Initiative Identity Assurance Framework (IAF) [16]. The Kantara Initiative is a continuation of the Liberty Alliance Identity Assurance Framework. The idea of the IAF is to propose guidelines to enable relying parties to evaluate if a specific identity can be trusted to be owned by a particular entity.

1.3 Trustworthiness of Context-Aware Service Providers

Figure 5 shows the interactions and trust relationships between the *Service User* and the *Service Provider* roles. The *Service User* accesses the context-aware service provided by the *Service Provider* and includes in the access request his/her digital identity. The service provider verifies the service user's identity and requests context information about the *Service User* or other *Context Owners* relevant to the service being requested.

In Fig. 5, we assume that the *Service User* is also the *Context Owner* of interest for the service being provided. In some scenarios, a service invocation might not require the context information of the *Service User*. In this case, the privacy enforcement trust relationship does not apply, since the *Service User* is only concerned with the reliability of the context-aware adaptation. One example of this scenario could be a *Friend Finder* service that does not require the location information of the friend, i.e. *Service User*, using the service. Only the location information of the target friend being located is relevant. However, users of a friend radar service will also be targets of their friends and we expect that subscribers of context-aware services will, in most cases, also provide their context information for the service they will be using.

We identify two types of trust relationships between context owners, service users, and service providers. The first trust relationship is related to the enforcement of the

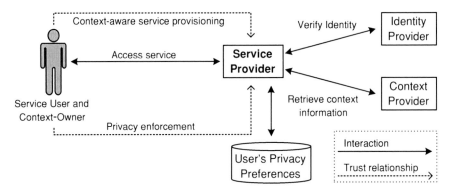

Fig. 5 Trust in service provider

privacy preferences of the *Context Owner* with respect to the context information that is being accessed by the *Service Provider*. Similar techniques to those described to evaluate the trustworthiness of context providers with respect to privacy enforcement can be used to assess trust values related to this trust relationship.

The second trust relationship is related to the competence of the *Service Provider* to reliably adapt to the context information. If the *Service Provider* is not competent, the result can be that the service provided is of less value. One example is a context-aware service that provides poor personalized tourist advice for a user in the sense that the advice is not correctly customized for the user's location.

With respect to the adaptation of the context-aware service, it is possible that the incorrect context-based adaptation is not the fault of the *Service Provider*. The *Service Provider* might be faulty due to untrustworthy context information retrieved from the context providers. Therefore, the context-aware service provider depends on his own competence to provide context-based adaptation and also depends on the reliability and competence of the *Context Provider* to provide context information about the context owners.

We assume that the service provider is always competent, and that an unsuccessful context-based adaptation is related to context information provided by untrustworthy context providers. We are aware that in reality this assumption cannot always be true: it is possible that a service provider does not reliably adapt despite receiving trustworthy context information at the required QoC level. This situation could be detected by analyzing the service implementation and the context information instances retrieved, and we consider it outside the scope of this chapter.

1.4 Trust Aspects and Dependencies Summary

Tables 1 and 2 summarize the roles, trust aspects, and trust dependencies. We do not present trust dependencies in the *Service User* and in the *Context Owner* roles because

Table 1 Summary of trust aspects for each role

Role → Trust aspect ↓	Identity provider	Context provider	Service provider
Identity provisioning (IDP)	●	–	–
Privacy enforcement (PE)	●	●	●
Context provisioning (CIP)	–	●	–
Context-aware adaptation (CA)	–	–	●

Table 2 Summary of trust dependencies

Depends on → Role ↓	Identity provider	Context provider	Service provider
Identity provider	IDP	–	–
Context provider	IDP	–	–
Service provider	IDP	CIP	–
Service user	IDP	CIP	CA
Context owner	IDP/PE	PE	PE

our initial set of trust aspects and mechanisms does not address any dependency on these roles. We focus on the trust aspects related to the trade-off between privacy and context-based service adaptation. An example of a dependency and trust aspect that could be considered is the dependency of the *Service Provider* on the *Service User* in the trust aspect of *paying for the service costs*.

In Table 1, the trust aspect of *identity provisioning* applies to the *Identity Provider* role. The *privacy enforcement* trust aspect applies to the *Identity Provider*, *Context Provider* and *Service Provider* roles, which manipulate the *Context Owner*'s context and identity profile information. The *context provisioning* trust aspect applies only to the *Context Provider* role, and the *context-aware adaptation* (a.k.a. *context-aware service provisioning*) applies only to the *Service Provider* role.

Table 2 shows the trust aspect dependency of each role with respect to all other roles, for example, the *Identity Provider* (row) depends on the *Identity Provider* role (column) for the *identity provisioning* (IDP) trust aspect. In fact, all roles depend on the *Identity Provider* role with respect to the IDP aspect.[2] The *Service Provider* depends on the *Context Provider* with respect to the *context information provisioning* (CIP) trust aspect. The *Service User* depends on the *Context Provider* with respect to the CIP trust aspect. The *Context Owner* depends on the *Context Provider*, *Service Provider*, and *Identity Provider* with respect to the *privacy enforcement* (PE) trust aspect.

---

[2] We address this circular trust dependency in our trust management model described in Sect. 2.

2 Trust Management Model

In this section we present our trust management model, which supports the specification of aspect-specific trust relationships. Our trust management model instantiates well-known concepts like direct trust establishment through personal experience or beliefs, and indirect trust establishment through recommendations. Our trust management model quantifies trust using Subjective Logic (in short, SL) [11, 14], which is a probabilistic logic capable of explicitly expressing uncertainty about the probability values.

We use our trust management model in two trust-based selection mechanisms to support users and service providers of context-aware services in managing their trust relationships and in selecting trustworthy entities to interact with. The first trust management mechanism we introduce supports service providers in selecting trustworthy context providers with respect to a QoC level. The second trust management mechanism supports service users in selecting context providers and service providers, taking into account the trade-off between privacy and context-based service adaptation.

Most of the existing trust management models refer to a specific application domain and, as such, propose special-purpose solutions that are not easily portable to other domains. Our context-aware service platform domain requires a specific formalism of combining trust relationships focusing on specific trust aspects we have not found treated appropriately in the literature. For reasons of simplicity, we specify our trust formalism using a simple set of rules; however, we do not exclude that existing formalisms for trust (e.g. Nielsen and Krukow [26]) could be specialized to express and combine multiple trust aspects as required by our domain.

2.1 Metamodels

We formalize trust as a relationship between two entities, the Trustor and the Trustee, as widely accepted by the literature [1, 6, 10, 29]. We define trust as *the measurement of the belief from a trusting party point of view (trustor) with respect to a trusted party (trustee) focused on a specific trust aspect that possibly implies a benefit or a risk.* For example, *Bob* (Trustor) may trust to a *high degree* (measurement) *Alice* (Trustee) concerning her *competence in coding in Java* (trust aspect). The risk implication is only present when Bob accepts to depend on Alice to code a Java program on his behalf or to use a Java code provided by Alice.

We refer to the terms *trust* and *trust relationship* interchangeably, always meaning the relationship between two entities. We use the term *trustworthiness* to refer to the amount, measurement, or degree of trust (by a trustor in a trustee) in a trust relationship. Furthermore, the entities in a trust relationship that we refer to as Trustor and Trustee are digitally represented in an information system using digital identities. With the expression

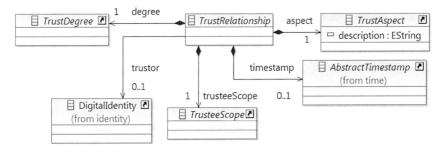

Fig. 6 Trust relationship metamodel

$$A \xrightarrow[v]{*;a} B \tag{1}$$

we represent a personal trust relation between A (the Trustor) and B (the Trustee) that tackles the trust aspect a and that has degree v. In our formalism, A and B are entities that belong to a set ID. Identities are assigned to different roles in different instances of our platform.

"$*$" is a placeholder for classes of trust relations. We consider two classes of trust relations: direct functional (df) and indirect functional (if) relations, so $* \in \{df, if\}$. Direct trust originates from A's direct experiences or evaluations of B. We distinguish two different sub-classes of direct trust: arbitrary and experience. Arbitrary trust is the trust determined based on personal beliefs without previous experience. Experience trust is trust determined based on A's direct evidence that contribute to belief or disbelief. Indirect trust originates when A's resorts to indirectly evaluating trust in B, for example, by combining trust values or asking for recommendations from other entities (see also [15]).

A *Trust Relationship* in our metamodel presented in Fig. 6 is a class that is composed of a *Trust Degree* of a trustor (*Digital Identity*) for a specific *Trust Aspect* specified at a specific moment (timestamp). A trust relationship also includes a *Trustee Scope* to allow flexibility in the specification of trust relationships considering the social trust concepts of *System Trust, Dispositional Trust, Situational Trust*, and *Trust Beliefs* [19]. The social trust concept of *Trust Beliefs* is mapped to an abstract *Trust Relationship* in our metamodel.

In Fig. 7 we show the metamodel of the different possible trustee scopes we support for a trust relationship. The concept of *Dispositional Trust* is the intrinsic/inherent disposition an entity has to trust any other given entity in the absence of evidence or previous experiences, which we believe can be used to support trust bootstrapping. The concept of *System Trust* is the impersonal trust perception an entity has regarding the set of regulations and safeguards of the system as a whole. The *Situational Trust* represents the personal trust a *Trustor* has in one particular or all *Context Situations*, and the *Personal Trust* represents the trust a trustor has in a specific trustee represented by a digital identity pattern. For example, a trustor may specify a trust relationship with all trustees owning a digital identity containing the attribute city with value

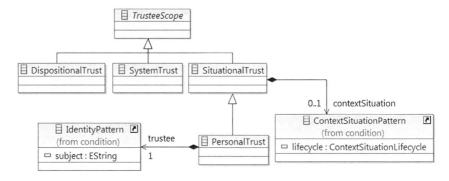

Fig. 7 Trustee scope metamodel

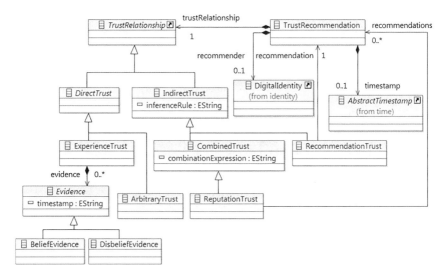

Fig. 8 Trust relationship types metamodel

Amsterdam issued by a government agency. Our formalism focus on the *Personal Trust* concept but for completeness we include in our metamodel all possible trustee scopes described in the work of Mayer et al. [19].

Figure 8 presents the trust types we support in our metamodel. The direct trust may be arbitrary or based on experience consisting of belief and disbelief evidences observed directly by the trustor. The indirect trust may be just an arbitrary domain specific combination of trust relationships, reputation trust based on recommendations of other trustors, or a recommendation based on a trust relationship of one specific trustor. Trust recommendations and trust evidence also have a timestamp attribute to support evaluation of trust relationships considering temporal parameters, for example, selecting more recent recommendations over old ones and observing the evolution in time of trust degrees in a community.

The aspect a of a trust relationship ranges over identity provisioning, privacy enforcement, and context information provisioning, that is $a \in \{idp, pe, cip\}$. We consider the set of roles $R = \{US, CO, IP, CP, SP\}$ from our context-aware service platform, namely, user (US), context owner (CO), identity provider (IP), context provider (CP), and service provider (SP). The function role: $ID \rightarrow R$ returns the role that, at the present moment, a given entity identified by an identity $id \in ID$ plays; initializing and updating this function is the exclusive competence of identity providers, but it can be invoked by any entity that has registered its identity. We assume that entities can access a set of functions that calculate the direct trust value from a Trustor to a Trustee based on the evaluation of its privacy enforcement (pe), identity provisioning (idp), and context information provisioning (cip) qualities. These functions receive as input the Trustor and Trustee identities ($ID \times ID$) and return the trust value for the specific trust aspect:

$$(trust\_PE : ID \times ID \rightarrow TValues)$$
$$(trust\_IDP : ID \times ID \rightarrow TValues) \qquad (2)$$
$$(trust\_CIP : ID \times ID \rightarrow TValues)$$

For example, $trust\_PE(Alice, Bob)$ is the evaluation of Bob's honesty, competence, and reliability in its privacy enforcement aspect from Alice's point of view. These functions are our starting point for trust evaluation; on their output we can establish the degree of trust between the Trustor and the Trustee. If we specify our reasoning in terms of an inference system, i.e., in terms of axioms and deductive rules of the form premises/conclusion, definitions (1) and (2) can be used, at a meta-level, to define our set of axioms. In all the following rules that express our algorithm, we assume that $role(A) = US$, that is, Trustor A is a service user.

$$\frac{[trust\_PE(A,B)=v]}{A \xrightarrow[v]{df:pe} B} \quad role(B) \in \{CP, IP, SP\}$$
$$\frac{[trust\_IDP(A,B)=v]}{A \xrightarrow[v]{df:idp} B} \quad role(B) = IP \qquad (3)$$
$$\frac{[trust\_CIP(A,B)=v]}{A \xrightarrow[v]{df:cip} B} \quad role(B) = CP$$

In the first rule when $trust\_PE$ is invoked with parameters A and B, it returns a value v, which states that A has degree v of (direct) trust in B, with respect to the aspect pe (privacy enforcement). This aspect is significant when Trustee B is a context provider, an identity provider or a service provider.

Trust aspects can also be understood as quality requirements related to the provision of a service, for example, the provision of context information implies a particular quality level of the context information provided and also requires privacy preferences to be respected. Trust relationships for different aspects could be specified for any type of service considering the quality requirements of service users.

Figure 9 shows the different trust aspects we propose in our trust metamodel also to represent trust on a specific trustee to provide trust recommendations and trust on

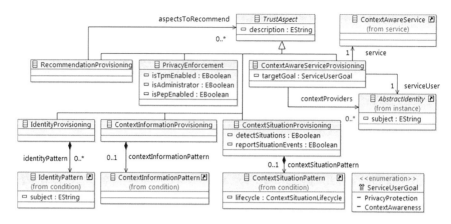

Fig. 9 Trust aspects metamodel

the provisioning of a context-aware service with two possible target goals: privacy protection and context-awareness. These goals are relevant in the mechanism for selection of context-aware service providers introduced in Sect. 5. We also model in Fig. 9 a trust aspect for context situation provisioning in addition to the context information provisioning aspect in order to support specification of trust relationships focusing on context situations and situation events (see Sect. 1.1).

Trustors can perceive or interpret the level of trust for the different aspects as an isolated or combined measurement of, for example, honesty, competency, reputation, usability, credibility, and reliability to perform a specific action. We assume this level of trust always refers to a combination of the concepts of "honesty, competence, and reliability for a certain aspect", because this is the most common interpretation of trust observed by an extensive survey conducted by Mcknight and Chervany [20]. Other trust concepts are also important and are considered future work. A list of possible trust concepts and their correlations based on user studies can also be found in [29].

Regarding the choice of the domain of *trust degrees*, existing trust management models have different proposals. Some authors quantify trust as a real numeric value (e.g. between -1 and 1), a discrete value (e.g. trust or distrust), or a combination of both, where each element in the discrete set has a numeric equivalent (e.g. values in $(0, 1]$ mean trust, values in $[-1, 0)$ denote distrust, and 0 means unknown). Our proposal is independent of any particular solution; we assume a generic domain *T Value*.

We instantiate *T Value* in the set of opinions of the Subjective Logic (in short, SL) [11]. SL is a probabilistic logic capable of explicitly expressing uncertainty about the probability values. The basic assumption is that there is always uncertainty and that the truth is always expressed from an individual perspective. The SL formalism has been proven to be an appropriate formalism for addressing trust calculations because it allows more realistic modeling of real-world situations that reflect ignorance and uncertainty [14] compared to traditional probabilistic approaches [2]. We describe in a following section the SL formalism, which is used in our trust mechanism to instantiate our $TValue$ domain.

2.2 Subjective Logic

In our trust model we adopt the formalism of Subjective Logic (SL) [11, 13, 14] to quantify the degree of trust in a trust relationship specified in our trust management model. SL has been proven to allow a more realistic representation of real-world situations because it supports the expression of uncertainty regarding propositions. Traditional probabilistic approaches limit themselves to expressing belief and disbelief mass only. The strongest point of SL is that it combines the structure of binary logic with the capacity of probabilities to express the degrees of truth and uncertainty about propositions.

The basic assumption of SL is that nobody can ever determine whether a proposition about the world is true or false. Furthermore, the truth about a proposition is always expressed from the point of view of a specific individual, in the sense that it is subjective and unique to the person experiencing it and does not represent a general or objective point of view. In SL, the probability regarding a proposition is expressed through a Subjective Logic Opinion.[3] An Opinion is represented using the symbol: ω_x^A, where A is the belief owner and x is the proposition. The opinion ω_x^A is a ordered quadruple $(b, d, u, a) \in [0, 1]^4$, where:

- b: is the belief mass supporting that the specified proposition is TRUE;
- d: is the belief mass supporting that the specified proposition is FALSE;
- u: is the uncertainty amount or uncommitted belief mass (neither TRUE or FALSE);
- a: is the base rate or atomicity that indicates the a priori probability that the specified proposition is TRUE in the absence of a committed belief mass.

To represent opinions graphically, SL adopts a two-dimensional equilateral triangle representation, presented in Fig. 10. A point inside this triangle represents an opinion. To clarify how to interpret the subjective logic triangle, we present on the left the reference axes and on the right the position of an example opinion.

In the left-side triangle of Fig. 10, the belief, disbelief, and uncertainty axes run from the opposite side of the edge with the respective label, assuming the maximum value of 1 (one) in the edge with the label and zero in the opposite side. The probability axis is the bottom axis of the triangle, which is equivalent to the traditional probability axis because it represents opinions with zero uncertainty.

In the right-side triangle, we omit the axes and present an example opinion $\omega_x^A = (0.6, 0.3, 0.1, 0.5)$. The opinion represents 60 % of belief mass, 30 % of disbelief mass, and 10 % of uncertainty. We also show the atomicity as a line dividing the

---

[3] We refer to *Opinion* from now on as meaning a *Subjective Logic Opinion*.

Fig. 10 Subjective logic
triangle of opinions

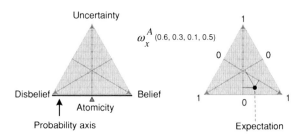

Table 3 Subjective logic
opinions equivalence

Subjective logic	Equivalent to
Belief = 1	TRUE of binary logic
Disbelief = 1	FALSE of binary logic
Belief + disbelief = 1	Traditional probability
Belief + disbelief < 1	Degrees of uncertainty
Belief + disbelief = 0	Total uncertainty

triangle in the middle, indicating an equal 50 % a priori probability of the proposition being TRUE or FALSE. Table 3 presents the equivalence between a subjective logic opinion and traditional probability theory.

The semantic of a subjective logic opinion can be better understood by means of an example. Imagine that the observer *Ricardo* wants to quantify his opinion about a proposition x related to his friend *Rodrigo*. The proposition x states that *"Rodrigo arrives on time for his appointments"*, meaning that *Rodrigo* is never late for any given appointment. In the absence of evidence, *Ricardo*'s opinion about proposition x is of complete uncertainty, and is represented in subjective logic as: $\omega_x^{Ricardo} = (0, 0, 1, 0.5)$. This opinion means that *Ricardo* has no belief mass to support that the proposition x is either TRUE or FALSE, and that from the complete uncertainty belief mass there is 50 % atomicity, or a priory probability, that the proposition is TRUE or FALSE.

Now let us imagine that *Ricardo*, after meeting with *Rodrigo* ten times, has experienced that in seven of these meetings, *Rodrigo* was on time. For the other three times, *Ricardo* himself was late, so he is uncertain about *Rodrigo*'s punctuality. Considering his experience, *Ricardo*'s opinion has changed, and his new opinion about proposition x now is $\omega_x^{Ricardo} = (0.7, 0, 0.3, 0.7)$. This opinion means that *Ricardo* has 70 % of belief mass about proposition x, and 30 % of uncertainty. From the uncertain belief mass it is possible for *Ricardo* to assume that, if *Rodrigo* continues with his behavior, the a priori probability (atomicity) that he will be on time is now 70 % for the uncommitted belief mass.

In SL, the probability expectation E is calculate using the following formula $E_x^A = b + ua$. The intuition is that the expectation of an observer is the sum of the committed belief mass b, and the uncommitted belief u mass multiplied by the a priori probability or atomicity a. Considering our previous example, *Ricardo*'s probability expectation about *Rodrigo*'s punctuality for the next appointment is $0.7 + 0.3 \times 0.7$, which is

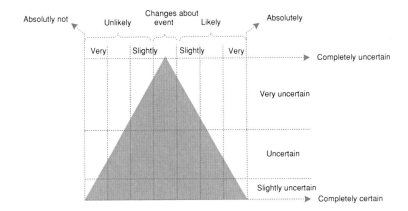

Fig. 11 Fuzzy interpretation of SL triangle

equal to 0.91. In summary, the atomicity represents how much of the uncertainty mass contributes to the probability expectation value that *Rodrigo* will be on time for the next appointment.

Subjective logic opinions can be also derived from the number of positive (r) and negative (s) previous experiences with an entity. The following formulas show how to calculate the belief (b), disbelief (d), and uncertainty (u) opinion parameters from previous observations. The weight (W) is usually instantiated to the value 2 and represents the impact of new experiences in the opinion parameters [12].

$$b = \frac{r}{r+s+W} \quad d = \frac{s}{r+s+W} \quad u = \frac{W}{r+s+W} \quad W = 2 \qquad (4)$$

The advantage of subjective logic is not only the capability to express uncertainty, but also the many operators to compute over sets of opinions. Imagine that more than one person has different opinions about proposition x—how can these opinions be combined for a final conclusion? The combination of the opinions could be done through the *consensus* operator, which provides a fair combination of opinions. SL also provides operators to perform subtraction and addition of opinions. Using a traditional probabilistic approach, the combination of a 100 % probability with a 0 % probability would result in a 50 % probability, whereas when the *consensus* operator is used, the result would be complete uncertainty due to the conflicting opinions.

The SL logic literature suggests different ways of interpreting subjective logic opinions, using a set of concepts and mappings. Figure 11 shows the proposed division by Jøsang [13]. We believe that the mapping proposed by [13] is too fine grained for users to understand, and for this reason we propose a different and simplified mapping strategy.

We have mapped the subjective opinion triple (b, d, u) to an ordered set {very untrustworthy, untrustworthy, unknown, trustworthy, very trustworthy} whose elements model the judgment of user perspectives [25]. An opinion *op* whose belief

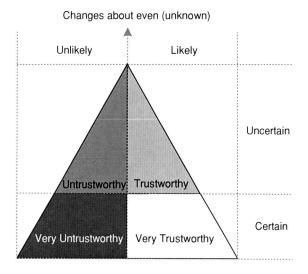

Fig. 12 The function: [0, 1]3'
VT, T, U, VU that maps an SL
opinion onto the set of user
judgments

is higher than its disbelief is considered trustworthy if it has an uncertainty of
not lower than 1/3 and is very trustworthy otherwise. An opinion *op* whose belief
is not higher than its disbelief is considered untrustworthy if it has an uncertainty
of not lower than 1/3 and is very trustworthy otherwise (see Fig. 12). The *unknown*
mapping represents opinions with even chances of equal belief and disbelief.

2.3 Evaluation of Trust Recommendations

A trust mechanism for supporting the evaluation of trust recommendations is required
because users and service providers may interact during context-aware service provisioning with entities that have unknown identities. If the identities are unknown,
there are no trust values associated with this entities and no record of direct previous
experiences exists. The approach we adopt for exchanging trust recommendations is
inspired by the approach adopted by Almenarez et al. [2]. However, our approach is
more complete because we take uncertainty into account and support the exchange
of trust recommendations for the different trust aspects supported by our trust management model.

In the approach of Almenarez et al., the recommendation requests are broadcasted and the recommendation responses are combined taking into account only the
responses of recommenders' with trustworthy identities. By using recommendations,
indirect trust relationships can be established based on information received from
other entities. Recommendation requests are only broadcasted when trustworthiness
values are required for entities that are not known from direct experience or belief.

Recommendations can be received from trustworthy or untrustworthy recommenders. For this reason, it is important to support in our trust management model

trust relationships related to the trust aspect of providing trust recommendations. Trust recommendations, despite being a trust aspect themselves, are also related to other specific trust aspects. For example, one entity might be trustworthy to give recommendations only about the privacy enforcement trust aspect about other entities. Recommendation trust degrees therefore state the amount of trust belief one trustor has in another trustee to provide recommendations about other entities with respect to a specific trust aspect.

In our trust management model, we merge the recommendations received from third parties using also the consensus operator from the Subjective Logic (SL), which has been proven to be a proper tool for this type of trust combinations [14]. The SL consensus operator is used to merge SL opinions in a "fair" way and, if conflicting opinions are received, the amount of uncertainty in the resulting trust degree is increased.

In contrast to the approach of Almenarez et al. [2] our proposal of combining recommendations using the SL consensus operator takes uncertainty into account. One major drawback of not considering uncertainty is a less accurate and less realistic trust result when conflicting recommendations are combined. Using the approach of Almenarez et al., when conflicting recommendations are received the result is an average of the belief probabilities. Using SL, when conflicting recommendations are received, there is an increase in the uncertainty.

Furthermore, Almenarez et al. [2] only consider the trustworthiness of the recommenders' identities and do not address trust values related to different trust aspects. In our approach, we subtract the trustworthiness of the identities from the trust recommendations, and we also support trustworthiness values for different trust aspects, including the trustworthiness of an entity to provide recommendations about specific trust aspects. For example, a trust recommendation received from a consumers' privacy protection organization will be influenced by the trustworthiness of the identity of the organization and by the trustworthiness value related to the organization's capability of providing recommendations about privacy enforcement trust. We do not propose more complex algorithms for calculating trust from indirect knowledge, for more complex mechanisms we refer to Toivonen et al. [30].

3 Mechanism for Selection of Context Providers

In this section, we present our mechanism for the trustworthiness evaluation of context providers. This evaluation is done by context consumers, which in our context-aware service platform are context-aware service providers. We do not position trustworthiness as a Quality-of-Context (QoC) parameter because it is not a quality attribute of the context information instance but a *degree of belief* from the point of view of the context consumer (e.g. the context-aware service provider) with respect to the context provider. The trust aspect of provisioning context information specifies the degree of belief of a context provider to provide context information about a context owner, and according to an advertised QoC level.

The bootstrapping of the trustworthiness values in our QoC model is done through pre-defined trustworthiness values or based on recommendations received from trusted third parties. Pre-defined trustworthiness values are usually defined based on the Dispositional trust [20], that is, the likelihood of trusting other entities in the absence of concrete trust evidence. A simple and optimistic strategy would be to consider a context provider trustworthy by default, in case no recommendations are received or no evidence exists to believe the opposite. If multiple recommendations are received, they are combined using a "fair" combination, which is supported by the consensus operator from SL [11].

After the bootstrapping, the trustworthiness values evolve based on the feedback evidence about the perception of users of the context-aware service regarding the reliability of its adaptation. When the users of the context-aware service notice wrong or inappropriate service adaptation, they can provide *negative* or positive feedback. Our feedback mechanism was inspired by the work of Huebscher et al. [7]. Positive feedback is mapped to a *trustworthy* opinion and negative feedback to an *untrustworthy* opinion. In case a positive feedback is received, the current trustworthiness value of the context provider for the specific context owner identity, context type, and QoC level is increased, and for the negative feedback, the trustworthiness value is decreased by the same amount.

The trustworthiness value decrease/increase is also computed by applying the consensus operator of the SL to the actual trustworthiness value of the context provider and to the feedback received. Negative feedback only affects the trustworthiness of a context provider for a specific context type, context owner, and QoC level; in other words, it is possible for a context provider to be very trustworthy for one context owner and very untrustworthy for another.

If the context-aware service adaptation is not satisfactory, the service users have the possibility to indicate positive or negative experiences, and also indicate which faulty context-based adaptation behavior they are experiencing. Based on the specific feedback from the users, the service provider is able to detect which context provider is not fulfilling his promises regarding the quality of context and, therefore, the trustworthiness value of the context provider is decreased. The context-aware service provider may be able to detect, depending on the granularity of the feedback from the users, the exact context provider and context owner that is causing the faulty context-based adaptation. This detection is useful if context information from multiple context providers and context owners is being used by the service.

We do not support in our trustworthiness evaluation mappings of QoC levels and trustworthiness value combinations. For instance, if a context provider is trustworthy to provide an *entity's location* with ± 1 m precision, nothing can be said about the trustworthiness of the same context provider to provide the same *entities' location* with ± 1.01 m precision. We acknowledge that mappings between QoC levels and trustworthiness values are possible and depend on the context type.

In order to be able to collect relevant feedback, the context-aware service provider needs to be able to map the positive and negative feedback regarding the context-based service adaptation to the context provider that influences that behavior. The mapping of feedback allows a service provider to identify the reasons for a faulty context-aware service adaptation. For example, a negative feedback for a context-aware meeting service stating that one meeting attendee arrived to a meeting after the time predicted by the service may indicate that the person's location information provided was of low quality.

The positive and negative feedback capture the situation where the context provider provides context at a lower quality than advertised because he was dishonest, incompetent, and/or unreliable. The mapping from the user feedback (positive/negative) to the reason for lower-quality context information (dishonesty, incompetence, and/or unreliability) can only be evaluated if the granularity of the feedback includes enough detail about the faulty context-based adaptation. We only consider in our examples discrete positive/negative feedback without refining the faulty behaviors because they depend on the type of context-aware adaptation executed by the service.

The assumption we make is that, if the context-aware service is not adapting properly to the context, then the context provider is the one to blame. This might happen if the context information provided by the context provider capabilities is not as good as the capabilities being advertised. We support with our mechanism the situation where the context provider advertises a certain quality level and always provides context at a quality level lower than that advertised. In this case, the trustworthiness value of the context provider will never increase for the advertised QoC level because the trustworthiness value is increased according to our mechanism taking into account the QoC level that is associated with the context information instance. The assumption we make is that the context provider is not being dishonest because he still provides context according to the correct QoC specification.

We acknowledge that in some scenarios with a high number of context types and QoC levels it is not operationally feasible nor usable for users to manage the trustworthiness values for each combination. In scenarios where the number of combinations is too numerous, simplification strategies can be adopted; for instance, trustworthiness values for context providers could be stored only considering the context types without distinguishing trustworthiness values for each possible QoC level.

4 Mechanism for Selection of Context-Aware Service Providers

In this section, we present our mechanism for the trustworthiness evaluation of the entities that collaborate in the provisioning of a context-aware service. This evaluation is done by service users, are also the context owners. In the description of this mechanism the assumption we make is that the context-aware service always uses the context information related to the service user in the context-based adaptation of the service provided.

We define trust as a relationship between entities that are represented through digital identities. We therefore conclude that trust in a digital identity is influenced by the trust (regarding the trust aspect idp) in the identity provider that has provided that identity. The trust value associated with the provider of the trustee identity influences all the trust values associated with that identity. Our assumption is that it is not possible to trust the trust values associated with an entity who's identity is not trusted. This inter-relationship between trust in identities and trust in identity providers is synthesized by the following inference rule for indirect trust:

$$\frac{A \xrightarrow[v]{df:a} B \quad A \xrightarrow[v']{df:idp} C}{A \xrightarrow[v\otimes v']{if:idp} B} \quad role(C) = IP \quad C \text{ provides B's identity} \quad (5)$$

The above rule expresses the following: If A's direct trust degree in B regarding the trust aspect a is v, and if the identity of B is provided by identity provider C, and if A's indirect trust in C for aspect identity provisioning is v', then A's indirect trust in B regarding aspect a is the result of $v' \otimes v$, which represents the value v subtracted by the value v' (e.g. $v \otimes v' \leq v$).In the Subjective Logic (SL) domain we map $\otimes$ onto the fair combination *discount operator*.

The identity provider himself also needs to be identified through a digital identity. Therefore, we introduce a circular problem if the identity provider identifies himself with a self certified identity. For this reason, the previous rule does not apply for the identity provisioning trust aspect in case the identity provider provides his own identity. The trust value of an identity provider has to be determined by different means, for instance, through a pre-defined list of trustworthy or untrustworthy identity providers. The trustworthiness of a self signed identity can be also directly mapped to the trustworthiness of the entity with respect to identity provisioning.

Once the service user has evaluated the trust relationships with all the entities that assume the context provider and service provider roles, the user deduces the combined trust value in the context provider (CP) role himself. The trust in the context provider and service provider roles has already been influenced by the trust the user has in the identity providers. This is a generalization step that allows the service user to evaluate his/her trust in the context provider role when the context-aware service retrieves context information from more than one context provider. The following rules express this generalization step.

$$\frac{A \xrightarrow[v]{if:a} C \quad [role(C)=CP]}{A \xrightarrow[v]{if:a} [CP,\{C\}]}$$
$$\frac{A \xrightarrow[v]{if:a} C \quad A \xrightarrow[v']{if:a} [CP,S] \quad [role(C)=CP]}{A \xrightarrow[v\otimes v']{if:a} [CP,S \cup \{C\}]} \quad C \notin S \quad (6)$$

Here $a \neq idp$, because identity provisioning has already been in place. The rule on the top says that A's trust in the CP role can be initiated with the trust A has in members of the CP role. The rule on the bottom says that new members can contribute to A's trust in the CP role; so if A's trust in the role CP is v, and if A's trust in member C is v', then the new A's trust in the role is $v \otimes v'$. Here $v \otimes v'$ expresses a fair combination of the two trust values as, for example, using the Subjective Logic consensus operator.

The same generalization step for the context provider role could be applied for the service provider role in a similar way. For the service provider role, this general-ization step would be required if more than one service provider were responsible for providing a context-aware service, for instance, in case of a more complex context-aware service composition. In our examples, we do not address these more complex cases; however, we acknowledge this possibility.

By using the consensus operator, we assume that all the entities that play the role of context provider have the same impact on the adaptation of the context-aware service and on the enforcement of the users' privacy. It is possible that two context providers contribute differently to the provisioning of the context-aware service and also that they have different impacts on the privacy of the context owners. We assume the impact to be always the same, and more advanced scenarios to be beyond the scope of our work.

The final step of our mechanism consists of evaluating the service user's trust in the context-aware service as a whole. This evaluation depends on the trust the user has in both the CP and the SP roles regarding the privacy enforcement and con-text provisioning aspects. The context provisioning aspect is only influenced by the members of the CP role. In this final step, we address two different user goal profiles derived from the trade-off between privacy enforcement and privacy adaptation. The first profile has higher priority in privacy enforcement and will accept less service adaptation. The second profile has higher priority in context-aware service adapta-tion even if privacy is not respected [3]. We name these two profiles privacy-focused and service-focused users.

The rule that expresses how to calculate A's (user) trust in a service provider B when context provider role is played by entities in S is formalized as follows:

$$
\frac{A \xrightarrow[v]{if;pe} B \quad A \xrightarrow[v']{if;cip} [CP, S]}{A \xrightarrow[f(v,v')]{if;pe \times cip} B \times [CP, S]} \quad role(B) = SP \tag{7}
$$

In this rule, the user combines his trust in the service provider role in the privacy enforcement aspect, and the trust he has in the context provider role in the context provisioning aspect. Function f expresses a particular way of aggregating trust, which depends on the two user profiles we address. In order to give an example of f, and for illustration purposes, we map TValues onto the ordered set $\{VT, T, U, VU\}$ whose elements model judgment of user perspectives: very untrustworthy (VU), untrustworthy (U), trustworthy (T), and very trustworthy (VT).

Informally, Fig. 13 shows the resulting trust in the context-aware service when the trust expectation in the service provider regarding the privacy enforcement aspect and the trust expectation in the context provider regarding the context information provisioning increase. The best case scenario for both user profiles is the one where

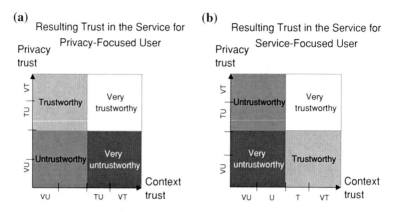

Fig. 13 Resulting trust in the service according to a user profile that focuses on privacy (**a**) and on a user profile that focuses on service adaptation (**b**)

the trust expectations for both the privacy enforcement and context information provisioning trust aspects are at least trustworthy.

According to Fig. 13a, for the privacy-focused profile the best cases are when privacy enforcement is at least trustworthy. The worst cases are when privacy enforcement is untrustworthy, because it is more likely that trustworthy context information about the user will be under a privacy risk. For the service-focused profile (Fig. 13b), the best cases are when context provisioning is at least trustworthy, and it is even better when privacy is also enforced. The worst cases are when the context information is not trustworthy, which results in bad service adaptation. However, in this case, it is preferable to have privacy enforcement, if possible. We assume here that a context-aware service receiving untrustworthy context information is more likely to adapt wrongly to the current user situation. From this discussion we support the conclusion that for both user profiles, the best case is when trust in context information and privacy enforcement is high; however, depending on the profile, the worst-case scenario is not the same.

An example of the function $\int$ can be obtained by first applying π to v and v', then applying one of the functions of Fig. 13, and then mapping back each user category onto a "representative" opinion of that category. For example a representative opinion of VT can be the triple $(0.75; 0.01; 0.24)$, of T it can be $(0.50; 0.01; 0.49)$, and so on. To the best of our knowledge, functions with the properties sketched in Fig. 13 cannot be obtained by composing existing SL operators with π.

It is also possible that the trust values for privacy enforcement and context provisioning are unknown. If both values are unknown, we assume that the resulting trust in the service provided will be unknown as well. If only the trust value for the privacy enforcement trust aspect is known, then the resulting trust for the privacy-focused user is the value of this aspect and for the service-focused user the resulting trust is unknown. If only the trust value for the context provisioning aspect is known, then the resulting trust for the service-focused user is mapped to the value of this aspect and for the privacy-focused user the resulting trust is unknown.

If the value is unknown, another possibility would be to inform the privacy-focused user of the risk (s)he is taking by using the service. For instance, if the trust value for the context provisioning trust aspect is high, and the value for privacy enforcement is unknown, then the risk the privacy-focused user is taking is high. For the service-focused user, privacy is considered secondary; therefore, this risk approach does not apply.

5 Prototype Implementation

In this section, we describe our prototype implementation that demonstrates the technical feasibility of our trust management model in a simulated context-aware service scenario. Our prototype implementation provides a graphical user interface to support users and service providers in the management of trust relationships and trust-based selection of identity providers, context providers, context situation providers, and context-aware service providers. The trustworthiness evaluation in our prototype is implemented using the Subjective Logic (SL) API [11] for trust calculations using SL opinions.

Our trust management model is integrated in the *Model-based Security Toolkit* (SecKit) developed at the European Commission Joint Research Center (JRC). The objective of the SecKit is to provide a collection of metamodels and components to support the engineering and management of secure computer systems. In this section we describe only the SecKit prototype implementation parts that implement the trust management model we propose for context-aware service platforms.

The SecKit user interface supports the specification of interrelated design and runtime models of a computer system. The design models includes specifications of the system structure, information, behavior, roles, context, identities, and security policies. This design models are used as a reference to support the specification of security-focused runtime models, which include the trust model consisting of the trust relationships proposed by us.

Figure 14 shows the SecKit graphical user interface (GUI) with the set of sub-tabs displaying the runtime models. The runtime models capture the entity instances of the running system, the known identities, the role hierarchy, and the policy rules. The first tab in Fig. 14 shows the *Entity Model* that can be filtered according to the type of service they provide. The first tab in Fig. 14 is configure to show in a tree only the entities that perform the role of *Identity Provider*, and for each entity the specified trust relationships for the *Identity Provisioning* trust aspect .

For each entity in the tree it is possible to visualize as a child node of the entity the trust relationships specified with the respective trust degree using a colored icon. The colors we adopt for the trust degrees are green, light green, gray, light red, and red to specify respectively a very trustworthy, a trustworthy, an uncertain, an untrustworthy, and a very untrustworthy trust degree. A double click in the respective

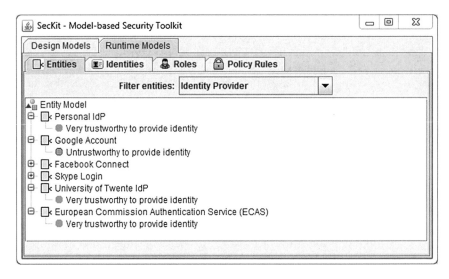

Fig. 14 Entity model, filter by identity providers

Fig. 15 Subjective logic triangle

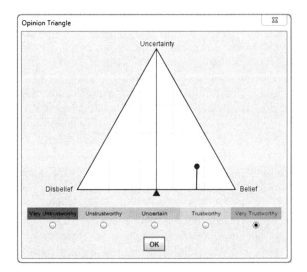

trust relationship shows the opinion triangle screen allowing the trust relationship details to be modified (see Fig. 15). The trust belief also shows the corresponding mapping of the trust degree to the subjective logic triangle following the mapping proposed by us in our trust management model.

Figure 16 shows the *Identities* sub-tab, which also displays in a tree the *Identity Model* with all the known identities configured in the system. Identities are grouped into packages to simplify the management and for each identity the trustworthiness of the associated identity provider can be visualized as a child node. Identities are

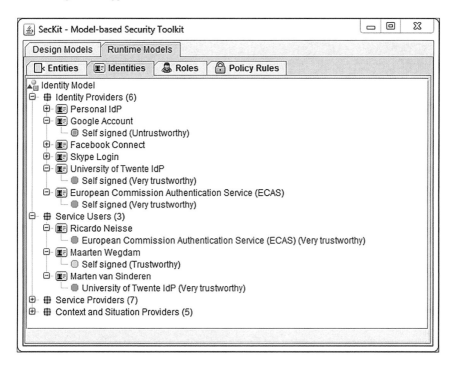

Fig. 16 Identities and trust degrees

assigned the trustworthiness of the identity provider that issued the identity, and this trustworthiness influences the trustworthiness of all the trust relationships specified with this identity according to the respective inference rule (5) specified in our trust management model.

Figure 17 shows the entity model configure to display only *Context Provider* entities specified in our simulated scenario. Context providers are capable of providing context information of a specific context type and only of a respective context owner of interest. In Fig. 17 the entity model tree shows all context providers available and for each context provider the respective trust relationships using the same icons already describe.

A context provider can be associated to trust relationships focusing on the privacy enforcement trust aspect or context provisioning trust aspect of our trust management model. For the privacy enforcement trust aspect we defined three parameters: the context provider is under the user control (Administrator), there is a Policy Enforcement Point (PEP) enabled, and there is a Trusted Platform Module (TPM) chip enabled. For the provisioning of context information the trust relationship can be parametrized with the respective context type, context owner, and quality-of-context (QoC) level supported by the provider.

In our prototype we also implemented the possibility of querying a context provider to retrieve simulated context information using a pop-up menu option in the

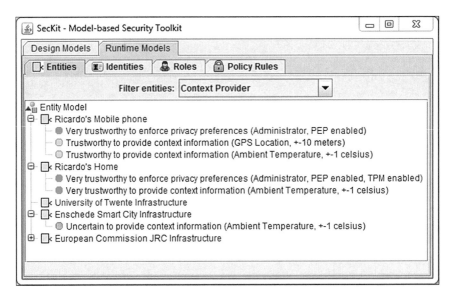

Fig. 17 Context information providers

entity tree. Figure 18 depicts the query result for the *Ambient Temperature* context provider. The query result shows the context value, QoC attributes, and a trustworthiness feedback option. After reviewing the context value a negative or positive feedback value can be provided and is used as evidence to increase/decrease the trustworthiness value of the context provider.

In addition to managing the trustworthiness of context providers we also support context situation providers. Situation providers are responsible for detecting situation events representing the moments a situation begins and ends to hold. The trustworthiness assessment of a situation provider is related to the signaling of situation events for a specific situation type and entities participating in the situation.

Figure 19 shows the entity model configured to display *Service Providers* entities in our simulated scenario. Service providers can be discovered considering the service user identity from the list of available identities and the context provider from the list of available context providers. The tree with the entities shows the list of service providers and respective trustworthiness values as child nodes in the tree for the privacy protection and context adaptation user goals. These trustworthiness values are calculated using the mechanism implementation for selection of context-aware service providers, taking as input the select user identity and context provider. The user is then able to selected based on his/her goal which service is more trustworthy considering his/her needs and goals.

The trustworthiness values in Fig. 19 indicate that *Google* and *University of Twente* are both very trustworthy to adapt their services to the user context considering the selected identity and context provider. However, *University of Twente* is very trustworthy to protect the user's privacy, and is therefore a better choice. This is

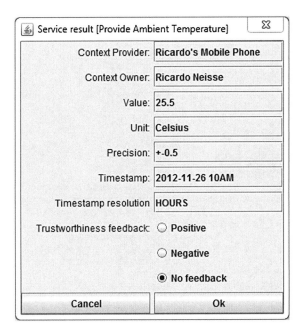

the case because the *University of Twente* is considered trustworthy to enforce the
privacy preferences in addition to its capabilities to adapt the service to the user
context.

The objective of our prototype implementation is to evaluate the technical feasibility of our trust management model and mechanisms to support users and service providers in selecting trustworthy entities to interact with. Service users are
interested in the assessment of the overall trustworthiness of a context-aware service
provider taking their primary goal into consideration. Service providers are interested
in assessing the trustworthiness of context information providers to maximized their
context-based adaptation capabilities.

With respect to the usability of our prototype implementation, we observe that
the selection of context providers may be difficult if many fine grained QoC levels
and trustworthiness values are specified. One important issue would be to allow
context consumers to specify their QoC and trustworthiness requirements by means
of ranges of trustworthiness values and QoC levels, and possibly by means of ranges
of individual QoC attributes. For example, a context consumer could state a minimum
and maximum precision and a minimum trustworthiness value required for a context
provider without specifying the absolute required values.

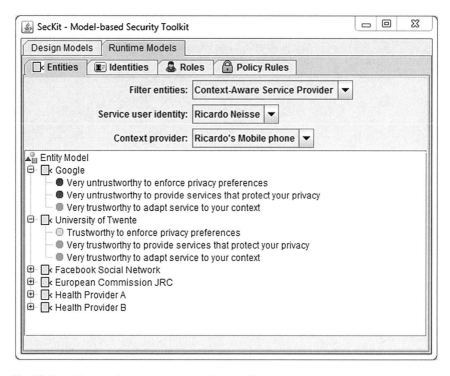

Fig. 19 Resulting trust in context-aware service providers

6 Conclusions and Future Work

We propose a new trust management model that supports the quantification of trust degrees for aspect-specific trust relationships that are relevant in our target context-aware service platform. Our model is extensible and considers trust aspects related to identity provisioning, privacy enforcement, context information provisioning, and context-aware service provisioning. We identified the inter-dependencies between these trust aspects and developed mechanisms based on a formalism for combining these trust aspects in order to evaluate the resulting trust users have in a context-aware service provider. We addressed two different resulting trust calculations considering the two different user goals we distinguish: privacy enforcement and context-based service adaptation. These goals are derived from the trade-off between privacy and context-based service adaptation.

Based on our trust model and mechanisms specification we have designed and implemented a proof-of-concept prototype that demonstrates the technical feasibility of our contributions and show how our model and mechanisms can be used to assist users and service providers in their trust decisions and in the selection of trustworthy entities to interact with. Even though we do not support in our trust management model and mechanisms all possible scenarios, to the best of our knowledge we are the first to provide a systematic analysis of the trust issues in context-aware service platforms for different trust aspects. This analysis contributes to a better understanding of the problems and future research in this area.

Our trust management model and mechanisms were designed to be applied in our target context-aware service platform. However, we foresee that our contributions could be easily applied to other service scenarios. Our contributions in the area of trust management are generic and extensible, and could be applied to other sets of trust relationships related to different trust aspects and trust management requirements. We have learned that in service oriented architectures trust relationships should focus on reliability issues of specific services, and may be combined in an overall trustworthiness assessment strategy that depends on the goals of the stakeholders considering the dependencies between the service providers.

The specification of trust relationship requirements and evaluation of each service provider individually may not be usable from a service user perspective. We plan to work further in the integration of our trust management model in the model-based security toolkit developed by us [22] and extend the policy refinement support with refinement of trust relationship requirements. Our plan is to specify high-level system trust requirements considering an abstract system design model and propose trust refinement rules to support the analysis of the personal service provider trust relationships considering the high-level trust requirements.

References

1. Abdul-Rahman A, Hailes S (2000) Supporting trust in virtual communities. In: HICSS'00: Proceedings of the 33rd Hawaii international conference on system sciences, vol 6. IEEE Computer Society, Washington, USA, p 6007
2. Almenárez F, Marín A, Campo C, García CR (2004) A pervasive trust management model for dynamic open environments. In: First workshop on pervasive security and trust in mobiQuitous 2004, Boston, USA
3. Berendt B, Günther O, Spiekermann S (2005) Privacy in e-commerce: stated preferences vs. actual behavior. Commun ACM 48(4):101–106
4. Daskapan S, Ali Eldin A, Wagenaar R (2005) Trust in mobile context aware systems. In: International business information management conference (5th IBIMA), Dec 2005, Cairo, Egypt
5. Dockhorn Costa P (2007) Architectural support for context-aware applications: from context models to services platforms. PhD thesis, University of Twente, December 2007
6. Grandison T, Sloman M (2000) A survey of trust in internet application. http://citeseer.ist.psu.edu/grandison00survey.html
7. Huebscher MC, McCann JA (2005) A learning model for trustworthiness of context-awareness services. In: PERCOMW'05: Proceedings of the third IEEE international conference on pervasive computing and communications workshops. IEEE Computer Society, Washington, USA, pp 120–124
8. Huebscher C, Mccann A (2005) An adaptive middleware framework for context-aware applications. Personal Ubiquitous Computing 10(1):12–20
9. Hulsebosch RJ, Salden AH, Bargh MS, Ebben PWG, Reitsma J (2005) Context sensitive access control. In: SACMAT'05: Proceedings of the tenth ACM symposium on access control models and technologies. ACM, New York, USA, pp 111–119

10. Jøsang A (1996) The right type of trust for distributed systems. In: NSPW'96: Proceedings of the 1996 workshop on new security paradigms, pp 119–131
11. Jøsang A (2001) A logic for uncertain probabilities. Int J Uncertainty Fuzziness Knowl Based Syst 9(3):279–311
12. Jøsang A (2010) Evidential reasoning with subjective logic. In: 13th international conference on information fusion 2010
13. Jøsang A (2010) Subjective logic book (draft). http://persons.unik.no/josang/papers/subjective_logic.pdf
14. Jøsang A, Hayward R, Pope S (2006) Trust network analysis with subjective logic. In: ACSC'06: Proceedings of the 29th Australasian computer science conference. Australian Computer Society Inc., Darlinghurst, Australia, pp 85–94
15. Jøsang A, Keser C, Dimitrakos T (2005) Can we manage trust? In: iTrust 2005, trust management, third international conference. Paris, France, pp 93–107
16. Kantara Initiative (2013) Identity assurance framework. http://kantarainitiative.org
17. Lachmund S, Walter T, Gomez L, Bussard L, Olk E (2006) Context-aware access control making access control decisions based on context information. In: 3rd annual international conference on mobile and ubiquitous systems—workshops, July 2006, pp 1–8
18. Li W, Kilander F, Jansson CG (2006) Toward a person-centric context aware system. In: Workshop on requirements and solutions for pervasive software infrastructures, PERVASIVE, 2006
19. Mayer RC, Davis JH, Schoorman DF (1995) An integrative model of organizational trust. The Acad Manage Rev 20(3):709–734
20. Mcknight DH, Chervany NL (1996) The meanings of trust. http://misrc.umn.edu/wpaper/WorkingPapers/9604.pdf
21. Neisse R (2012) Trust and privacy management support for context-aware service platforms. PhD thesis, University of Twente, March 2012
22. Neisse R, Doerr J (2013) Model-based specification and refinement of usage control policies. In: Eleventh international conference on privacy, security and trust (PST), 2013
23. Neisse R, Holling D, Pretschner A (2011) Implementing trust in cloud infrastructures. In: Proceedings of the 11th IEEE/ACM international symposium on cluster, cloud and grid computing (CCGRID), 2011
24. Neisse R, Wegdam M, van Sinderen M (2008) Trustworthiness and quality of context information. In: Proceedings of the 2008 international symposium on trusted computing (TrustCom), Nov 2008
25. Neisse R, Wegdam M, van Sinderen M, Lenzini G (2007) Trust management model and architecture for context-aware service platforms. In: Proceedings of the 2nd international symposium on information security (IS07), Nov 26–27. Vilamoura, Portugal, pp 1803–1820
26. Nielsen M, Krukow K (2003) Towards a formal notion of trust. In: PPDP'03: Proceedings of the 5th ACM SIGPLAN international conference on principles and practice of declarative programming. ACM, New York, USA, pp 4–7
27. Pons AP (2006) Biometric marketing: targeting the online consumer. Commun ACM 49(8): 60–66
28. Pranam K, Li D, Lalana K, Shashidhara G, Anupam J, Tim J (2004) Enhancing P3P framework through policies and trust. Technical report, UMBC
29. Quinn K, Lewis D, O'Sullivan D, Wade VP (2006) Trust meta-policies for flexible and dynamic policy based trust management. In: POLICY'06: proceedings of the seventh IEEE international workshop on policies for distributed systems and networks (POLICY'06). IEEE Computer Society, Washington, USA, pp 145–148
30. Toivonen S, Lenzini G, Uusitalo I (2006) Context-aware trust evaluation functions for dynamic reconfigurable systems. In: Proceedings of the WWW'06 workshop on models of trust for the web (MTW'06), Edinburgh, Scotland

Trust Management in ULOOP

**Carlos Ballester Lafuente, Jean-Marc Seigneur, Rute Sofia,
Christian Silva and Waldir Moreira**

Abstract This White Paper provides an insight on Trust Management within the context of the User-centric Wireless Local Loop (ULOOP) project, depicting the main principles and the overall trust management framework, and also describing its main individual components. It has as motivation to disseminate ULOOP concepts and to raise awareness towards trust management in user-centric wireless networks.

Keywords Crypto-id · Trust management · Metrics · Cooperation · Incentives · Rewarding

1 Introduction

The flexibility inherent to wireless technologies is giving rise to new types of access networks and allowing the Internet to expand in a user-centric way. This is particularly relevant if one considers that wireless technologies such as Wireless Fidelity (Wi-Fi) currently complement Internet access broadband technologies, forming the last hop

C. Ballester Lafuente (✉) · J.-M. Seigneur
Institute of Services Science, University of Geneva, Carouge, Switzerland
e-mail: carlos.ballester@unige.ch

J.-M. Seigneur
e-mail: jean-marc.seigneur@unige.ch

R. Sofia · C. Silva · W. Moreira
COPELABS, University Lusófona, Lisbon, Portugal
e-mail: rute.sofia@ulusofona.pt

C. Silva
e-mail: christian.silva@ulusofona.pt

W. Moreira
e-mail: waldir.moreira@ulusofona.pt

A. Aldini and A. Bogliolo (eds.), *User-Centric Networking*,
Lecture Notes in Social Networks, DOI: 10.1007/978-3-319-05218-2_5,
© Springer International Publishing Switzerland 2014

to the end-user. This fact becomes even more significant due to the dense deployment of Wi-Fi Access Points that is common today in urban environments.

Due to such density, a relevant aspect that can be worked upon is leveraging such "wireless local-loop" by developing networking mechanisms that allow adequate resource management and a future Internet architecture to scale in an autonomic way. Such wireless local-loop could then reach rates closer to the ones provided by current access technologies, while using a lighter management infrastructure.

The EU ULOOP project [8] is investigating and implementing technology to overcome the limitation of today's broadband access technologies, expanding the backbone infrastructure by means of low-cost wireless technologies that embody a multi-operator model, i.e., a local-loop based upon what a specific community of individuals (end-users) is willing to share, backed up by specific cooperation incentives and "good behaviour" rules.

This represents a paradigm shift in the Internet evolution, as the user may be in control of parts of the network, in a way that is acknowledged (or not) by Internet stakeholders. In such scenarios where several strangers are expected to interact for the sake of robust data transmission, trust is of vital importance as it establishes a way for the nodes involved in the system to communicate with each other in a safe manner, to share services and information, and above all, to form communities that assist in sustaining robust connectivity models.

This whitepaper describes the role of trust management in the context of the ULOOP project, and how it is used to achieve security in a flexible way, without necessarily implying the use of strong security associations.

2 Trust Management Principles

In ULOOP, trust management and incentives for cooperation are related to under-standing how to define and build circles of trust on-the-fly. Such circles of trust are capable of sustaining an environment for allowing devices to share resources in order to support the dynamic behavior of user-centric networks [11]. Trust management is based on reputation mechanisms able to identify end-user misbehavior and to address social aspects, e.g., the different types of levels of trust users may have in different communities (e.g., family, affiliation). In situations where the created network of trust is not enough to allow resources to be shared, ULOOP devices are able to use a cooperation incentive scheme based on the transfer of credits directly proportional to the amount of shared resources.

Another key aspect relates to the development and validation of a set of methods and techniques that make it possible to optimize network resources in regards to social behavior, i.e., exploiting shared interests or OSN information to create/optimize/add trust to ULOOP communities.

2.1 Notions

In ULOOP, there are two fundamental roles: *ULOOP node* and ULOOP *gateway* [15].

- A **ULOOP node** concerns a role (software functionality) that a wireless capable device takes. Concrete examples of nodes can be specific user-equipment, access points, or even some management server.
- A **ULOOP gateway** is a role (software functionality) that reflects an operational behavior making a ULOOP node capable of acting as a mediator between ULOOP systems and non-ULOOP systems—the outside world. This gateway role may or may not be owned and controlled by a ULOOP user, it may also be controlled by an access operator. The key differentiating factor of the role of gateway, in contrast to a regular ULOOP node, is the operational intelligence and mediation capability.

Similarly to ULOOP nodes, the ULOOP gateway functionality may reside in the user-equipment, in Access Points, or even in the access network. Hence, they exhibit a feature that is key in user-centric environments: their behavior as part of the network is expected to be highly variable. Gateways will be active or inactive based on several conditions such as users' wishes and network load.

From a trust perspective, we consider two additional definitions: a *requestee* and a *requester* in a trust negotiation process. A requestee in ULOOP can only be a ULOOP gateway, while the requester role can be assumed by both, a node and a gateway: nodes perform trust negotiation towards gateways; gateways perform trust negotiation among themselves.

2.2 Requestee: Requesting Trust

In ULOOP, a requester goes through four stages: (i) boot up; (ii) requestee discovery; (iii) data transfer; (iv) dispositional trust adjustment.

The boot-up phase is present in any ULOOP node, be it a requestee or a requester, since it aims to establish the initial set of conditions for the participation in a ULOOP community.

From a requester perspective, this implies generation of its virtual identity. Based on this virtual identity the requester initiates the creation of a set of trust parameters, process that we name as dispositional trust setup. This process will influence the way the requester is willing to cooperate with other ULOOP nodes.

Since the ULOOP trust environment may not be enough as an incentive for cooperation, the boot up phase ends up with the assignment of a set of credits that the requester may use to access shared resources.

While in idle mode, the requester behaves as any Wi-Fi node, listening passively to wireless messages (beacons) sent by nearby Access Points (AP)—ULOOP gateways, which may take the role of requestee. Based on collected information the requester will try to establish trust associations with one of the available ULOOP gateways via the regular MAC Layer attachment process. In other words, the requester contacts gateways available by sending, in the MAC association frames, a few parameters required to try to establish a trust association and hence, to get access to a specific service, e.g. Internet access.

2.3 Requester: Providing Trust

As previously mentioned, a requestee in ULOOP is always a gateway that may offer resources to a ULOOP node or to another gateway. The functionality of a requester has four major blocks: (i) boot up, (ii) cooperation request process, (iii) data reception, (iv) monetization and dispositional trust adjustment.

After boot-up and following the regular process of an AP, the requestee emits periodic wireless messages (beacons) announcing its presence to nodes around. As a response, the requester will send a tuple providing indications about its dispositional trust and resource thresholds. This information will allow a requester to attempt a connection with the different gateways around as part of the regular MAC Layer attachment process.

Data transmission will start with the reception of a request to send from the requester, and will proceed only if the requestee has incentives to cooperate with such requester based on the established trust association only [12].

In case the requestee is not motivated to cooperate only based on the trust association with the requester, the latter will have to send an explicit request for cooperation. After the reception of such request, the requestee will trigger a cooperation incentive scheme, which includes the negotiation of the number of credits that the requester should transfer to the requestee at the end of the transmission, in case the requester has enough credits for the requested resources [6].

2.4 Operational Example, Untrusted Environment

We start by providing a description of what occurs if an unknown node (requester) arrives to a community and wants to use some shared resources. Then the requester will send beacons aiming to setup a trust association with some ULOOP gateway (requestee), provided that: (i) the requestee allows such interaction to occur (dispositional trust aspects) and; (ii) there are gateways in the community which are willing to accept requested from unknown nodes to the community (dispositional trust aspects from the gateway).

This process triggers social trust computation. For the case of an unknown node, this process implies that for some period of time the node will have a basic trust level which could simply relate to the provisioning of basic services, e.g. Internet access, HTTP port 80 only, and 1 % of bandwidth. As this is an unknown node to the community, most likely developing trust associations will not be enough to motivate requestees to cooperate with such a new element. **Hence, it is most likely that in a scenario with unknown nodes, the cooperation incentive scheme will be triggered for any data transfer requested by a new node, until an adequate level of trustworthiness can be reached**—just looking from a trust framework perspective, this implies that the new node has to be able to somehow share resources that imply that the new element must share resources in order to see a positive change in its trust level. Another way to achieve this is through cooperation incentives [12].

2.5 Operational Example, Trustworthy Nodes

Assuming we have a well-established ULOOP community, where some nodes have reached some level of trustworthiness, the dynamic conditions of a ULOOP network may lead to changes in the trust level among any pair of ULOOP devices over time. For instance, it may happen that the requestee does not have the required dispositional trust (which is adjusted over time), or the requester did not yet reach a trust level required by a particular requestee. This process triggers social trust computation, leading to an adjustment of the trust association value between requester and requestee.

After the adjustment of the trust association between requester and requestee, it may still happen that the requestee is not willing to cooperate. This may happen, for instance, if the amount of resources that the requestee has available is low, and extra incentives may be needed to associate some of those resources for the transmission requested by the requester. In this case the cooperation incentive scheme will be triggered, in order to allow requester and requestee to negotiate the credit value of the needed resources.

3 ULOOP Trust and Cooperation Framework

In ULOOP, trust management and cooperation incentives are related to the under-standing of how to define and build circles of trust on-the-fly. Such circles of trust aim at sustaining an environment for allowing devices to share resources and to support the dynamic behavior of user-centric networks [5]. Trust management is based on reputation mechanisms able to identify end-user misbehavior and to address social aspects, e.g., the different levels of trust users may have in different communities (e.g., family, affiliation). In situations where the created network of trust is not enough to allow resources to be shared, ULOOP devices are able to use a cooperation incentive scheme that allows a node to gain credits in an amount directly proportional to the amount of shared resources: such credits can then be used to gain access to other resources.

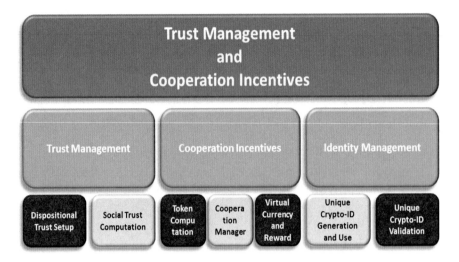

Fig. 1 ULOOP trust framework

Hence, trust management and cooperation incentives framework is split into three main blocks: (i) Trust management; (ii) Cooperation Incentives; (iii) Identity management. These major blocks are illustrated in Fig. 1.

As illustrated, we have further split each of the blocks into sub-blocks, which correspond to different object-oriented modules as has been explained in D2.3 [13], D3 [14] and finalized lately in D3.1 [4].

The ULOOP Trust Framework can be deeply exploited to provide the cooperation incentives required to induce users equipped with Wi-Fi-enabled devices to become members of the ULOOP community and to adopt a pro-social behavior. First of all, the existence of a trust-management system that allows end-users to set their own dispositional trust level lowers the access barrier for reluctant users. Once users have become ULOOP members, the Trust Framework triggers a positive feedback by inducing each member to share his/her own resources with other members in order to increase his/her reputation that is essential, in its turn, to gain access to someone else's resources at better conditions.

In the paradigmatic case of a ULOOP node requesting Internet connectivity to a ULOOP gateway, both the willingness of the gateway to cooperate and the service level granted to the node can depend on the trust of the gateway on the node. On the other hand, in case of multiple gateways offering the same service, the choice of the node can depend on his/her trust on the gateways. As a consequence, higher trust levels lead to more opportunities of cooperation, which, in their turn, provide the chance of further increasing the reputation of the parties involved.

Another key aspect relates to the development and validation of a set of methods and techniques that make it possible to optimize network resources in regards to social behaviour, i.e., exploiting shared interests or On-line Social Networks (OSN) information to create/optimize/add trust to ULOOP communities.

4 Trust Components in ULOOP

4.1 Virtual Identities: The Unique Crypto-ID

Virtual identities of ULOOP nodes and gateways are used in the process of computing and managing the set of trust associations among any pair of ULOOP devices. The goal is to mitigate the impact of impersonation and non-repudiation, while insuring the right level of privacy, e.g. by relying on PET. ULOOP considers identity-based cryptography to generate virtual identities, i.e., *crypto-IDs*. Such identity is associated to a single user, who may own more than one end-user device.

Once in place, it is more familiar for the user to manage one virtual identity and to avoid attacks based on the use of different virtual identities per user, for example, preventing voting twice. Moreover, using a unique crypto-ID avoids a potential complex process of identity disambiguation.

After the verification of the identification of a ULOOP user, the crypto-ID generated based on such identification is expected to be used across any country. Nowadays, most countries are implementing already digital identity systems in order to automate most of the national services. One such example is Portugal, where the citizen card has an embedded chip with a one-time generated crypto-ID that is used to authenticate the holder of the card against different services. Moreover, some of those services, such as changing the address or some other personal information, require an extra secret key that is provided with the card. This process will allow unique crypto-IDs to be generated based on any system that EU countries will decide to implement in the future to identify their citizens electronically.

In ULOOP, the crypto-id of a new node that is introduced into the system is calculated using the SHA256 cryptographic hash function producing a message digest of 256 bits over the public key of the user.

Unique crypto-IDs are generated based on a set of information extracted from the user's device, namely a public key. Such information is used to generate a unique crypto-ID based on a hash function, taken over the previous piece of information, which is implemented in any ULOOP node or gateway. The local generated crypto-ID will need to be verified by an authorized entity in order to allow the ULOOP node/gateway to gain full access to the ULOOP community. While such verification does not happen, the ULOOP device gets a minimum trust level in the community, allowing it to use a predefined set of minimum resources.

In ULOOP, owners (users) are likely to be responsible for more than one active device as previously mentioned. One of such devices is considered to be a primary device, and the remainder equipment shares the same crypto-ID generated by the first personal device, as well as the reputation level and trust associations associated to the unique crypto-ID. This is possible by using secure in range wireless or wired communications. Synchronizing the reputation levels and trust associations among personal devices will allow the user to always make use of the earned reputation level, trust associations and credits that resulted from the usage of the unique crypto-ID in another personal device. Synchronization of trust information can be done by using prior-art on file and data synchronization.

The validation of the unique crypto-ID can be done by making use of any opportunity to access the Internet (limited Internet access should be allowed by the minimum trust level). This may create some problem in extreme cases, in which Internet access is not possible for a long time. However, such scenarios are more related to delay-tolerant networks than to ULOOP. In the latter case it is expected trust management and cooperation incentives to create the conditions to make Internet access more pervasive than today.

Some user data is required for the validation: first name, last name and mobile phone number in order to be able to perform the SMS validation. Moreover the user is asked to choose a nickname that will be linked to the crypto-id. Verification near the Identity Management System will ensure the uniqueness of the nickname. The Identity Management System, owned by the Identity Validator, proves the ownership of the provided mobile phone number sending a SMS with a secret to that mobile phone number.

The ULOOP node intercepts incoming SMS messages (with a predefined format) and when recognizes the message sent by the Identity Management System, it sends back the secret received in the SMS via http together with the chosen nickname and other additional information.

If the Identity Management System recognizes that the replied data and secret are the same stored for that Crypto-Id in its database, then the validation can be considered completed and the confirmation is sent to the node together with an X.509 [9] certificate. The purpose of the X.509 certificate is to bind the public key of the node to a particular distinguished name or to an alternative name such as an e-mail address, or in ULOOP case, a *nickname*. The node marks the Crypto-Id as validated and stores the X.509 certificate.

4.2 Dispositional Trust

Dispositional Trust (DT) is defined in ULOOP as the general willingness of a given user to trust others. As such, in a first implementation of DT, the owner of a ULOOP node will set up this value manually. The DT setup is done in the boot-up phase as explained in Sect. 2 and it may or may not remain the same during the node's lifetime. However, as it might provide a better protection of the ULOOP owner, depending on the surrounding environment of the node and other external factors, an adaptation process may be carried out to readjust DT automatically [10] or after asking the user in order to protect the node's integrity. For the first implementation of ULOOP, we consider a single set-up, without changes during the trust negotiation aspect. Adaptation is an aspect that we expect to address during year 3 of ULOOP.

The dispositional trust module allows the user to configure a personal device with his/her disposition to trust other devices. If the user has multiple personal, he/she only has to set his/her dispositional trust for one device: the others will get that information from the first one when in direct contact. For the first device, the owner is prompted to set its DT, e.g. being able to select from a list of predefined values, which range from 0 to 1, being 0 "paranoid", which means that a priori the node will not trust anyone, and being 1 "blind trust", which means that the node will trust no matter what.

If the device is not the first one being configured, the user is presented with two options: (i) to clone the dispositional trust level assigned to other devices that are already in ULOOP and that she/he owns, as described in D3 [14, Sect. 2.1.4.1.1] for the usage of unique crypto-IDs in different personal devices: (ii) to assign a new DT level for the node being introduced, as explained in the previous paragraph.

Since some nodes are carried by Internet end-users, their networking composition, surrounding environment and organization can rapidly change. As such, the dispositional trust level on a given node might not be appropriated in all circumstances and should be able to be adapted and changed over time, in order to protect the node's integrity. The process of dispositional trust adaptation might occur in two different cases:

- The node has a dispositional trust level that is inappropriate and leaves it too open to attacks.
- The node joins a different community to the initial one in which the dispositional trust level had been setup.

In the first case, every time a timeout occurs a social interaction analysis process is triggered, and the evidence collected together with other evidence such as QoS and attack evidence is used to determine if the dispositional trust level should be re-adjusted. This level can be readjusted manually by the user if so specified, or automatically by the system if not.

In the second case, every time the node joins a different community, the functionality checks if it is the first time that the node joins or not, and according to that, he user is prompted to specify a new dispositional trust level to use while in that community. This value could be as well automatically re-adjusted according perhaps to the social relationship with the nodes that are present (already known nodes, unknown nodes, their trust values, etc.).

4.3 Trust Computation

ULOOP considers the use of computational trust management as a complementary approach to security where a level of trust in the requesting entity is automatically computed based on different types of evidence.

Nodes are associated to other nodes by means of *trust associations,* as illustrated in Fig. 2. A trust association T_{ijk} is the k-th directed association between nodes i and j, and is related to the respective owner's interests and social networking perspective. A trust association holds a cost which we name as *trust level.* The trust level provides

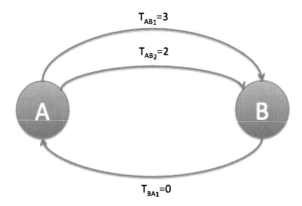

Fig. 2 Social trust association examples

a measure of previous trust behavior, of *Quality of Experience (QoE)* of nodes, etc. Hence, two nodes may in fact hold more than one trust association among them, as represented in Fig. 1, where nodes *A* and *B* hold three different trust associations: A has two trust association to node B, T_{AB_1}, which relates to the exchange of data owned by A (where A is the originator), and T_{AB_2}, which relates to the exchange of information which A is relaying (A is not the source). B has one trust association to A, T_{BA_1} with a cost of zero which e.g. could mean that B still does not trust A to relay his data. As shown in Fig. 2 the different trust associations have a specific cost, and the computation of such cost is based upon the nodes expectations and beliefs.

The weight of a specific trust association considers *local* and *external* influences. Examples of local influences are the degree of connectivity and reputation level of node B. External influences are influences that do not relate to the nature of each node but to external networking conditions (e.g. too much overhearing probability around node B).

For instance, if Alice has the choice to connect to two nearby ULOOP gateways, Charles' gateway and Bob's gateway but she has never interacted with Bob's gateway before. Bob's ULOOP gateway is the gateway that would give her the quality of service she requires. Fortunately, she has already interacted with Charles' gateway. As she has no direct observation of Bob's gateway, therefore, she asks Charles for a recommendation. Charles has already used Bob's gateway and sends his recommendation to Alice. A third type of evidence used in ULOOP concerns reputation, which is the aggregation of different recommendations from different recommenders that are not exactly known. That reputation value may come from the aggregation of evidence external to ULOOP, for example, from the mining of existing online social networks.

The computation of trust is provided by a function implemented in ULOOP nodes and gateways. Trust computation is a dynamic cost function that has to be sufficiently strong to provide, based on a local perspective, attack resistance. It comprises therefore the dispositional trust of a node, as well as evidence concerning contacts with other nodes.

One of the main challenges of the final ULOOP trust metric will be to make it attack-resistant such as resistant to the Sybil attack. To be able to bootstrap the ULOOP community, it will also be important to have a good number of users who are generally disposed to trust others. For this reason, cooperation management depicted in the next section will try to reward those users who are essential to sustain a high level of cooperation.

5 Incentives and Rewarding

As mentioned earlier, trust allows users to seamlessly exchange services and resources without worrying about security issues. However, trust also can prevent the ULOOP communities to grow (i.e., scalability), added to the fact that it can also create disjoint communities (i.e., egoistic, better-than-others behaviour). And clearly such characteristics go against the dynamic communication found in user-centric networks.

In order to avoid ULOOP communities not to scale and the appearance of disjoint groups of users—which indeed have a negative impact in exchange of services/resources, cooperation incentives [1–3, 7] are used to allow new (and untrusted) users to join these communities and encourage users already in the trusted communities to interact with untrusted ones.

Incentives are given in the form of cooperation credits which, once earned, can be used to obtain more services/resources in other untrusted environments (or even exchanged for virtual currency as later explained in the Rewarding process in this section).

The Cooperation Manager (CM) is the entity responsible for coordinating the cooperation. We briefly explain the function of the CM and direct the reader to a more detailed view in [12]. First, CM assigns an initial amount of cooperation credits to new ULOOP nodes. Then, it controls the cooperation process among nodes (where credits are used for encouraging cooperation). CM also periodic evaluates cooperation to check whether or not cooperation took place as negotiated.

By providing new ULOOP users with an initial amount of cooperation credits, will allow them to interact with others who will see them as untrusted users. The more users cooperate, the more cooperation credits will be earned. This consequently allows already established ULOOP communities to scale and prevent the formation of disjoint communities. Additionally, the evaluation process allows for penalizing cooperation misconduct, which guarantees seamless future interactions between ULOOP users. Cooperation can take place in two different modes, namely volunteer or retailer. In the former, cooperation happens for the 'greater good' of the users and a symbolic amount of cooperation credits is exchanged between the involved ULOOP users. As for the retailer cooperation mode, services/resources are sold for an agreed amount of cooperation credits. At this point the cooperation process is further handled by the Reward Manager, which will decide whether the Requestee accepts or refuses to engage in the cooperation considering the amount of credits offered by the Requester.

This last rewarding mechanism is implemented in ULOOP to provide further cooperation incentives and to allow the ULOOP community to become part of the Internet value chain. Such a mechanism is based on a custom virtual currency system the details of which are provided in a separate document [6]. As long as ULOOP credits are used only within the ULOOP community, they work as an additional guarantee of reciprocity: the user who provides a service/resource earns ULOOP credits that can be spent later on to pay for other services/services. ULOOP credits can eventually be traded for money at the boundaries of the ULOOP community (discussing this case is outside the scope of this white paper).

Although the virtual currency system is orthogonal to the trust management system, the two systems interact with each other is three main ways. First, the trust of the requester on the Requestee acts as a discount factor, which allows trusted Requestees to pay less. Second, the trust of the requester on the Requestee increases the earning opportunities of members acting in retailer mode. Finally, the Trust Framework of ULOOP is deeply exploited in the implementation of the virtual currency system to reduce discourage cheating behaviors.

6 Conclusions

Trust Management in ULOOP is used to obtain security in a flexible way without the use of strong security associations. This is achieved through:

- The use of a solid trust management framework including a combination of dispositional trust levels and social trust metrics and computation.
- The use of virtual identities represented by unique crypto-IDs (one crypto-ID per user).
- Introducing cooperation incentives to complement and strengthen the trust framework and metrics.
- Introduction of rewarding mechanisms.

References

1. Aldini A (2013) Formal approach to design and automatic verification of cooperation-based networks. IARIA Int J Adv Internet Technol 6(1&2):42–56
2. Aldini A, Bogliolo A (2012) Model checking of trust-based user-centric cooperative networks (AFIN), IARIA, August 2012. Rome, Italy, pp 32–41
3. Aldini A, Bogliolo A (2013) Modeling and verification of cooperation incentive mechanisms in user-centric wireless communications. In: Rawat DB, Bista B, Yan G (eds) Security, privacy, trust, and resource management in mobile and wireless communications. IGI Global, Hershey

4. Ballester C (ed) (2012) ULOOP consortium, D3.1: trust management and cooperation incentives pre-prototype software. EU FP7 IST ULOOP project (grant number 257418) deliverable, June 2012
5. Ballester C, Seigneur J-M, di Francesco P, Moreno V, Sofia RC, Bogliolo A, Martins N, Moreira W Jr (2013) A user-centric approach to trust management in Wi-Fi networks. In: IEEE INFOCOM—Demos Track. Turin, Italy
6. Bogliolo A (ed) (2012) Crediting aspects in ULOOP, EU FP7 IST ULOOP project (grant number 257418) white paper, August 2012
7. Bogliolo A, Polidori P, Aldini A, Moreira W, Mendes P, Yildiz M, Ballester C, Seigneur J-M (2012) Virtual currency and reputation-based cooperation incentives in user-centric networks (IWCMC), August 2012. Cyprus, pp 895–900
8. EU FP7 ULOOP Project http://www.uloop.eu
9. Housley R et al (1999) Network working group, RFC 2459: internet X.509 public key infrastructure certificate and CRL profile. http://www.ietf.org/rfc/rfc2459.txt. Accessed Jan 1999
10. Lafuente CB, Seigneur J-M (2013) Dispositional trust self-adaptation in user-centric environments. In: AINA 2013, 26–28th March 2013. Barcelona, Spain
11. Lafuente CB, Seigneur J-M, Moreira W, Mendes P, Maknavicius L, Bogliolo A, di Francesco P (2012) Trust and cooperation incentives for wireless user-centric environments. In: IADIS e-Society 2012, 10–13 March 2012. Berlin, Germany
12. Mendes P (ed) (2012) Cooperative networking in ULOOP, EU FP7 IST ULOOP project (grant number 257418) white paper, August 2012
13. Sofia R (ed) (2011) ULOOP consortium, D2.3: ULOOP overall specification. EU FP7 IST ULOOP project (grant number 257418) deliverable, Sept 2011
14. Sofia R (ed) (2011) ULOOP consortium, D3: ULOOP high level architecture specification. EU FP7 IST ULOOP project (grant number 257418) deliverable, Dec 2011
15. Sofia R, Marce O (eds) (2011) User-centric wireless local loop framework. EU FP7 IST ULOOP project (grant number 257418) white paper

Designing a Trust-Based Virtual Currency System for User-Centric Networks: The ULOOP Case

Lorenz Cuno Klopfenstein and Saverio Delpriori

Abstract The functioning of mobile networks, and especially mobile ad-hoc networks (MANETs), is based on the assumption that a subset of the nodes that compose the network are able and willing to cooperate, for instance forwarding packets on the behalf of others. In particular, in user-centric networks based on local-loop forwarding, where resources might be constrained, there is no direct gain in altruistic behavior. Incentives are needed to overcome the selfishness of every participant of the network. The ULOOP project envisions a user-centric network system where nodes are persuaded into cooperation by the presence of both trust-based and monetary incentives. The transfer of monetary units in exchange for resources requires a credit interchange protocol between nodes that satisfies the particular criteria of an ULOOP network. Such a protocol is presented in the following paper and is then discussed and evaluated on its particular properties and constraints.

Keywords User-centric network · Virtual currency · Reward · Cooperation incentive

1 Introduction

During the last years, modern communication devices started becoming widespread, to the point of being ubiquitous, and are relied upon throughout the whole day, never leaving our sides. While traditional computing devices still serve specific needs in specific places, many devices follow their owners during their everyday lives and serve a variety of needs and tasks. Most of these devices, including smartphones,

L. C. Klopfenstein (✉) · S. Delpriori
University of Urbino, Urbino, Italy
e-mail: cuno.klopfenstein@uniurb.it; lck@klopfenstein.net

S. Delpriori
e-mail: saverio.delpriori@uniurb.it

A. Aldini and A. Bogliolo (eds.), *User-Centric Networking*,
Lecture Notes in Social Networks, DOI: 10.1007/978-3-319-05218-2_6,
© Springer International Publishing Switzerland 2014

multimedia players and gaming consoles are capable both of accessing the Internet and of creating local ad-hoc networks with other gadgets. Devices are increasingly powerful and are becoming capable of providing services to other participants on the network. The potential of *User-Centric Networks* (UCN) is starting to be explored in this context, where not only hardware, but also services and resources are provided by the users themselves, and can be shared in a peer-to-peer fashion.

This "willingness to share" is the foundation for cooperation-based networks such as UCNs: operation and utility of an UCN depends more on social aspects than on technological ones. For instance, a wireless UCN may not be based on any fixed infrastructure, but may be built on heterogeneous and opportunistic links among nodes, which spontaneously form a network. A wireless UCN that has no users, not being defined by a physical part, in fact does not exist. Moreover, it must be taken into account that an UCN has to deal with a fragile infrastructure (if any) and parts of the network which may be heavily mobile. These characteristics make an UCN complex to control and potentially vulnerable to malicious users.

Nevertheless, a wireless UCN relies exclusively on the cooperation among nodes in a network. Since nodes are usually constrained by their limited resources, some nodes may also decide to act selfishly in order to gain services without sharing theirs.

Reputation and monetization are two possible solutions to the non-cooperation problem: a *reputation system* evaluates each node's behavior and makes it possible to distinguish between a trustworthy and a malicious or selfish node. On the other hand, a *reward system* uses a virtual currency unit to control transactions among nodes and thus provide service access against a form of virtual payment.

ULOOP proposes an integrated system based on both reputation and reward, which aims to mitigate the limitations of both approaches. Analytical studies based on game-theory [7, 11] have demonstrated that mixed incentive strategies, combining reputation and price-based mechanism, are effective in inducing pro-social behaviors, while also isolating selfish nodes in an open community. In [1–3] a formal study of the benefits of the joint usage of these mechanisms was developed and determined that cooperation incentives work properly for both parties of a service exchange, motivating honest behavior with a higher number of accepted requests and a lower average cost obtained when compared with the results obtained by a cheating party.

In the next section, the overall resource sharing system of ULOOP will be discussed, detailing the trust and rewarding mechanism and the service sharing protocol adopted by the nodes of a network.

Section 3 will describe the virtual credits transfer protocol in detail, including the overall design goals and the current technical implementation. Finally, a short discussion about potential improvements to the protocol design will be given in Sect. 4.

2 Resource Sharing in ULOOP

An ad-hoc network is composed of a number of nodes interacting with each other and cooperating in order to guarantee a degree of functionality and utility of the network. Network utility for participating nodes is usually dependent on the ability of each node to send and receive information from and to other participants. Even from a selfish point of view, guaranteeing a fair share of network utilization from each other node can be seen as useful.

However, selfishness must be taken into account because the impact of a non-cooperative node can be detrimental to the overall quality of the service. The degree with which a node is willing to grant services to another node depends on its "willingness to cooperate". In ULOOP, this parameter is based on the so-called *dispositional trust value*, which is a fundamental setting of the node itself (set by the owner or the user of the node, determining the node's disposition to cooperate), and is then relied upon by the trust mechanism of the system. The dispositional trust and trust mechanism are described in depth in the previous paper of this book [4].

Furthermore, while nodes can rely only on others *dispositional trust* settings, a node can improve the chance that another node may be willing to invest resources and grant access by using *incentives* during the cooperation request (see [5]). As mentioned before, ULOOP is based on a double system of incentives: *reputation* and *monetary rewards*.

The incentive scheme of ULOOP relies on the following principles:

- Reputation, expressed as the collective perception of a user trustworthiness within a community, can affect cooperation decisions and opportunities, insofar as more trustworthy nodes are picked with higher likelihood as cooperation partners.
- Reputation cannot be traded and cannot be manipulated, except through cooperation. By cooperating successfully, nodes can build a reciprocal trust bond.
- Monetary rewards are based around a virtual currency, called *credits*, mainly used to facilitate cooperation and to guarantee reciprocity. A node, even if untrustworthy, can prompt another node to cooperate if the payment is deemed sufficient.
- Virtual currency flows freely through the network, but in order to keep inflation at bay, it is only created *ex-nihilo* when a new node joins the ULOOP network. Each node obtains an amount of initial credits proportional to its dispositional trust: credits are assigned to new nodes to compensate the benefit the network at large gains from the node because of cooperation opportunities and positive externalities. Thus, the more trusting the node is, the more credits it obtains on start-up. The value of the network increases as new users join, in a controlled way.

Interactions in an ULOOP network occur between nodes assuming two complementary roles: the *requester* and the *requestee*. Each node may assume both roles at different times, depending on its goals and capabilities.

- A **requester** has the need to acquire some resources which the node lacks (usually, Internet access) and will request those resources from a node in communication range.
- A **requestee** receives a cooperation request from a requester and may evaluate whether to accept the cooperation (thus assigning some of its resources to the requester node) or to refuse it.

Of course, nodes that cannot provide resources will always operate as requesters. Gateways on a wireless network will instead usually operate as requestees for Internet access.

The cooperation process is managed by an ULOOP component named *cooperation manager*. This component will attempt to negotiate for resources when needed. Monetary incentives are always dispatched along with the request, in order to increase the chances of success if needed.

The cooperation request is then evaluated by another component: the *reward manager*. This component determines whether the incentives offered are sufficient to grant access to the resources asked for. Approval conditions can depend on a certain number of factors and could potentially be different for each ULOOP node (for instance, a node might ask for a higher share of monetary incentives while another one might base its decisions mainly on trust).

Reward managers can operate in two main rewarding modes, which affect how they react to incoming cooperation requests: they can act as a *retailer*, or as a *volunteer*, depending on the settings chosen by the owner of the node. Both modes can contribute to an ULOOP network and selecting a mode over another does not influence the node's capability to ask for cooperation on its own. *Retailer* nodes will always exchange their services for monetary incentives, effectively forcing requesters to pay for their cooperation. This mode of operation allows infrastructural nodes, for instance, to sell Internet access or other networking services. Nodes operating in *volunteer* mode will accept any monetary incentive, but do not restrict access: the presence of an incentive does not influence the likelihood of the requestee accepting the request.

The requestee has the option of accepting or rejecting any cooperation request, based on trust and monetary principles. When rejected, the cooperation is terminated and can be attempted again at the requester's leisure. However, notice that any credits sent along with the cooperation request are considered to be lost and cannot be reclaimed by the requester for successive attempts.

Since each node introduces new value by obtaining new credits on start-up, which are occasionally lost through exchange as seen above, the incentives mechanism works like a *fixed circulation* stamp-trading scheme as described in [9].

2.1 ULOOP Node Lifetime Phases

During its lifetime, an ULOOP node goes through the following phases:

- **Bootstrap phase**: performed once for each network node. The initial *dispositional trust* of the node is set and the initial amount of credits is assigned to each node.
- **Cooperation phase**: one of the users acts as a *requester* and looks for an ULOOP member providing the required service or the required resource. The requester decides the amount of credits to be offered to the selected requestee.

- If the requestee is operating in *volunteer* mode, it will always accept any credit transfer and refuse the request only when lacking resources.
- If the requestee is operating in *retailer* mode, it may decide whether to accept the request or not based on the the requester's reputation and on the monetary incentives provided.

When a cooperation is accepted the transaction terminates, credits are transferred and the requestee provides the service or the resources to the requestee.

- **Evaluation phase**: at any time, the requester may evaluate the requestee by comparing the agreed service levels and those actually provided. This periodic evaluation will influence the trust ratings among nodes and is important to discourage misbehaviours.

3 Credit Transfer Protocol

3.1 Design Principles

The protocol envisioned for ULOOP must be reasonably *secure*, it can rely on the actions of an external authority (the *Bank*), but must be designed to support temporary *offline operations*. Payments must be undeniable, reasonably traceable and allow the authority to prevent fraud and cheating (see [12]). Payments do not hold an intrinsic value, so monetary loss prevention is not of paramount importance.

Any payment-based credit transfer protocol demands some sort of guarantee that the credits circulating are legal and cannot be spent twice for different transactions. This can be achieved in different ways, for instance using a centralized authority or a tamper-proof secure module in each node, as discussed in [6]. In our crediting system, a centralized trusted authority, namely the Bank, will generate credits. The Bank generates all credits and assigns them to each user of the system. No user can spend more credits than he can afford, in principle.

Credits are assigned when a node joins the ULOOP network. The virtual currency is injected into the economic system and then relies on the traditional money economy in order to circulate, instead of being regenerated or recharged by the authority itself. This is done mainly not to require a tight attachment to the Bank during operations.

Payments are based around the concept of *Promises of Payment*: these are the messages passed between nodes, which stand for a credit transfer. The generator of such a Promise essentially binds a subset of its credits to the promise and sends it to the receiver. The receiver can already count the received credits as part of his funds, but eventually needs to *cash-in* the Promise by contacting the Bank. The Bank will check whether the Promise is valid and will then confirm the credit transfer.

Promises of Payment can be transferred without contacting the Bank beforehand. Credits received from a payment can also be used for another payment without first cashing-in, if the new payment is covered by the credit balance last known to the Bank (i.e. a node should not overdraw its Bank account when generating payments).

They need to be undeniable, meaning that only the owner can generate a Promise using its own credits; traceable, meaning that sender and receiver of a Promise are clearly identified for each payment; and secure.

While payments can be transferred between nodes without contacting the Bank, cheating detection requires the Bank to operate: the central authority has the ability to deny a fraudulent payment and inform the payee. When cheating behavior is detected, the invalid payment is ignored and no credits are transferred. The nodes involved can react to the situation by adapting their trust metrics in order to punish misbehaving nodes.

Payments do not require any confirmation, thus credits invested in a Promise of Payment are deemed lost both if the payment is confirmed and if it is rejected.

3.2 Payment Protocol Overview

ULOOP requires that new nodes, when first joining an ULOOP community, must register their identity and may optionally validate it in order to be legally identified. The payment protocol piggybacks this initial step and, as soon as the identity of the user is ready (in the form of a *CryptoID*, see [4]) the Bank is contacted. This process creates a *bank account* for the user which will contain its initial credits assigned by the system.

Available credits are synchronized and kept in a local wallet. From that point on, Bank access is needed only to eventually cash-in payments.

When a payment between nodes is required, the payer can simply extract credits from its wallet and create a so-called Promise of Payment object. This object is signed by the payer and thus constitutes proof of the intent of payment from payer to payee. The signed object is transferred to the payee (usually inside a service request) and the payer gets access to the payment, while being able to verify its author.

Credits are considered to be transferred automatically and local wallets can be updated accordingly.

However, nodes that have received a payment may wish to periodically contact the Bank in order to check for fraudulent payments and also to synchronize the status of the bank accounts.

3.3 Technical Protocol Specification

3.3.1 The Bank

The Bank is the central authority of the virtual currency system in ULOOP and is implemented as a RESTful web service. Each action the Bank offers (mainly opening accounts, checking payments and synchronizing bank accounts) is exposed as an HTTPS endpoint. Data is usually encoded using the JSON format.

3.3.2 Registration

A client willing to participate in a ULOOP Network, needs to go through the registration process, in order to be acknowledged by the central authority and to obtain his own initial amount of credits.

During the bootstrap phase of each ULOOP node, the client will contact the central authority with a registration message, containing his personal CryptoID and the initial amount of credits for its account. The initial amount of credits is determined by the *cooperation manager* [8], based on the amount of dispositional trust set by the user. The amount is also checked by the Bank for correctness.

The registration message is signed using the node's private key (generated at the same time the CryptoID is generated during bootstrapping). The public key is appended to the message and sent to the bank through a HTTPS request.

The Bank will validate the registration (in order to prevent double registrations and requests of too many initial credits). If it is correct, the Bank will return a confirmation and the bank account is now open: the node can update its wallet with the amount of credits requested and payments can be generated to pay for services. The Bank will store the node's public key inside the bank account in order to be able to identify the node in further exchanges.

On the other hand, if the registration is not valid, the Bank will return a failure code and no bank account will be opened. The node must attempt a new registration before being able to submit any payment.

$$reg = < sign_{PvtK}(CryptoID, c), PubK >$$

Where CryptoID is ID of node, c is number of credits, *sign* the RSA signing function, PvtK private key and PubK public key.

3.3.3 Payment Generation

When node A (requester) makes a service request to node B (requestee) on a ULOOP network, the service request message will contain several parameters, among which the payment itself, used as an incentive for the requestee to provide its service.

As described before, the payment represents the commitment by A to provide B with an amount c of credits, which will be transferred by the Bank when the cooperation is accepted and the Promise of Payment is cashed-in by B.

The Promise of Payment (*proof$_a$* in Fig. 1) contains an *unique ID* generated for that specific payment. ULOOP nodes generate a fresh UUID identifier which should guarantee a reasonable lack of collision between IDs generated in the scope of the project. The ID uniquely identifies the payment and also serves as a nonce. Additionally, the object also contains the ID of A, the ID of B (in the form of CryptoIDs) and the actual amount of credits c to be payed.

The final payload to be transferred from A to B contains the Promise of Payment signed by A's private key and the full public key of A. The public key must be

Fig. 1 Registration and
payment messages between
ULOOP components

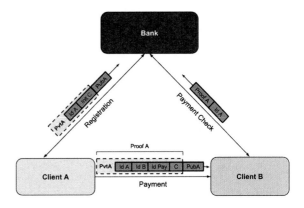

transmitted at each payment because B might not be able to reach the Bank in order
to get knowledge of the public keys of other nodes.

$$payment_{promise} = < proof_a, PubK > = < sign_{PvtK}(ID_{payment}, ID_a, ID_b, c), PubK >$$

Where *PvtK* and *PubK* are private and public keys of node A. $D_{payment}$ is the GUID
identifier of the payment (generated by node A, also used as nonce).

The signed part of the payload is called *proof$_a$*. That payload is kept by the receiver
of the payment.

On receiving a payment, node B is able to verify the validity of the payment: if
the payment is signed correctly, B gains an amount of c credits and keeps track of
the payment in a storage of pending payments to be cashed-in through the Bank as
soon as possible. B will also keep the signed proof of the payment.

3.3.4 Checking Payments

Pending payments can be *cashed-in* as soon as there is a connection to the central
Bank: this cash-in mechanism signals the payment to the Bank, which will check
whether the payment is valid or not. This action requires a single HTTPS request to
the Bank by node B, containing the signed payment by A (*proof$_a$*) and the identifier
of A (which is used by the Bank to determine the public key necessary to check the
payment's authenticity).

$$paymentcheck = < proof_a ID_a >$$

The bank knows $PubK_a$ from the registration phase. Once the proof is authen-
ticated and the payment is determined to be valid, the payment is registered as
cashed-in and cannot be altered or denied. Credits that were removed from A and
added to B's wallet are reflected in the bank accounts of both nodes.

If a payment cannot be confirmed (either because it isn't valid or node A in fact did not own the credits necessary to perform the payment) it is marked as failed and node B does not obtain the credits.

3.3.5 Bank Synchronization

The ULOOP credits system is conceived lo let the nodes continue their transactions even if the central authority is out of reach, therefore a node can build up a set of transactions not checked by the Bank. In order to avoid fraudulent payments and to have his local wallet synchronized with the account kept by the Bank, each node periodically tries to go through the synchronization process which is composed by two sub parts.

First, the node tries to contact the Bank sending a *payment check request* for each one of the pending payments it has received and stored locally. If the Bank can be reached, it will respond by confirming or rejecting the payments. Then a further *get saldo* request is issued to the Bank in order to get confirmation of the actual number of credits available to the node and to synchronize the node's wallet to its authoritative bank account.

3.3.6 Payment Cash-in Race Condition

Having payments checked by the Bank is the only way, for a client A, to be sure to actually collect credits earned from a transaction with a potentially dishonest client B. Say that node B's balance counts 300 credits, and B buys services from client A for 200 credits. Now imagine that client B also needs services from client C for 200 credits while owning only 100 credits. B will have to wait to earn another 100 credits selling services. Because the ULOOP credits system is partially asynchronous there is no way to client C to know how many credits another client has, so client B decides to behave dishonestly and purchase services from C sending him a Promise of Payment of 200 credits, which we know is only partially covered by his current budget.

In a case like this, the principle followed by the ULOOP credit system is that the first client who has the payments checked by the Bank obtains the credits. As outlined in Fig. 2, if a second client tries to obtain credits, verifying a payment received from a client who does not own as much credits as he spent (e.g. the client B), he cannot have those credits recognized by the bank.

In our example, although client C performed the transaction after client A, if C checks the payment from B before A, client A will not be able to check the Promise of Payment obtained from B and will not obtain any credits for the services he might already have provided. While the payment system can work offline from the Bank, this mechanism naturally urges all nodes to perform synchronization with the Bank as soon as possible, in order to protect their own gains, which on its turn ensures that the Bank is kept in a coherent state whenever possible.

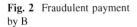 **Fig. 2** Fraudulent payment by B

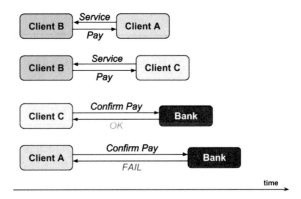

4 Potential Improvements

The current protocol requires frequent public key transmission in order to let nodes authenticate payments, but this of course has a cost in terms of bits transmitted for each payment. This could be circumvented using ID-Cryptography when generating the promise of payment: by regenerating the public key from the payer's CryptoID itself, the size of the payment payload could be reduced significantly (see [10]).

Monetization, i.e. the process of transforming *fiat money* into virtual credits and vice versa, was not implemented in the current protocol. However, in order to provide means for users to acquire new credits in an unbalanced environment and also to provide means for service providers to monetize their service offering, a monetization scheme would be required. This could be reasonably implemented through the central Bank or via a brokerage system.

As mentioned in Sect. 2, credits that are sent along with the cooperation request are considered to be spent and cannot be reclaimed by the requester for successive attempts because of the overall request protocol used in ULOOP. Nodes can eventually recollect credits when synchronizing with the Bank, given that the requestee has signaled the payment as being refused in the meantime. However, this depends on the actions of other nodes and is therefore neither immediate nor guaranteed. More sophisticated negotiation policies are under analysis.

Another option would be to implement a self-recharging virtual currency as described by [6] that would eliminate reliance on fragile mechanism for recycling currency through the economy. The Bank would issue virtual budgets, which could be used by nodes until depletion: no node would run out of credits and no node would be able to hoard credits, possibly manipulating the price of resources.

Acknowledgments The research leading to these results has received funding from the EU IST Seventh Framework Programme ([FP7/2007–2013]) under grant agreement no. 257418, project ULOOP (User-centric Wireless Local Loop).

References

1. Aldini A (2013) Formal approach to design and automatic verification of cooperation-based networks. Int J Adv Internet Tech 6(1–2):42–56
2. Aldini A, Alessandro B (2012) Model checking of trust-based user-centric cooperative networks. In: The 4th international conference on advances in future internet (AFIN 2012)
3. Aldini A, Bogliolo A (2013) Modeling and verification of cooperation incentive mechanisms in user-centric wireless communications. In: Security, privacy, trust, and resource management in mobile and wireless communications IGI Global, 2013
4. Ballester C, Seigneur J-M, Sofia R, Silva C, Moreira W (2014) Trust management in ULOOP. In: User-centric networking—future perspectives. LNSN, Springer, London
5. Bogliolo A et al (2012) Virtual currency and reputation-based cooperation incentives in user-centric networks. In: IEEE 8th international wireless communications and mobile computing conference (IWCMC 2012)
6. Irwin D et al (2005) Self-recharging virtual currency. In: Proceedings of the 2005 ACM SIGCOMM workshop on economics of peer-to-peer systems, 2005
7. Li Ze, Shen Haiying (2012) Game-theoretic analysis of cooperation incentive strategies in mobile ad hoc networks. IEEE Trans Mobile Comput 11(8):1287–303
8. Mendes P, Moreira W, Jamal T, Haci H, Zhu H (2014) Cooperative networking in user-centric wireless networks. In: User-centric networking—future perspectives, LNSN, Springer, London
9. Moreton T, Andrew T (2003) Trading in trust, tokens, and stamps. In: Proceedings of the 1st workshop on economics of peer-to-peer systems, 2003
10. Shamir A (1985) Identity-based cryptosystems and signature schemes. In :Advances in cryptology. Springer, Berlin
11. Srivastava V et al (2005) Using game theory to analyze wireless ad hoc networks. IEEE Commun Surv Tutor 7(4):46–56
12. Tanenbaum AS (1989) Computer networks. vol 4. Prentice-Hall, Englewood Cliffs

Part III
Resource Management in User-Centric Environments

Cooperative Networking in User-Centric Wireless Networks

Paulo Mendes, Waldir Moreira, Tauseef Jamal, Huseyin Haci
and Huiling Zhu

Abstract Mobile social networking is becoming a new trend allowing users to have instant and real-time access from their devices to any Internet based networking platform. This requirement is not fulfilled by current wireless networks, which provide a limited number of wireless access points with instant Internet access. However, the flexibility of wireless technologies is giving rise to new types of wireless access networks, allowing the Internet to expand in a user-centric way. This new user-centric wireless networking trend aims to mitigate such problems, by allowing sharing of wireless resources for a faster and ubiquitous Internet access. The European project ULOOP—User-centric Wireless Local Loop is investigating new ways of deploying wireless local loops by means of low-cost technologies, where resource sharing is backed up by specific cooperation incentives and mechanisms. This paper aims to introduce the new concept of user-centric networks, an analysis of the most relevant incentive models, and a cooperative networking model to sustain the deployment of low-cost user-centric wireless networks based on cooperation incentives and mechanisms.

P. Mendes (✉) · W. Moreira · T. Jamal
COPELABS, University Lusófona, Building U First Floor, Campo Grande 388, 1749-024
Lisboa, Portugal
e-mail: paulo.mendes@ulusofona.pt

W. Moreira
e-mail: waldir.junior@ulusofona.pt

T. Jamal
e-mail: tauseef.jamal@ulusofona.pt

H. Haci · H. Zhu
University of Kent, Canterbury, UK
e-mail: hh219@kent.ac.uk

H. Zhu
e-mail: h.zhu@kent.ac.uk

A. Aldini and A. Bogliolo (eds.), *User-Centric Networking*,
Lecture Notes in Social Networks, DOI: 10.1007/978-3-319-05218-2_7,
© Springer International Publishing Switzerland 2014

Keywords Mobile social networking · User-centric networks · Cooperation model · Resource management

1 Introduction

The technical convergence of the Internet and mobile devices is opening new opportunities for the development of truly pervasive Internet access, supporting social networking activity anytime and anywhere. For instance, although much of the user-generated content available on online social networks is created with mobile devices, few users benefit from broadband mobile Internet services, due to the lack of pervasive wireless Internet access: lack of global deployment of access points, and the existence of access points that are closed behind a strong authentication methods that requires too much intervention from the end-user.

Nevertheless, the flexibility of wireless technologies is giving rise to new types of access networks, allowing the Internet to expand in a user-centric way. This is particularly relevant if we consider that wireless technologies such as Wireless Fidelity (Wi-Fi) complement Internet access broadband technologies, forming the last hop to the end-user in a global way.

A relevant aspect of such density is the possibility to leverage such "wireless local-loop" by developing networking mechanisms that allow adequate resource sharing, as well as an autonomic management of future Internet architectures. Such wireless local-loop could then reach data rates closer to the ones provided by current access technologies, while using a lighter management infrastructure.

The European project ULOOP—User-centric Wireless Local Loop—(Reference 257418) is investigating concepts and developing new technology to overcome the limitation of today's broadband access technologies, by developing models and technology suitable for the deployment of User-Centric Networks (UCNs). Such technology is expected to assist in expanding the backbone infrastructure by means of low-cost wireless technologies that embody a multi-operator model, i.e., a local-loop that a specific community of individuals is willing to share, backed up by specific cooperation incentives and mechanisms.

The purpose of this paper is three-fold: (i) to introduce the concept of user-centric networks in the context of the evolution of the Internet wholesale model; (ii) to present a cooperation model to sustain the deployment of low-cost user-centric wireless networks; (iii) to explain the fundamental properties of the solutions needed to implement such cooperation model.

To achieve the enumerated goals, the remainder of this paper is structured as follows. Section 2 presents prior art related to user-centric networks and models for the creation of cooperation incentives. In Sect. 3 we provide a characterization of user-centric networks, including an applicability study and an analysis of the impact that the new concept may have on the Internet wholesale models. Section 4 describes the proposed cooperative networking model for the sustainability of user-centric networks. In Sect. 5 we detail the incentives framework encompassed in the proposed model, while Sect. 6 presents cooperation aspects that are fundamental to the success of user-centric networks, namely cooperation at the level of resource management. Finally, Sect. 7 provides a general discussion on the proposed framework and future directions.

2 Related Work

2.1 User-Centric Networks

Cooperation is a concept that has been for long applied to several OSI Layers, being the main principle behind successful P2P applications. In more recent years, special focus has been put in systems and mechanisms that allow networks to self-organize and to automatically establish connectivity among involved entities, in order to accommodate future service needs.

This is the core belief of BIONETS, an EU-funded project (Reference 027748) that got inspiration from biological models. BIONETS aimed to develop autonomic network models that mould to social environments they serve. The ultimate goal was to develop truly user-centric Internet models, thus allowing networks to naturally evolve and to become autonomous in order to better accommodate new services and societal needs.

The EU project HAGGLE (Reference 027918) made an approach to the user-centric and autonomic perspective going beyond current network paradigms by exploring *application-driven* message forwarding (overlays), as well as the impact of human communication on network functionality, and opportunism in communication. The concept of *Pocket Switched Networks (PSNs)*, developed int he HAGGLE project, is described as related to mobile networks providing isolated connectivity graphs (*islands of connectivity*). PSN mechanisms are based upon the pragmatic fact that today end-users have, for a specific period of time and due to a specific schedule, intermittent connectivity between different islands of connectivity.

The Eu project SOCIALNETS (Reference 217141) followed the same line of thought but aiming to model networks according to human-behavior. The concepts to be developed rely on an interdisciplinary approach that involves social anthropology and networking, as well as computer science.

From a commercial perspective, FON explores user-centric connectivity models based upon the wireless local-loop and existing Internet broadband access, but without any intervention from access providers, such os which are operating as *Mobile Virtual Network Operators (MVNOs)*, i.e., a legal entity that provides wireless services under a specific brand name, but does not own licensed frequency spectrum.

As another commercial example, Whisher goes a step further into user-centric models, being all the functionality placed in software integrated into the end-user device. Such choice allows Whisher to rely on ad-hoc networking between communities of pre-registered users, thus allowing these communities to share Internet connectivity. The ad-hoc network then obtains Internet access from a specific registered user willing to relay connectivity.

User-centric networks [25] go a step beyond the provided examples, incorporating any of the mentioned models, but always having in mind an end-user centric model for sharing Internet connectivity based on a set of cooperation incentives. The proposed developed in the EU ULOOP project, and described in this article follows the user-centric networking concept based on a reciprocity-rewarding cooperation model that encompasses trust and reward based incentives as well as cooperative mechanisms for wireless resource management (allocation and relaying).

2.2 Cooperation Models

Cooperation incentives [20, 28] are essential requisites of any community, the success of which strongly depends on the willingness of its members to cooperate and can be impaired by selfishness. This is particularly true in user-centric networks, where even the underlying communication infrastructure is dynamically built by users who share their wireless connections, although the inherent limitations of mobile devices (in terms of battery, CPU, and bandwidth) can keep users from adopting social behaviors. When inherent motivations (including fairness and sense of community) provide no sufficient cooperation incentives [6], they need to be complemented by extrinsic motivations, such as virtual currency and community enforcement based on reputation, which provide the means for implementing the so-called soft security, which is characterized by relaxation of the security policies and enforcement of common ethical norms for the community [18].

The development of a virtual currency [9] allows nodes to be remunerated for resource sharing (e.g. packet relaying or Internet access). Nodes accumulate credit through cooperative behavior and use this credit to purchase resources from other nodes. The idea of virtual currency is intuitively appealing in many scenarios, but the analysis of these techniques has been mostly based on heuristics. Moreover, the use of virtual currency requires at least some increase in network traffic overhead, node complexity, and may also have the potential for fraud and/or collusion [1, 2, 4, 10].

Another approach for encouraging cooperation in user-centric networks is community enforcement through the dissemination of reputation information about each node. The basic idea is that nodes monitor the behavior of other nodes to see if they are sharing resources or if they are misbehaving. A self-learning repeated-game framework for community enforcement of cooperation under different assumptions was also proposed recently [24]. Although community enforcement has been shown to encourage cooperation, there are several potential drawbacks including the potential for misinterpreting the behavior of nodes [22], increasing node complexity, needed to monitor the behavior of other nodes, increasing network overhead for reputation propagation messages, as well as the potential for collusion and/ or spoofing.

A game-theoretic analytical study [21] has recently revealed that reputation-based and price-based strategies should be integrated in order to optimize the effects of cooperation incentives. Game theory has been widely used to conduct a mathematical analysis of the complex interactions among nodes of wireless ad-hoc networks [21, 26]: the results of the analytical study are consolidated by simulation results showing the fast convergence towards cooperative behaviors in the case of mixed incentive strategies.

This paper provides an orthogonal view of the benefits of mixed cooperation incentives by employing an extrinsic cooperation incentive framework in user-centric networks, by the joint application of reputation management and virtual currency systems [3, 5]. By looking at the state-of-the-art for cooperative solutions, one can find works that force the cooperative behavior with specific propose, such as to provide data rate rewards [29], exchange power for profits, and increase revenue and spectrum efficiency [19]. However, these frameworks differ from the one developed in the ULOOP Eu project in the sense that ours is much broader (i.e., more than just encouraging data forwarding or maximizing a specific type of resource) than existing mechanisms.

3 User-Centric Networks

The potential of such UCNs has been exploited by means of different scenarios, ranging from basic functionality (e.g. wireless ad-hoc networks activated on-the-fly), to more elaborate commercial cases (e.g. FON, MIFI), where specific hardware and software is provided. A common aspect to all these scenarios is the incentives that users may have to share network resources based on their profile and behavior, namely his/her willingness to share (reciprocity), and roaming habits.

Figure 1 shows an example of a potential deployment of a wireless user-centric network in which spontaneous wireless access points and mobile devices have incentives to cooperate within trustful communities by (i) sharing Internet connectivity and wireless resources based on incentives; (ii) profiting from extended coverage by cooperative relaying; (iii) augmented mobility.

Albeit often associated to resource management, these networking structures—UCNs—are also advantageous in terms of social interaction, an aspect that the Internet is often accused of disrupting. The mentioned benefit is a direct result from the underlying principle of this type of networking structure: Internet services are shared among end-users that may or may not directly relate in real environments, and such sharing has the potential to strengthen social interaction in reality. Going a step further, in relation to the current notion of virtual operator, UCNs integrate the notion of micro-provider [25]: the network is "spread" by means of the end-user willingness to share his/her subscribed Internet access and management can be de-centralized, or there can be a central coordinator (virtual operator) in charge of management.

This section aims to provide a description of the major properties of UCNs, and to show their importance by providing an applicability analysis and by describing the

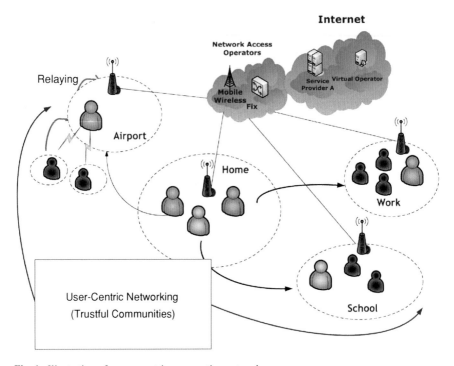

Fig. 1 Illustration of a user-centric cooperative network

increasing flexibility given to the end-user in the context of the Internet wholesale model.

3.1 General Characterization of UCNs

In this initial study we assume that UCNs rely on any form of *radio* technology, be it cellular or wireless. Moreover, we argue that UCNs rely on four basic characteristics, namely, *connectivity sharing*, *cooperation*, *trust*, and *self-organization*.

3.1.1 Willingness to Share

The first feature considered is the willingness of end-users to openly share connectivity, i.e., the willingness of end-users to become micro-providers within autonomic communities. In order to ensure widespread usage, connectivity sharing has to be deployed based upon software or hardware that is available or easily deployable in end-user equipment, independently of radio technology, operating system, and device.

Connectivity sharing can be performed either actively or passively, i.e., the end-user owning the device may or may not be aware that he/she is sharing connectivity in a precise moment in time. However, in either case, it is necessary to ensure that the end-user allows such sharing to happen. Furthermore, being based upon end-users' willingness to share connectivity, UCNs naturally follow human living patterns having a higher probability of providing better coverage in densely populated areas.

3.1.2 Cooperation

A second main feature of UCNs is cooperation. Users must rely upon some form of incentive to share connectivity that they own, let it be trust boundaries, or some form of compensation (e.g. wider Internet connectivity). Cooperation aspects in UCNs also relate to the need to behave properly, according to specific community's criteria.

Cooperation mechanisms should ensure that there are both incentives and rewards for end-users to behave adequately, thus reducing the probability of having malicious users entering the network. In what concerns current models, such incentive is simply being able to take advantage of a broader and free coverage.

But for the future and in more sophisticated versions of UCNs, such incentive could relate to the amount of time (or throughput) that is shared by a specific end-user: the more (service) the end-user shares, the more the end-user can benefit from. In addition, *good* behavior can be rewarded, e.g., to provide more connectivity to end-users that behave according to some pre-established criteria.

However, it may also be necessary to implement adequate misbehavior detection and rule-out mechanisms to avoid network operation disruption.

3.1.3 Trust

Trust is the third pillar, and has strong impact on cooperation, as well as on connectivity sharing. UCNs are completely user-centric and consequently, building a network of trust is essential when considering widespread usage.

Consequently, trust models must consider social interaction and human interests as the basis for building trust. In addition, reputation mechanisms (e.g., similar to the E-Bay model) should be integrated. With these mechanisms as well as incentives to behave adequately, misbehavior will be reduced.

3.1.4 Self-Organization

Self-organization, the fourth pillar, relates to the autonomic facet of UCNs. It is not possible to predict how many users will be hanging from a specific micro-provider at a time, nor is it possible to predict the dynamic morphology of a specific UCN and

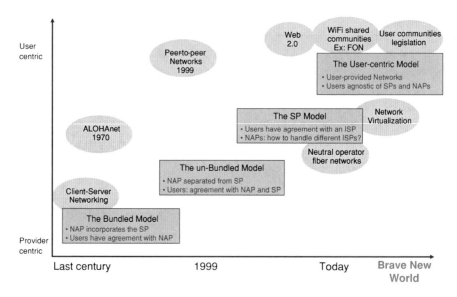

Fig. 2 Wholesale model evolution

how it will spread. *Spreading* is here defined as the method that the infrastructure, as well as the connectivity model, are built upon.

Spreading can be characterized as a whole by the number of nodes sharing (or profiting from) connectivity, the way that nodes interconnect (e.g., node degree), the average number of end-to-end hops, as well as the average bandwidth that nodes can take advantage from.

The resulting topologies are therefore strongly dependent upon human behavior, particularly, *human mobility patterns* and *social interaction* models. For instance, Watts and Strogatz [27] define an association among individuals according to their common professional interests. Other studies focus on the aggregation of different variables [7], which are part of the users' profile and behavior. In networks that operates based on users' behavior, such as UCNs, users' profiles are used to control connectivity among users, as well as the sharing of resource through the creation of wireless virtual communities.

3.2 Impact on the Internet Wholesale Models

The traditional Internet wholesale model is the bundled model (c.f. Fig. 2), where a Network Access Provider (NAP) also incorporates the role of Service Provider (SP). This model imposes several restrictions to the implementation of a cooperation framework, since the end-user can only set a communication relationship with one NAP, restricting the choice of SPs.

The traditional wholesale model started to evolve in 1999, when the FCC regulated in the USA the unbundled model, where a NAP owns the infrastructure, being SPs separate entities. This model brings more flexibility to the development of cooperative networking, since the end-user keeps subscription agreements both related to the access and to services being provided.

A third and current wholesale model is the SP-centric model, where SPs have a direct relationship with end-users in what concerns Internet access. The end-user has the possibility to bypass NAP limitations, keeping an agreement only with the SP. Although this brings in more flexibility, it still restricts users in what concerns choice of specific SPs.

A step further into the virtualization of the Internet architecture is the role of virtual operator of which the most common example today is a Mobile Network Operator (MNO). A virtual operator is an entity that provides some form of service (e.g., connectivity coordination) but that does not have its own infrastructure to provide the service.

UCNs go a step further concerning the notion of virtual operator, integrating the notion of micro-provider: the network is "spread" by means of the end-user willingness to share his/her subscribed Internet access and management is de-centralized, or there is a central coordinator (virtual operator) in charge of management.

This means that any cooperative model aiming to sustain the deployment of user-centric wireless networks need to consider the role of virtual operators and micro-providers.

3.3 User-Centric Wireless Local Loop: UCN Instantiation

The main motivation for the ULOOP project relates to the need to assist an autonomic deployment of user-centric wireless local loops. Such support is provided by developing software functionality that sustains a robust, trustful, and autonomic network growth in a user-friendly way, thus becoming the basis for generating new services and consequently, new business models for current access and Internet stakeholders.

In the case of user-centric wireless local loops, the term user-centric refers to a community model that extends the reach of a high rate, multi-access broadband backbone by means of communication opportunities provided by end-users, based upon cooperation incentives. Such incentives may relate to an individual or a community of individuals, as well as to access stakeholders. Moreover, user-centricity can be discussed from two different perspectives.

To better explain where the boundaries of a user-centric wireless local loop reside and which type of outcome is expected, Fig. 3 provides a generic illustration of the Internet, being highlighted the location of SPs (ISPs, VO, ASPs), of Network Access Providers (NAP) as well as of the Internet end-user Customer Premises (CP). In such scenario, two different NAPs provide Wi-Fi coverage to different scenarios ranging from residential complementary Wi-Fi Internet access (privately owned WLANs) to Wi-Fi municipalities, or commercial hotspots.

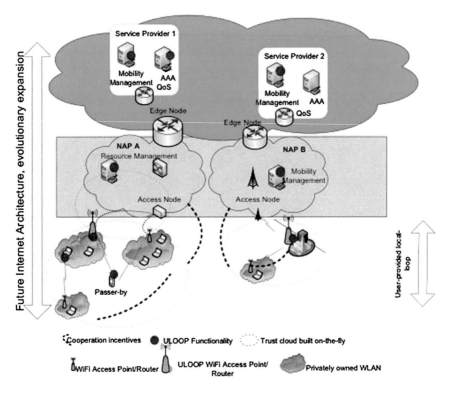

Fig. 3 User-centric wireless local loops

As illustrated in Fig. 3 the development of user-centric wireless local loops assume that an infrastructure providing Internet access to specific locations is available, and users are simply willing to expand such infrastructure in a way that is user-friendly and plug & play. It also considers that within specific trust spheres, specific cooperation incentives can be provided in order for both the access and the end-user to cooperate and assist in further expanding the Internet.

In order for that to happen, there are a fundamental aspect that is often disregarded and which is crucial to sustain the development of user-centric networks, and in particular user-centric wireless local loops: cooperation model based on incentives able to sustain a shared management of wireless resources.

3.3.1 Applicability Study

In order to illustrate the applicability of UCNs, an particular of user-centric wireless local loops, this section describes a scenario aiming to support service sharing, monitoring and user traceability aspects.

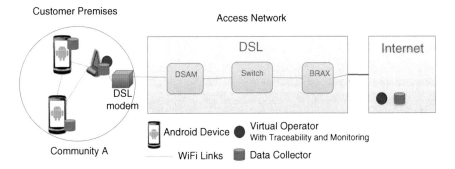

Fig. 4 Data monitoring and user traceability

Figure 4 shows a *Community A* visited by users that (i) have direct access to a gateway (e.g. a DSL access point); (ii) have access by means of few-hop wireless sharing.

End-user devices within the community are able to share Internet connectivity based on specific cooperation incentives, which also allow devices to track user expectations and service response within the community. Here, tracking relates to information that is not personal.

This can be a use-case in an airport, where a local coffee shop gets a reward for sharing its Internet access with users that have a boarding pass. Users that are far away from the coffee shop can also benefit from Internet sharing, since users getting Internet access directly from the coffee-shop gains credits by relaying traffic. Such credits can be used to access other shared services. For instance, the coffee-shop gateway will perform resource allocation based on the amount of credits earned by contending users.

4 Reciprocity-Rewarding Cooperation Model

In user-centric wireless networks, cooperation incentives are used to motivate nodes to share resources and connectivity in order to augment their networking experience. Cooperation is an inherently positive interaction between parties, which can improve the worth of the network they live in. More and more networks depend on the individuals' willingness to cooperate, especially when sharing resources or working with peer-to-peer protocols. Since individual nodes are usually strongly constrained in terms of bandwidth and computing power, cooperation and resource sharing must be motivated, especially considering the presence of selfish nodes that might impair the functioning of the entire network.

The Prisoners' Dilemma (PD) game is perhaps the most popular model that has been used aiming to give an explanation of how and why cooperation can emerge in social, biological and economic systems. The PD game model is very simple in showing how a dominant individualistic action leads to the most inefficient collective outcome when all individuals adopt it. The most famous example of this type of strategies is the "Tit-for-Tat" in which individuals cooperate on the first move and then cooperate or defect exactly as the other player did on the preceding move. This direct reciprocity based on retaliation and reputation plays a fundamental goal in promoting cooperative behavior in user-centric networks.

Although reputation and reciprocity are good models for wireless user-centric networks, selfish nodes may refuse to be cooperative, since nodes are usually constrained by limited computation and network resources. Therefore a combination of reputation systems and reward-based systems seems to be a solution to solve the node noncooperation problem: the former evaluates node behavior by reputation values and uses a reputation threshold to distinguish trustworthy nodes and untrustworthy nodes; a reward-based system uses virtual cash to control the transactions of a Internet connectivity service. Although these two type of systems have been widely used, very little research has been devoted to investigate the effectiveness of node cooperation incentives provided by networking systems.

We propose an integrated reciprocity-reward system [5], aiming to mitigate the limitations of both systems: by applying strategies of using a threshold to determine the trustworthiness of a node in the reputation system, and by compensating cooperative nodes in a system potentially manipulated by selfish nodes. Incentives for cooperation are created by adopting policies that are based on the threat of retaliation for non-cooperating nodes; rewards are provided as a virtual payment for cooperation.

In this section, we provide an introduction to reciprocity and reward based incentives, as well as a description of the proposed reciprocity-reward incentive model.

4.1 Reciprocity-Based Incentives

When operating based only on reciprocity, cooperation is fully based on the behavior of the end-user: the cooperation process between a requester and a requestee is essentially a function of the trust level between them.

Based on reciprocity, the requestee acts based upon a set of reputation recommendations sent by neighbors, aiming to provide a service/resource only based on reciprocity. The reputation of a node can be based on direct observation and recommendations: in the proposed model reputation is distributed and kept locally for each node. Nodes which are unknown are assigned a base reputation value, depending entirely on the basic attitude that each node has to agree upon the reputation of others. Therefore, depending on its base reputation value, a node can be distrustful or cooperative in nature. Reputation values are then adjusted as nodes interact and cooperate.

During a reciprocity based cooperation, the requestee accepts service requests with the goal to increase its reputation in the community for latter usage of shared resources. For a requestee, high reputation increases the probability of being chosen by a requester, while for a requester it increases the probability of being served by the requestee of choice. Nevertheless, the requestee may give different quality of service levels according to requester's reputation or received credits. Reciprocity-based incentives trigger a positive feedback loop within community members, since trust increases the cooperation opportunities for both requesters and requestees. Furthermore, trust is also used as a discount factor by requestees acting in retailer mode, as detailed in next subsection.

Reciprocity-based incentives have a similar impact to what occurs in social trust schemes: based on its own reading of the trust level towards the requester, a requestee accepts or denies a cooperation request offer (instantiated by a set of cooperation credits and the trust level of the requestee towards the requester) since his/her main motivation is to be part of a community, and to provide a service/resource just based on reciprocity.

4.2 Reward-Based Incentives

Reciprocity-based incentives may not be enough to encourage nodes to engage in cooperation. Therefore we include the notion of rewarding in the proposed cooperation model: rewards are provided in terms of a negotiation of cooperation credits used to stimulate the participation and interaction between users.

Rewards in the form of virtual credits are monetary units of a virtual currency that can be exchanged between individuals. Like fiat currency, virtual currencies have several requirements, among which forgery and cheating avoidance. The proposed model is based on a partially distributed virtual currency that allows individuals to exchange credits even in absence of a trusted third party. However, control of all payments is eventually performed by a central authority for robustness and security.

The amount of credits needed by a requestee to provide a given service/resource usually depends at first from his/her trust on the requester, as occurs with reciprocity-based incentives: the higher the trust, the lower the price. Similarly, the amount of credits offered by a requester usually depends on his/her trust on the requestee: the higher the trust, the higher the offer.

This operation guarantees a tight relationship between reciprocity-based and reward-based incentives: the requester offers credits to the requestee independent of the operation mode of the requestee (c.f. Sect. 4.3).

4.3 Cooperation Model

Figure 5 illustrates the proposed reciprocity-reward based cooperative networking model in which services (e.g. Internet access, few-hop relay) are shared based on reward incentives (e.g. Internet sharing) and behavior incentives (e.g. few-hop relay). For the specific service "Internet access" the illustrated gateway takes into account

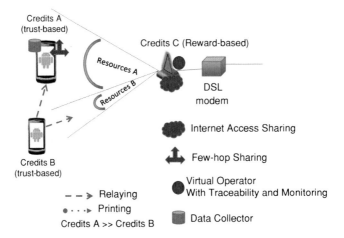

Fig. 5 Cooperative networking model

the higher cooperation level (reputation) of the user with credits A, allowing it to use more wireless resources (spectrum) than the user with credits B. The proposed model can be applied independently of the considered resources (e.g. printing, monitoring, cooperative sensing).

Both mechanisms provide a form of cooperative incentive and they motivate members of the network to adopt a positive behavior: reputation provides a reciprocal reward for cooperating individuals, which allows them to keep track of pro-social behavior and to benefit from it. Virtual currency, on the other hand, provides a reward mechanism even in the absence of reciprocity.

Monetary incentives offset part of the implications of a reputation-only based system. A scenario where only individuals with high reputation foster mutual cooperation at the loss of other network nodes of lower reputation could be a barrier for new nodes entering an unknown network where they have no existing reputation. Credits allow users to surmount this obstacle by providing an external, direct incentive, promoting the establishment of trust relations.

On the other hand, reputation can encourage the exchange of credits: the list of potential requests for a service request could be sorted by trustworthiness, thus giving better visibility to individuals with high reputation. Achieving high reputation through cooperation and pro-social behavior is also a viable strategy to maximize the earning opportunities and the amount of credits received as direct cooperation incentives.

As mentioned before, the chance of obtaining access to a service/resource and the quality of service reserved by the requestee can be parameterized both on the reputation of the requester and on the direct incentive offered at service request. Even in typically uncooperative scenarios, such as networks where nodes have very short presence, when access is very dynamic, or when connectivity is intermittent, a mixed incentive model provides motivations for entering the network while discouraging cheating and selfish behaviors.

In the proposed model, each request is complemented by the transfer of virtual credits: direct monetary incentives are an intrinsic part of the service request. The amount of credits passed by a requester along with the service request depends on the reputation relationship between nodes and the level of service requested. Requestees may decide to accept or refuse the cooperation depending on reputation and credits obtained. Thresholds for both parameters can be defined. Essentially, in the proposed model, a requestee may operate in two possible modes:

- Retailer: cooperation is intended to provide a positive return to the node itself, both in terms of acquired reputation and credits gain. The node may define minimum credits required to provide specific services and to grant a specific quality of service. If the threshold is not reached, the request is rejected. The value may depend on the requester's reputation value, thus lowering prices for known nodes. Similarly, the amount of credits offered by a requester usually depends on the requestee's reputation: the higher the reputation, the higher the offer.
- Volunteer: cooperation occurs based on reciprocity. The requestee acts based upon a set of reputation recommendations sent by neighbors, aiming to provide a service/resource only based on reciprocity. The requestee accepts service requests with the goal to increase its reputation in the community for latter usage of shared resources. For a requestee, high reputation increases the probability of being chosen by a requester, while for a requester it increases the probability of being served by the requestee of choice. Nevertheless, the requestee may give different quality of service levels according to requester's reputation or received credits.

Intertwining reciprocity and reward based incentives makes reputation and credits an integral part of a service request, where credits can compensate an eventual lack of existing reputation relations between requester and requestee. The only difference between volunteer and retailer operation at the requestee is that the offer of cooperation credits is always accepted by a volunteer, while it can be refused by a retailer in case it is lower than the reward he/she needs to provide his/her service.

In the mixed approach, we assume that the trust level is a variable to be put in relation with the cost variable as follows: definition of a minimum reward (cost) asked by the requestee regardless of the requester's reputation; a maximum reward asked to serve requesters with low reputation; and a reputation threshold above which the minimum cost is applied to the requester. In the proposed framework, the behavior of the reciprocity-reward curve is imposed to all members, in order to be used as tuning parameters to adjust the behavior of the community as a whole. The minimum and maximum cost values can be set by any member and for each category of service (e.g., relaying, information sharing).

Notice however that the mixed approach might induce an undesired effect: the unfair underestimation of reputation due to the avidity of the requestee. In fact, the higher the reputation the lower the direct reward obtained by the requestee. To avoid this effect, the value of reputation can be limited to the component of the metric that accounts for community-scale evidence of requester's reputation.

Fig. 6 Cooperation framework

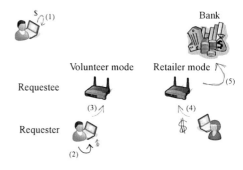

5 Cooperation Framework

As described in the previous section, the sustainability of the proposed cooperation model is achieved based on a framework that combines reciprocity-based and reward-based cooperation incentives. The former provides reciprocity based rewarding by enabling a rigorous definition of reputation as a metric positively affected by pro-social behaviors and positively affecting the opportunity of benefiting from the community. Reward-based incentives provide a motivation based on a virtual currency that can be used to motivate cooperation even in absence of reciprocity, possibly enabling monetization of pro-social behavior.

The presupposed cooperation model adopts four driving principles while handling reciprocity and reward based incentives: (i) reputation can affect individual decisions and opportunities; (ii) reputation cannot be traded for virtual currency or money; (iii) virtual currency is a commodity money mainly used to facilitate cooperation among community members; (iv) virtual currency can eventually be traded for fiat money at the only purpose of allowing the community to benefit from third-party services and to provide services to non-member end-users.

Depending on the type of adopted incentives, a requestee can operate in two different modes, volunteer and retailer, as mentioned in the previous section. In the proposed framework, the requestee can switch between such modes according to his/her attitude and to his/her need for credits. In this way, a network node keeps earned credits, which can in a later stage in time be used to obtain additional services/resources. Incentive credits can eventually be monetized (i.e., traded for real money) at the boundaries of network communities.

The proposed cooperation incentive framework comprises three distinct phases, as illustrated in Fig. 6:

- Boot-up phase, performed once for each network node, where an initial reputation level that a node is willing to accept from any other nodes is set and the initial amount of credits is assigned to each node taking into account its disposition to trust/cooperate with others (step 1 in Fig. 6): the amount of credits is higher for nodes with an average dispositional trust level, aiming to compensate the lacking of trust when motivating the node to cooperate in a community network (nodes with low dispositional trust have low probability of cooperate with others; nodes with high dispositional trust do not require extra motivation to cooperate with others). To avoid inflation, in the proposed framework the amount of credits awarded to each new member conventionally corresponds to the incremental value brought to the community by such member, due to positive externalities and network effects.

- Cooperation phase, in which a user, hereafter called requester, looks for a member sharing/providing the required service/resource, hereafter called requestee. The requester decides the amount of credits (step 2 in Fig. 6) to be offered to a specific requestee to encourage cooperation based on his/her consideration about the requestee's reputation.

  - If the requestee is operating in a volunteer mode, it can accept the amount of credits offered by the requester (step 3 in Fig. 6), increasing its reputation, or it may either refuse the request because of the lack of resources or willingness to cooperate;
  - If the requestee is operating in a retailer mode, it enters a negotiation process to decide whether or not to accept the request based on his/her considerations about the requester's reputation and about the adopted incentive mode (step 4 in Fig. 6). If the offer is accepted, the requestee starts a transaction process encompassing both the provision of the service/resource and the payment of the reward in terms of virtual currency (step 5 in Fig. 6) encompassing a trustful third party (Bank entity).

- Evaluation phase, in which the requester evaluates the requestee by comparing the provided and agreed service levels. This periodic evaluation is important to discourage misbehaviors.

Within the cooperation phase, reputation (of the requestee) can represent an incentive whenever the topological network organization assigns a more important position to nodes with higher reputation. In particular, the list of potential requestees available to negotiate a request should be provided/considered in descending order with respect to the related reputation and the choice of the requester can be driven by his/her trust on the candidate requestees. Hence, high reputation helps to achieve more preferences and, possibly, to increase the reputation itself. This strategy is effective especially whenever combined with direct incentives that allow the requestee to receive a reward for the delivered service: the higher the reputation the higher the earning opportunities. On the other hand, the reputation of the requester is a key factor during the cooperation phase if the requestee is operating in retailer mode. First, the quality of service that is negotiated (e.g., amount of bandwidth, CPU, storage) can be parameterized to be proportional to the requester's reputation (c.f. Sect. 6).

During the cooperation phase, the requestee provides the negotiated service /resource, while the requester pays the negotiated amount of credits. Notice that, in this phase, payment and service provision are not conditioned to each other. Rather, both the requester and the requestee are bound to the negotiated terms. The non-observance of the agreement possibly made by one of the peers does not authorize the other to do as much. This lack of control facilitates transactions, while the risk of abuses is mitigated both by the final evaluation, which will affect reputation in case of misbehavior, and by the small granularity of atomic transactions, which makes the risk of each of them almost negligible for the parties.

The evaluation phase is fundamental to make the whole reputation system reliable. In fact, explicit and/or implicit mechanisms for evaluation of services have an impact over the quality of the information used to estimate the reputation gain/loss after a transaction. Explicit mechanisms involve the direct participation of the requester who might be asked to vote explicitly the quality of the received service. Implicit mechanisms are based on transparent evaluation systems that compare the result of the transaction with the negotiated parameters. In any case, the result of the application of such mechanisms represents a feedback that affects both future local recommendations provided by the requester and the global reputation of the requestee stored in public repositories. Note that whenever such repositories are available, the provision of (honest) feedback is worth receiving a reward in its turn (e.g., in terms of reputation), as shown in literature [23]. The same benefits can be obtained in a distributed reputation system by rewarding (honest) recommendations in terms of reputation gain that the requester locally assigns to the recommender. The lack of feedback could also affect reputation.

6 Incentives Framework Applied to Resource Management

In the proposed cooperation model reputation of the requester is a key factor during the cooperation phase, independently of the operation mode of the requestee. In what concerns the management of resources in UCNs, this means that the quality of service that is established (e.g., amount of bandwidth, CPU, storage) between requester and requestee can be parameterized to be proportional to the requester's reputation.

In a simplest setting reputation thresholds can be applied to ensure access to the requested service in a binary way, i.e., a given service is available only for reputation metrics higher than a certain threshold. In a more complex setting where services have a cost, the level of reputation can represent a variable contributing to the calculation of such a cost. Second, multiple transactions (that are negotiated by different peers with the same requestee) competing for the same resource are queued based on a priority that depends on peer's reputation. In these cases, high reputation represents an enabling condition to access better services within stricter deadlines.

This section provides a description of the way wireless resources can be controlled in UCNs based on the proposed reciprocity-reward based cooperation model.

6.1 Cooperation-Based Resource Management

With the proposed cooperation model, the management of resources takes advantage of the willingness that users have in cooperating, based on the mentioned two types of incentives: reciprocity-based and reward-based. Cooperative-based resource management has three components: call admission control, resource allocation, and load-balancing.

Call admission control (CAC) decides whether a new request from an end-user can be accepted or not. In a UCN framework, as the one developed in the ULOOP project, resource allocation assigns resources among all accepted and active end-users. Load balancing can be used to react to congestion or to prevent congestion in the presence of a prediction mechanism.

The innovative aspects of a cooperation-based CAC relate to the capability of a Internet gateway to take into consideration not only available resources, but also its trust level towards the different requesters, and also the credits such requesters are considering to use in the cooperation process.

Based on the proposed cooperation model, the allocation of resources is done based upon the type of requested service (e.g. real-time vs non real-time); the amount of credits that a requester considers in a specific negotiation, and also based upon the trust level that the requestee has on the requester. Based on these three criteria, resources (e.g. bandwidth, power and bit rates) are allocated fairly among requesters aiming to guarantee the quality level required by them, but having in mind their willingness to share resources within communities—their reputation as sharers. For instance, a user with a high quality demand, but with a cooperation deficit shall earn lesser credits than others and as a consequence may get fewer resources than another user with a lower quality demand, but with a higher cooperation index (which translates into more earned credits) in case of scarcity of resources. That is to say that resource allocation does not consider only the service quality obtained by each individual end-user, but also the whole performance of the community: network capacity can be maximized by increasing the incentives for nodes to share services and/or resources by providing them higher priority to access shared services/resources.

In the presence of congestion, load-balancing aims to provide a more efficient usage of resources by different gateways. A congested gateway may decide to shift traffic towards another gateway based on local measurements and by analyzing the measurements of neighbor gateways. The latter cooperate with the congested gateway by deciding to receive some of its traffic, since they will get extra credits for that. The decision to help balancing the load of a congested gateway is done based on a cooperation index, which is locally computed based on the analysis of the trade-off between accepting traffic from neighbors and the advantage of having more credits. Based on the list of neighbor gateways that are willing to cooperate with the congested gateway, the later will select a neighbor based on the trust level that the users, trying to access the congested gateway, have on the potential new gateway.

6.2 Cooperation-Based Relaying

Wireless channel conditions in wireless networks are subjected to interference and multi-path propagation, creating fading channels and decreasing the overall network

Fig. 7 Cooperative relaying

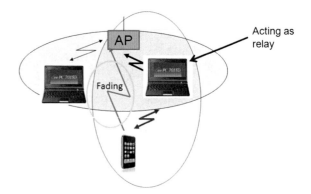

performance. While fast fading can be mitigated by having the source retransmitting packets, slow fading, caused by obstruction of the main signal path, makes retransmission useless, since periods of low signal power last for the entire duration of the transmission.

The impact of slow fading can be mitigated based on wireless cooperative relaying. Cooperation, as illustrated in Fig. 7, occurs when overhearing nodes assist the transmission from source to destination by relaying different copies of the same signal from different locations, allowing the destination to get independently faded versions of the signal that can be combined to obtain an error-free signal [8].

With the introduction of cooperative relaying, the relay selection procedure requires special attention, since it has a strong impact on network and transmission performance. Cooperation procedures may be very unstable when relay selection relies on current channel conditions and is agnostic of interference induced by simultaneous transmissions and local CPU load at the potential relay. As a result, the performance of cooperative relaying in realistic scenarios may be very different from interference-free scenarios that have been the focus of most of the existing proposals so far.

Aiming to support stable and scalable relaying solutions for UCNs, we propose a cooperative relaying protocol, called RelaySpot [11–13, 15], which is aware of interference [16, 17] and agnostic of channel state information, in order to increase stability. With the proposed scheme, potential relays opportunistically configure their contention window based on an on-the-fly analysis of interference, mobility patterns, and history of successful transmissions toward the desired destination.

With RelaySpot, relay selection is performed in three steps: First, each node checks if it is eligible to be a relay by verifying two conditions, overheard a good frame sent by the source and be positioned within the so called cooperation area; Second, eligible nodes start the self-election process by computing their selection factor; Third, self-elected relays set their contention windows based on their selection factor, and send a qualification message towards the destination after the contention window expires, as shown in Fig. 8.

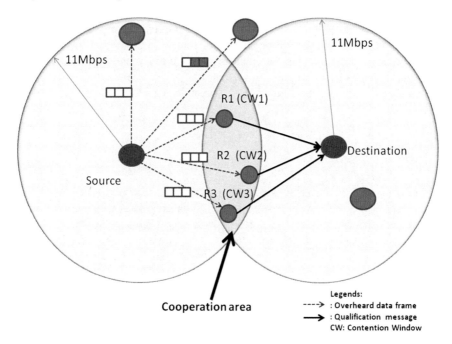

Fig. 8 Opportunistic relay selection

As mentioned, to be eligible as a relay nodes need to verify if they are inside the cooperation area. This is done in two steps: First by analyzing if the are willing to cooperate with the source, to relay its data to a destination; Second by computing their Cooperation Factor (CF).

With RelaySpot, a potential relay decides whether to help a source or not based on the trust level that it has towards the different sources, and also the credits such sources are considering to use in the cooperation process. This is a similar process as the one used by gateway to admit a new communication session from a end-user devices, as described in Sect. 6.1.

If a node decides that we has incentives to help the source, its cooperation factor, as eligible relay, is computed based on the rate of source-relay channel (R_{sr}) and the relay-destination channel rate (R_{rd}), as shown in Eq. 1. The rate of the source-relay and relay-destination channel is computed by overhearing RTS and CTS frames exchanged between source and destination. The CF ensures that potential relays are closely bounded with the source while having good channel towards the destination: an eligible relay must have a CF that ensures a higher data-rate than over the direct link from source to destination.

$$CF = (R_{sr} * R_{rd}) / (R_{sr} + R_{rd}), \quad CF \in [0, \infty] \tag{1}$$

RelaySpot implements a cooperative relaying solution for UCNs, by combining the opportunistic relay selection procedures described in this section with cooperative relay scheduling and switching [14, 17], aiming to improve the utilization of spatial diversity, minimizing outage and increasing reliability. Simulation results show that opportunistic relay selection based on the proposed interference-aware and history-aware metric can provide higher throughput and lower latency than interference agnostic and CSI dependent solutions based on node degree and distance toward the destination.

7 Summary

In a pervasive Internet, users expect to have instant and real-time access from their devices to any kind of Internet based platform, such as online social networks. This requirement is fulfilled by User-Centric Networks (UCN), which are today starting to be exploited by means of different scenarios with a common aspect: the incentives that users may have to share network resources based on the user's profile and behavior.

Mixed incentive strategies, combining reputation and reward-based mechanisms, have proved effective in inducing pro-social behavior while isolating selfish or cheating nodes in a community. Hence, this paper describes a cooperation model and framework entailing both reciprocity and reward based incentives to support cooperation in UCNs. The formal approach adopted in this work is being developed in the European project ULOOP.

In the ULOOP project the proposed reciprocity-reward based cooperation model is being used to control admission to the network, to manage wireless resources, and to provide cooperative relaying.

References

1. Aldini A (2013) Formal approach to design and automatic verification of cooperation-based networks. IARIA Int J Adv Internet Technol 6(1, 2):42–56
2. Aldini A, Bogliolo A (2012) Model checking of trust-based user-centric cooperative networks. In: 4th international conference on advances in future internet (AFIN), IARIA, pp 32–41
3. Aldini A, Bogliolo A (2012) Trading performance and cooperation incentives in user-centric networks. In: International workshop on quantitative aspects in security assurance (QASA'12), Pisa
4. Aldini A, Bogliolo A (2014) Modeling and verification of cooperation incentive mechanisms in user-centric wireless communications. In: Rawat D, Bista B, Yan G (eds) Security, privacy, trust, and resource management in mobile and wireless communications, IGI Global, pp 432–461
5. Bogliolo A, Polidori P, Aldini A, Moreira W, Mendes P, Yildiz M, Ballester C, Seigneur J-M (2012) Virtual currency and reputation-based cooperation incentives in user-centric networks. In: Proceedings of IEEE wireless communications and mobile computing conference (IWCMC-2012), Cyprus, pp 895–900, Aug 2012
6. Declerck CH, Boone C, Emonds G (2013) When do people cooperate? the neuroeconomics of prosocial decision making. Elsevier Brain Cogn 81(1):95–117

7. Eagle N, Pentland A (2011) Eigenbehaviors: identifying structure in routine. In: Proceedings of international conference on ubiquitous computing, Orange County, USA
8. Elmenreich W, Marchenko N, Adam H, Hofbauer C, Brandner G, Bettstetter C, Huemer M (2008) Building blocks of cooperative relaying in wireless systems. Electr Comput Eng 125(10):353–359 (Springer)
9. Greengard S (2011) Social games, virtual goods. Commun ACM 54(4):19–22
10. Jakobsson M, Hubaux J-P, Buttyan L (2003) A micro-payment scheme encouraging collaboration in multi-hop cellular networks. In: Proceedings of international conference on financial crypto, Guadeloupe, French West Indies, Jan 2003
11. Jamal T, Mendes P (2010) Relay Selection Approaches for Wireless Cooperative Networks. In: Proceedings of international conference on wireless and mobile computing, networking and communications, Niagara Falls, Oct 2010
12. Jamal T, Mendes P (2011) Interference-aware opportunistic relay selection. In: Proceedings of ACM CoNext—student workshop, Tokyo, Dec 2011
13. T Jamal, P Mendes (2011) RelaySpot: a framework for opportunistic cooperative relaying. In: Proceedings of international conference on access networks, Luxembourg, June 2011
14. Jamal T, Mendes P (2014) Cooperative relaying for dynamic networks. Models, approaches, techniques, protocols, architectures, tools, applications and services,Springer lecture notes in computer science, wireless networking for moving objects
15. Jamal T, Mendes P, Zuquete A (2012) Opportunistic relay selection for wireless cooperative network. In: Proceedings of IFIP NTMS, Istanbul, Turkey, May 2012
16. T Jamal, P Mendes, A Zuquete (2012) Wireless cooperative relaying based on opportunistic relay selection. Int J Adv Netw Serv 5(1):116–127
17. Jamal T, Mendes P, Zuquete A (2013) Analysis of hybrid relaying in cooperative wlan. In: Proceedings of wireless days, Valencia, Spain, Nov 2013
18. Josang A (2012) Robustness of trust and reputation systems: does it matter? In: Proceedings of IFIPTM international conference on trust management, Surat, May 2012
19. Kandeepan S, Jayaweera S, Fedrizzi R (2012) Power-trading in wireless communications: a cooperative networking business model. IEEE Trans Wireless Commun 11(5):1872–1880
20. Lafuentea CB, Seigneur J-M, Moreira W, Mendes P, Maknavicius L, Bogliolo A, Di Francesco P (2012) Survey of trust and cooperation incentives for wireless user-centric environments. In: Proceedings of IADIS e-society, Berlin, Germany, Mar 2012
21. Li Z, Shen H (2012) Game-theoretic analysis of cooperation incentives strategies in mobile ad hoc networks. IEEE Trans Mobile Comput 11(8):1287–1303
22. Liu Y, Yang Y (2003) Reputation propagation and agreement in mobile adhoc networks. In: Proceedings of IEEE wireless communication and networking, New Orleans, Mar 2003
23. Dai Y, Yang M, Feng Q, Zhang Z (2007) A multi-dimensional reputation system combined with trust and incentive mechanisms in p2p file sharing systems. In: Proceedings of IEEE distributed computing systems workshops, Toronto, Canada, June 2007
24. Pandana C, Han Z, Liu KR (2008) Cooperation enforcement and learning for optimizing packet forwarding in autonomous wireless networks. IEEE Trans Wireless Commun 7(8):3150–3163
25. Sofia R, Mendes P (2008) User-provided networks: consumer as provider. IEEE Commun Mag, Feature Topic Consum Commun Networking—Gaming Entertainment 46(12):86–91
26. Srivastava V, Neel J, MacKenzie A, Menon R, DaSilva L, Hicks J, Reed J, Gilles R (2006) Using game theory to analyze wireless ad hoc networks. IEEE Commun Surveys Tutorials 7(4):46–56
27. Watts D, Strogatz S (1998) Collective dynamics of 'small-world' networks. Nature 393(6684):409–410
28. Yildiz M (2012) Cooperation incentives based load balancing in ucn: a probabilistic approach. In: Proceedings of Globecom, July 2012
29. Zhang G, Yang K, Qingsong H, Liu P, Enjie D (2012) Bargaining game theoretic framework for stimulating cooperation in wireless cooperative multicast networks. IEEE Commun Lett 16(2):208–211

Trust as a Fairness Parameter for Quality of Experience in Wireless Networks

Rute Sofia and Luis Amaral Lopes

Abstract This paper addresses the applicability of trust metrics as a fairness parameter in call admission control within the context of user-centric networks. The paper explains how trust can assist in improving user Quality of Experience in wireless networks, by taking into consideration not only channel conditions, but also trust levels derived from the interaction that users have in the context of Internet shared services.

Keywords QoE · Trust · Resource management · Wireless

1 Introduction

Wireless revolutionizes local area communications allowing citizens to provide communication services as well as to become micro-providers in *User-centric Networks* (*UCNs*). This emerging networking paradigm relies in the user's willingness to share connectivity and resources. In comparison to traditional Internet routing scenarios (be it based on wireless or fix line technologies), UCNs bring in forwarding challenges, due to their underlying assumptions, namely: (i) end-user device nodes may behave as networking nodes, (ii) nodes have a highly nomadic behavior, (iii) data is exchanged based on individual user interests and expectations.

Furthermore, emerging trends such as UCNs adding to the development of faster, more reliable wireless standards, miniaturization of devices, and reduced costs of hardware and services, is leading to a fast evolution of technological as well as

R. Sofia (✉) · L. A. Lopes
COPELABS, University Lusófona, Building U First Floor, Campo Grande 388, 1749-024
Lisbon, Portugal
e-mail: rute.sofia@ulusofona.pt

L. A. Lopes
e-mail: luis.amaral@ulusofona.pt

A. Aldini and A. Bogliolo (eds.), *User-Centric Networking*,
Lecture Notes in Social Networks, DOI: 10.1007/978-3-319-05218-2_8,
© Springer International Publishing Switzerland 2014

societal aspects in the way that people communicate. For instance, people expect to be able to send and retrieve information whenever and wherever they want. Yet, there are technological limitations which may affect this anytime-anywhere communication paradigm, e.g. gray areas (i.e., areas where the wireless signal strength is not enough to sustain connectivity); physical obstructions; limited battery devices; environmental aspects; limited resources and security issues. Related literature has been addressing aspects to mitigate wireless interference and to take advantage of cooperative diversity which may mitigate some of the problems posed by physical obstructions and coverage problems due to node mobility. However, it is imperative to say that, since information is relayed among nodes and these nodes can be highly dynamic, communication may experience delay, varying from short to long periods, as isolated areas (e.g., intermittent connected networks) may form in the case of node failure (e.g., damaged *Access Points, APs*) or mobility (e.g., user changes position). Thus, to increase the performance of multi-hop communication, several improvements can be made, by taking advantage of transmission opportunities provided by moving nodes and accessible APs, for instance. An example for such an occurrence is a citizen at a public location without Internet access. If Internet users are in the vicinity, and such users are part of a UCN, then some of them may share Internet access and data can be relayed until it reaches the closest Internet gateway. Another situation may occur when information is simply carried by users that happen to be moving towards the place where the destination is located. Nowadays, this is possible thanks to the size of devices which are making them easier to carry around, and also to the resource capabilities they have. For instance, the HAGGLE EU project [11] exploits store-carry-and-forward capabilities (i.e., devices' powerful features, user willingness, trust among users, opportunistic contacts) aiming to provide communication in scenarios with intermittent connectivity. HAGGLE considered human mobility and the power of users' devices to perform forwarding of information independently of the network layer. So it is easy to see that the way people communicate is arriving at a point where such communication must happen independently of the infrastructure available, and depending on the capabilities of intermediary devices as well as their mobility pattern, interests and social ties.

In what concerns the network layers, this new communication paradigm demands more reliable and efficient protocols, as today we have areas where spectrum abounds and creates interference—dense networks, e.g. residential households, shopping malls) as well as areas where communication is only possible through the formation of clusters of users (e.g., intermittently connected networks). Even in a metropolitan area, intermittent connected networks exist due to wireless environments, unexpected disruptions, and areas where the networking infrastructure is sparse (e.g., city parks).

A key factor that has assisted so far the expansion of UCNs in an indirect way, is trust as perceived by humans in social networking. Living examples of UCNs have expanded solely by the willingness of the end-user to become part of these communities, i.e., to *trust* unknown users and to allow others to rely on privately owned APs. Hence, these networks and their scalability is basically growing due to the Internet end-user's belief that the benefit of relying on UCNs is higher than the risk, which in itself is a pure social belief that can be applied in networking to improve the network operation [9].

UCNs integrate the notion of social trust schemes thus allowing users and operators to develop connectivity between devices based on trust circles. Such trust does not necessarily imply that users know each other; instead, it relates to *social interaction* and to the interests shared by familiar strangers, i.e., users that knowingly or unknowingly share some aspects of their daily routines (e.g. visiting the same coffee shop every Saturday morning). Hence, the user anonymity is kept, while social interaction metrics related to direct and non direct recommendations of nodes around, as well as to the trust openness of a user towards strangers assists in developing more robust connectivity links, in the sense that connectivity becomes intertwined with circles of trust that are built on-the-fly.

In addition to assist in creating, under specific circumstances robust connectivity on-the-fly, trust is also a *Quality of Experience (QoE)* parameter which can be applied in networking to improve the satisfaction of users involved in UCNs, and hence contribute to a better, in the sense of fairer, network usage.

This paper explains such notions, namely, how can trust, a social parameter based on individual beliefs, be applied in the context of resource management in wireless networks, in particular in environments such as UCNs where the end-user is a stakeholder of Internet connectivity. The paper is organized as follows. Section 2 describes work that shares our motivation. Section 3 goes over the trust application in the context of QoE and network fairness, while Sect. 4 gives insight into how trust can be applied to *Call Admission Control (CAC)* in the context of wireless networks, based on a specific proof-of-concept that has been developed for UCNs. The paper concludes in Sect. 5, providing a few guidelines for related research.

2 Related Work

Resource management in wireless networks and in particular call admission control is a topic that has been central to *Quality of Service (QoS)* research in wireless networks, for the last decade. Several schemes have considered ways to ensure fairness, being usually the intention to allow the network to serve more users at an instant in time [1]. Initial approaches considered static or dynamic threshold models [4] and priorities to provide fairness in terms of network utility, e.g. throughput. Pong et al. provided an analysis of the trade-off between fairness and capacity in the context of *Wireless Local Area Networks (WLANs)*, for scenarios with interference. This work explores fairness in terms of throughput as a measure of network utility, and allowed transmission time, explaining how different fairness parameters impact on the capacity of the link. Pricing model approaches [5, 12] were applied to ensure fairness, again in terms of network utility, but considering all of the potential network stakeholders. Game theory has also been considered as a way to assist a better notion of fairness in wireless networks [2].

More recently and due to the user-centric networking trend in wireless networks, the need to consider *Quality of Experience (QoE)* metrics that could assist a more dynamic behavior where the network can serve better more users as well as to increase user satisfaction emerged. Neely provides QoE networking contextualization in terms of QoS, for which parameters are more easily understandable, from a networking perspective [3]. Piamrat et al. consider mean opinion score (MOS) without interaction from real humans [6] and provide a simulation based performance evaluation showing that not only was the network better used in terms of throughput, but user satisfaction was also increased. Still in the context of translation between QoE and QoS, Zhang et al. explain the challenges and a possible solution to optimize QoE in next generation networks [14].

Our work builds on the need to integrate QoE, namely, the intention to allow more fairness in the way users are served by the network, while at the same time achieving better network usage. To achieve this we follow a dynamic approach, by considering the trust level that users have on thirds, to provide fairness.

3 Trust as a Fairness Parameter

In this section we start by providing a few notions related to *social trust* modeling, as trust as a QoE parameter has roots in social sciences paradigms.

In UCNs, a node *i* is defined to be a (wireless) device that belongs to an entity, its *owner*. An owner can be a specific user, or a group of users. In terms of trust and assuming that a networked device can be shared by different users, only the owner is responsible for such device. In other words: the trust computation, negotiation, and establishment is always associated to the owner identifier, and not to the device. Moreover, an owner may be responsible for more than one node. Nodes are associated to other nodes by means of trust associations.

A *trust association* is the k-th directed association between two nodes and is related to the respective owner's interests and social networking perspective. A trust association holds a cost, the *trust level*. The trust level provides a measure of previous trust behavior which can be considered as a QoE parameter. The rationale for this assumption is that a user is more willing to share resources within its trust circles. The computation of a trust level derives from each owner trust expectations and beliefs. Furthermore, such computation takes into consideration local and external influences. Examples of local influences are the degree of connectivity and reputation level of a node. External influences are influences that do not relate to the nature of each node but to external networking conditions (e.g. too much overhearing probability around a node).

Two nodes may hold more than one trust association among them, for instance, one per specific service. In this paper we consider a trust association to be unique and unidirectional between two nodes. Hence, a trust association from node A to node B may or may not have a different trust level than a trust association from node B to node A.

Table 1 Illustrative trust level categorization

Trust level	Meaning	Action
−0.1	Misbehaved user	Penalty
0	Not trusted	No data exchange
0.1	Recent acquaintance, quarantine mode	Only data that requires no confidentiality
0.3	Sufficient trust level	Node trusts enough to send data that is not confidential but requires reliability
0.6	Good trust level	Confidential data can be exchanged
1	Fully trusted	Closest list of trust

3.1 Examples of Operation

To better explain the notions behind trust as a social parameter that can be applied to networking this section provides a description of a few operational examples. For the given examples we consider that trust levels are deterministic and based on the values provided in Table 1. Such trust level representation is here provided for illustration purposes only. In a real environment the trust levels can be computed dynamically based on common reputation or recommendation schemes, as explained in Sect. 4.

Table 1, exemplifies a potential mapping of trust levels to service levels to be provided, i.e., a direct translation between QoE and QoS levels. A trust level of zero from node A to B ($T_{AB} = 0$) means that there is no trust association, which is reflected in the absence of data from A to B. When the level is set to 0.1, it means that a trust association has been recently established and that node B is in quarantine mode, i.e., node A will only transmit data to B that does not require confidentiality. A trust level of 0.3 is sufficient to allow node A to send data to node B. Such data although not confidential requires reliable treatment by node B. A trust level of 0.3 allows node A to send confidential data to node B, while a level of 0.6 means that nodes belong to a closest list of trusted devices bringing extra guarantees to node A beside confidentiality, such as non-repudiation and privacy by node B. When the trust level is negative it means that there is good evidence that node B has been misbehaved, which means that such node may be subjected to a penalty, such as getting lower priority when accessing the Internet.

Let us now consider the scenario illustrated in Fig. 1 where it is assumed that there is an already established community composed of three nodes (A, B, MP). Each node holds a trust table where each entry includes an index, trust level, as well as an aging value in seconds. The node MP holds two trust associations (1 and 2) towards node A (T_{MPA_1} and T_{MPA_2}), and one trust association towards node B (T_{MPB}). Both A and B are fully trusted by the MP that allows them to access the Internet with guarantees of non-repudiation. However, the MP only allows data originated from nodes adjacent to A to reach the Internet within a trust level of three, which means that the MP may impose some extra security mechanism to such traffic, such as tagging packets with

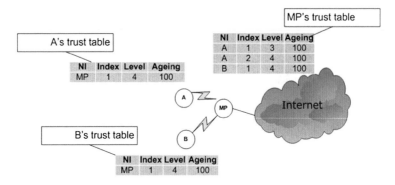

Fig. 1 Operational example for a community with three nodes: A, B, the MP

an alert label. Each trust association entry is refreshed after 100 s, as illustrated in the figure.

Let us now consider that a new node C recently registered in the community wants to access Internet. C is known to B—has a trust association to B—but not to A. When it tries to connect to the MP node C triggers a request for trust recommendations. This request has two purposes: (i) to check the reputation of the MP in the community; (ii) to ask neighboring nodes to recommend MP.

Both nodes A and B reply to C stating that they have good trust level associations towards node MP by sending the respective trust levels and hence node C places a new trust association towards the MP in its trust table, weighting its own beliefs (e.g. a weight) and the answers from the neighboring nodes. The computation of such weight would e.g. result in a trust level of 4.

At the same time, the MP broadcasts a request the network to collect recommendations about C. Let us first assume that node B knows C (has a trust association with C on level 3). Node B replies with this level to the MP, which accepts node C, creating a trust association of level 1, based on its own expectations and the provided answer. Node A claims that it does not know the node (answers with a trust level of 0). Depending on the answer of other nodes and on the node C behavior, the MP may or may not change the trust association level towards C.

Let us now consider that node C is not known by any node on the network. The MP gets no answer but based on its own willingness to share (belief concerning how to deal with unknown nodes) accepts the connection but places node C on trust level 0.1—quarantine.

3.2 Background on Trust Management

To provide a perspective on how the global trust framework works on UCN this section provides a description of the general functionality of the trust management scheme. For the remaining sections we shall consider the role of a *requestee* or of a *requester* in trust negotiation. A *requestee* in UCN corresponds to the notion of gateway (normally, an Acess Point). While the *requester* role can be assumed by both a node and a gateway: nodes perform trust negotiation towards gateways; gateways perform trust negotiation among themselves.

During boot up of the nodes, there are a few steps related to the initial setup of trust parameters, i.e., the way that nodes perceive others around, when they are in untrusted environments. Since the UCN trust environment may not be enough as an incentive for cooperation, the boot up phase ends up with the assignment of a set of *credits* that the requester may use to access shared resources. Based on the collected information the requester will try to establish trust associations with one of the responsive gateways and the MAC association part is established upon a successful end of trust negotiation.

Therefore, trust negotiation and initial credit assignment [8] are crucial to allow a node to associate with a gateway in UCNs. Full details concerning a model for trust management in UCN is out of the scope of this paper. The reader can find more information in Chaps. 1 and 2.

3.3 Applying Trust to QoS, Tokens as a Translation Currency Unit

In the previous section we have discussed and explained notions concerning how trust, which is a social parameter derived from human beliefs and interests, can be used as a QoE parameter to increase resource management fairness. As it is a social notion, trust is too subjective to be applied directly to networking. We consider trust weights to be values between 0 and 1, following the recent trends in distributed trust schemes. Trust, however, as a QoE parameter, cannot be used directly to improve the network performance and in particular fairness, as by simply considering trust levels, users that are more trusted would become greedy users and consume all of the resources in the network. Hence, to apply trust in the context of QoS it is necessary to consider incentives which motivate a good behavior, together with trust levels. In current UCNs users share Internet access in exchange of broader roaming. The network utility considered here is connectivity. However, there are additional networking resources that can be shared in exchange of other benefits. For instance, a user that cooperates frequently by opening its access points can be rewarded later not necessarily with the same type of service, but with a local service (e.g. access to a local printer in an airport; profiting from a relay in an area where the device does not have direct communication to any access point).

To make such exchange truly fair, we consider that resource assignment is based on the combination of trust associations and *credits* a node is willing to spend to get a service. Based on these two parameters, we consider a unique and virtual currency in the form of a *token*: a token is therefore a virtual measure (unit) of resources. Such a virtual currency allows for a direct exchange of different goods even in different instants in time.

Tokens are, as mentioned, the result of a utility function which has as input both a trust level of an association between two nodes, and a set of credits that a node is willing to spend to get a specific service. Such a function should take into consideration higher trust levels but also larger sets of credits. However, we expect it also to vary slowly and to depend more on the trust level, than on credits, as credits are an item that can be acquired and could lead easily to greedy situations. In other words, if the trust weight for a specific association between two nodes (*requestee* and *requester*) is low then even if the node requesting credits has a high credit level, the resulting token value should progress slowly. While if a requester—Requestee association shows a good trust level, then if it uses a high level of credits, the resulting tokens should also not increase linearly, as this would make the node consume all of the available resources.

To exemplify this line of thought, we consider two different functions to consider as provided in Eqs. (1 and 2).

$$tk(i, j) = tl(i, j) * \sqrt{c}, tl \in [0, 1]; c \in [0, \infty] \tag{1}$$

$$tk(i, j) = tl(i, j)^{\log(c)}, tl \in [0, 1]; c \in [0, \infty] \tag{2}$$

The main differences between Eqs. (1 and 2) are illustrated via Fig. 2. This figure shows two charts. (a) corresponds to Eq. (1, and (b) to Eq. 2), where values for the trust level *tl* and credits *c* have been varied to exemplify the impact on the computation of tokens.

Equation (1) (cf. Fig. 2a) is a utility function that results in a token progression that follows both credit and trust level growth. If by some reason the trust level is decreased (the user is penalized) then so are tokens, independently of whether or not the user has a large amount of credits. This function prevents greedy and misbehaving users to get a hold on all of the resources of the network, as tokens are to be exchanged by resources.

Equation (2) (cf. Fig. 2b) results in a larger number of tokens when credits and trust level is low, but with the increase on both these parameters, the resulting tokens also increase. The function is not so sensible to misbehaving users.

The functions here provided are discussed to explain how tokens can be defined to dynamically allow the translation between a social parameter such as trust, and networking resources.

The next section describes an example on how trust, and tokens, can be applied in the specific context of resource management, to call admission control, with the motivation to increase fairness in wireless networks and as a consequence, to improve user satisfaction.

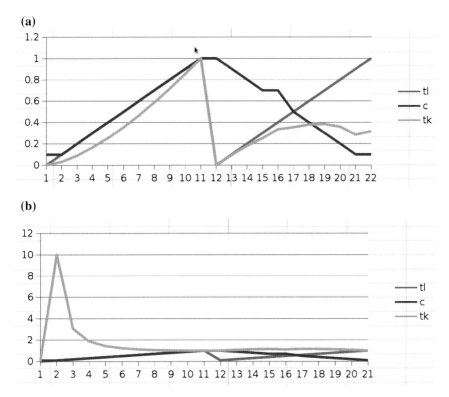

Fig. 2 Examples for token definition based on trust levels and credits. **a** Equation (1) variation based on credits and trust levels. **b** Examples for token defination based on trust levels and credits

4 Call Admission Control Based on Trust Circles

The description provided in this section is based on the UCN model provided by the European project ULOOP. The main blocks of a UCN have been described in Chap. 1 [8]. In ULOOP, resource management relates to self-organization and cooperative aspects, which are addressed from an OSI Layers 2 and 3 perspective. Resource management aspects in ULOOP comprise a block which is divided into four main sub-blocks which are described in this chapter. In what concerns *Call Admission Control* (CAC), this is a functional block which takes care of prioritizing requests based on both network conditions as well as on trust levels that the users (and consequently their owned devices) have in the network, towards networking devices. CAC starts prioritizing and queuing incoming requests so that contending requests can be treated based not only on the network conditions, but also on the trust association that a gateway (its owner) has towards a device (its owner). Queued requests are handled based on another thread, again related with prioritization as well as feedback from a resource allocation sub-module.

Incoming requests are classified by ULOOP gateways as "known" (e.g. flag Known = 1) or new (e.g. flag known = 0) as a way to prioritize requests from nodes that the gateway has recently authorized. A "Known" request example can be a request from a node that has just been served by this gateway e.g. 30 s ago, or a node that has been accepted to be transferred to this gateway by a gateway in the vicinities.

CAC then takes care of prioritizing requests according to a specific utility function that considers the trust level of the gateway towards the node, as well as the number of tokens that the node wants to exchange for a specific service. An example for such a function is provided in Eq. (3), where p corresponds to the priority which is proportionally dependent on the tokens that node i provides to j, to apply for a specific service, and dependent on the trust level that node j has on node i.

Moreover, the gateway checks whether or not it is a suitable gateway to handle the request. This is done based on local information that CAC can periodically collect from feedback of neighboring gateways. Assuming a case where the gateway decides it cannot serve the request or that there is a more suitable gateway, then the gateway redirects the request to that gateway, directly to the node, by providing the MAC of the best gateway.

Assuming a gateway that can serve a specific request, then CAC simply redirects the request to the resource allocation module.

$$p = tk(i,j) * tl(j,i) \tag{3}$$

5 Summary and Future Work

This paper addresses the applicability of trust metrics as a fairness parameter in call admission control within the context of user-centric networks. The paper explains how trust can assist in improving user Quality of Experience in wireless networks, by taking into consideration not only channel conditions, but also trust levels derived from the interaction that users have in the context of Internet shared services.

References

1. Ahmed MH (2005) Call admission control in wireless networks: a comprehensive survey. IEEE Commun Surv Tutorials 7(1–4):50–69
2. Fang Z, Bensaou B (2004) Fair bandwidth sharing algorithms based on game theory frameworks for wireless ad-hoc networks. INFOCOM 2004, 23rd annual joint conference of the IEEE computer and communications societies, vol 2
3. Ivanovici M, Beuran R (2010) Correlating quality of experience and quality of service for network applications. Quality of service architectures for wireless networks: performance metrics and management, pp 326–351
4. Kwon T, Kim S, Choi Y, Naghshineh M (2000) Threshold-type call admission control in wireless/mobile multimedia networks using prioritised adaptive framework. Electron Lett 36(9):852–854

5. Neely MJ (2009) Optimal pricing in a free market wireless network. Wirel Netw 15(7):901–915
6. Piamrat K, Ksentini A, Viho C, Bonnin J-M (2008) Qoe-aware admission control for multimedia applications in ieee 802.11 wireless networks. In: IEEE 68th vehicular technology conference, VTC 2008-Fall, pp 1–5
7. Pong D, Moors T (2004) Fairness and capacity trade-off in IEEE 802.11 WLANs. In: 29th annual IEEE international conference on local computer networks 2004
8. Sofia R (2014) User-centric networking: bringing the home network to the core. In: Aldini A, Bogliolo A (eds) User-centric networking—future perspectives. LNSN, Springer, London
9. Sofia R, Mendes P, Damásio MJ, Henriques S, Giglietto F, Giambitto E, Bogliolo A (2012) Moving towards a socially-driven internet architectural design. ACM SIGCOMM CCR Newslett 42(3):39–46
10. Sofia R, Mendes P (2008) User-provided networks: consumer as provider. In: IEEE communication magazines feature topic on consumer communications and networking—gaming and entertainment 2008
11. The Haggle Project. Code available http://code.google.com/p/haggle/
12. Xue Y, Baochun L, Nahrstedt K (2003) Price-based resource allocation in wireless ad hoc networks. Quality of service—IWQoS 2003. Springer, Monterey, pp 79–96
13. Yu F, Leung V (2002) Mobility-based predictive call admission control and bandwidth reservation in wireless cellular networks. Comput Netw 38(5):577–589
14. Zhang J, Ansari N (2011) On assuring end-to-end qoe in next generation networks: challenges and a possible solution. IEEE Commun Mag 49(7):185–191

Cooperative Relaying in User-Centric Wireless Networks

Tauseef Jamal and Paulo Mendes

Abstract An ever-growing demand for pervasive Internet access has boosted the deployment of wireless local networks in the past decades. Nevertheless, wireless technologies face performance limitations due to unstable propagation conditions and mobility of devices. In face of multi-path propagation and low data rate stations, cooperative relaying promises gains in performance and reliability. However, cooperation procedures are unstable (dependency upon current channel conditions) and introduce overhead that can endanger performance, especially when nodes are mobile. In this paper we describe a novel protocol, called RelaySpot, able to implement cooperative relaying in dynamic networks, based upon opportunistic relay selection, and cooperative relay scheduling and switching. RelaySpot removes the need for estimation and broadcast of channel conditions, and is expected to improve the utilization of spatial diversity, minimizing outage and increasing the transmission capacity of wireless local networks.

Keywords Opportunistic relay selection · Cooperative relay scheduling · Cooperative relay switching · Wireless resource management · Space-time diversity

1 Introduction

Over the past decade, Internet access became essentially wireless, with 802.11 technologies providing a low cost broadband support for a flexible and easy deployment. However, channel conditions in wireless networks are subjected to interference and multi-path propagation, creating fading channels and decreasing the overall

T. Jamal · P. Mendes (✉)
Copelabs, University Lusofona, Lisbon, Portugal
e-mail: paulo.mendes@ulusofona.pt

T. Jamal
e-mail: tauseef.jamal@ulusofona.pt

A. Aldini and A. Bogliolo (eds.), *User-Centric Networking*,
Lecture Notes in Social Networks, DOI: 10.1007/978-3-319-05218-2_9,
© Springer International Publishing Switzerland 2014

Fig. 1 802.11 rate adaptation
and cooperative relaying

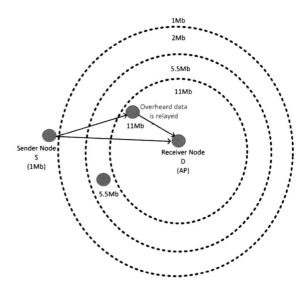

network performance [2]. While fast fading can be mitigated by having the source retransmitting packets, slow fading, caused by obstruction of the main signal path, makes retransmission useless, since periods of low signal power last for the entire duration of the transmission.

Cooperative networking can mitigate wireless performance constraints at the MAC layer. Cooperation occurs when overhearing nodes assist the communication between source and destination, by relaying different copies of the same frame from different locations, generating spatial diversity that allows the destination to get independently versions of the frame sent by the source. Hence, cooperative relaying allows a destination to get good frames with high probability. If all copies are corrupted, they can still be combined at the destination in order to obtain an error-free frame.

The development of cooperative relaying raises several research issues, including the performance impact on the relay itself, and on the overall network, leading to a potential decrease in network capacity and transmission fairness. Such research issues can be influenced not only by fading, but also by other performance constrains in wireless networks, such as the distance at which wireless nodes are from Access Points (APs), as well as the mobility of such nodes [8].

Due to the distance to an AP, a wireless node can observe a bad channel as compared to other nodes that are closer to such AP, leading to the use of 802.11 rate adaptation schemes. Figure 1 illustrates the transmission characteristics of wireless nodes, as a result of the rate adaptation functionality of IEEE 802.11: nodes closer to the AP transmit at high data rates, while nodes far away from the AP decrease their data rate after detecting missing frames. Figure 1 also illustrates the role that cooperative relaying may have increasing the performance of the overall wireless network, helping low data rate nodes to release the wireless medium sooner, helping high data rate nodes to keep the desirable performance, and allowing the network to achieve a good overall capacity.

The problems posed by the presence of low data rate nodes are magnified in mobile scenarios, since the data rate of nodes change over time, due to their mobility, requiring faster and more accurate selection of potential relays. However, the selection of a helper can be problematic, since potential relays are also mobile, which means that their relaying conditions may change over time. This requires relaying methods able to switch among relays to make a better use of spatial diversity.

In this paper we present a cooperative relaying MAC protocol, called RelaySpot, that is able to mitigate the problems posed by shadowing and by the presence of low data rate nodes. In a clear breakthrough in relation to prior art, the proposed relaying protocol is aware of the mobility of potential relays, leading to an increase in the performance of dynamic wireless local networks.

With RelaySpot, wireless networks do not need complicated distributed routing algorithms, as in ad-hoc network to extend the coverage of wireless local networks, due to its capability to switch among relays as mobility patterns change over time. With RelaySpot, standard 802.11 networks are able to offer ubiquitous high data rate coverage and throughput, with reduced latencies.

The paper is organized as follows: Section 2 describes related work. In Sect. 3 we describe RelaySpot, while Sect. 4 illustrates its operation. Section 5 presents the performance evaluation. Section 6 presents a summary of our findings and our conclusions.

2 Related Work

With the purpose of offering effective and efficient interaction between the physical and higher protocol layers, research on cooperative communication has been exploring the MAC layer [4]. Cooperative relaying, at the MAC layer, comprises two phases: relay selection and cooperative transmission. In the first phase a relay or group of relays are selected, while in the latter phase the communication via relay(s) takes place. The relays can be selected either by the source (source-based), the destination (destination-based), or by the relay itself (relay-based). Node mobility can greatly affect the cooperative transmissions and the selection of the relays due to additional overhead (due to relay failure). The mobility of wireless nodes should be taken into consideration when developing cooperative MAC protocols, in order to minimize such additional overhead and to maximize cooperation gain. However, the impact of node mobility on cooperative relaying has not been addressed in the literature.

For a better analysis of cooperative relaying, we can split current proposals into two classes of cooperative MAC protocols: proactive and reactive. In proactive relaying the source, the destination or a potential relay replaces a slow direct communication with a faster relaying communication, aiming to improve the transmission performance. In the case of reactive relaying, relays forward data to the destination when the direct communication fails, avoiding retransmissions.

Source-based relaying approaches require the source to maintain a table of Channel State Information (CSI) that is updated by potential relays based upon periodic broadcasts. As an example, with CoopMAC [13], the source can use an intermediate node (called helper) that experiences a good channel with the source and the destination. Based on the CSI broadcasted by potential helpers, the source updates a local table (cooptable) used to select the best relay for each transmission. Another example of source-based relaying is CODE [15], which uses multiple relays based on network coding. If nodes find that they can transmit data faster than the source, they add the identity of the source and the destination to their willingness list. Once the source finds its address on the willingness list of potential relay(s), it adds those relay(s) into its cooperative table. In general source-based approaches undergo two main problems: channel estimation and periodic broadcasts, which introduce overhead that is problematic in mobile scenarios. These problems are mitigated with RelaySpot, whose behavior was previously compared to CoopMAC, highlighting its operational advantages [12].

The relay enabled DCF (rDCF) protocol developed by Zhu and Cao [16] based on DCF is an example of destination based proactive relaying. In this protocol relays maintain a willingness list that contains the identifiers of the source-destination pairs that a relay can help. Periodically, each potential relay advertises its willingness list, and the source picks a relay from such list. rDCF proposes a triangular handshake mechanism used by the destination to select a relay based on a set of *Relay Requests to Send* messages sent by the source and the relay: these messages allow the destination to get information about the quality of the source-relay and relay-destination channels. rDCF has poor performance in the presence of messages with small payload, due to its relatively high overhead, and increases the probability of collisions, situation that does not occur with RelaySpot. Enhanced relay-enabled DCF (ErDCF) [1] inherits some characteristics of rDCF such as triangular handshake. But it uses short Physical Layer Convergence Protocol (PLCP) preamble for dual-hop cooperative transmission, which provides higher throughput and reduced blocking time. In ErDCF the data frame forwarded by a relay includes the duration field, which can minimize the collision risks. However, it increases the frequency of periodic broadcast, which increases overhead.

While the mentioned proactive approaches rely upon broadcast, Opportunistic Relay Protocol (ORP) [3] does not. It also does not rely on CSI for relay selection: the source opportunistically makes a frame available for relaying and all potential relays try to forward that frame. ORP is similar to RelaySpot in the sense that both do not rely on CSI for relay selection. However with ORP the source does not know the availability of relays, and therefore, it does not know the rate of the source-relay and relay-destination channels, leading to poor relay selection. With RelaySpot the destination is aware of the relay diversity, as well as the rates of the used links.

Reactive cooperative methods such as PRO [14] rely on relays to decide to retransmit on behalf of the source when the direct transmission fails. With PRO [14], relays are selected among a set of overhearing nodes in two phases: the first phase leads to the identification of qualified relays; in the second phase, qualification information is broadcasted, allowing qualified relays to set scheduling priorities. Contrary to RelaySpot, PRO presents a high probability of collision, as well as low efficiency in mobile scenarios due to CSI measurements.

In general, our analysis of related work shows that proactive and reactive approaches have their pros and cons. RelaySpot combines reactive and proactive mechanisms to avoid CSI estimation and broadcasts, and to reach low probability of collision, and good selection of relay communication (due to awareness about the data rate of the source-relay and relay-destination links). Moreover most of the prior art only considers relaying in static wireless scenarios. There are some approaches (e.g. CoopMAC and PRO) whose performance is evaluated also in mobile networks, but they are still agnostic of the mobility patterns of the involved nodes. To the best of our knowledge, RelaySpot is the first cooperative MAC protocol that is able to augment the performance of dynamic wireless networks, by being aware of the level of mobility of potential relays, combining it with an analysis of interference and transmission success rate.

3 RelaySpot MAC Protocol

RelaySpot is a hybrid cooperative relaying protocol where relays self-elected under cooperation conditions are used to increase the performance of active transmissions (proactive behavior) or to replace failed transmissions (reactive behavior). RelaySpot comprises three building blocks: opportunistic relay selection; cooperative relay scheduling; cooperative relay switching [7, 10].

In order to be applied to dynamic scenarios, and unlike previous work, RelaySpot does not require the maintenance of CSI tables, avoiding periodic updates and consequent broadcasts. The reason to avoid CSI metrics is that accurate CSI is hard to estimate in dynamic networks, and periodic broadcasts would need to be very fast to guarantee accurate reaction to channel conditions in such scenarios.

Moreover, relay selection faces several optimization problems, meaning that the best relay may be difficult to find. Hence, for dynamic scenarios, the approach followed by RelaySpot is to make use of the best possible relaying opportunity, and to switch between relays qualified within the cooperation area, if necessary.

With RelaySpot each session competing for the wireless medium is identified by a source-destination pair. The destination node is assumed to be reachable via the direct link or via one-hop relay. During the lifetime of a session, the destination dynamically selects the best qualified relay, from the set of self-elected relays, aiming to maximize the network utility in terms of throughput and latency.

We make the following assumptions during development of RelaySpot protocol:

- Single-channel: to keep RelaySpot simple, we consider that all nodes are using the same wireless channel. Interference is not handle by switching wireless channels, which would bring more delay and performance uncertainty, but by relay switching, selecting relays with a low interference factor.

- One-hop relaying: the current usage scenario is of a wireless local network with one access point and several mobile stations, which can act as source of data, as a relay, or both simultaneously. The usage of multiple chained relays, as a simpler alternative to layer-3 routing will be investigated in the future, as a solution to expand wireless coverage and increase the utility of a relay.
- Simultaneous access to a channel: Multiple sessions are allowed to access the channel at any given time, including the ones that are being relayed, which increases the wireless interference. To mitigate the effect of interference, dynamic switching among qualified relays is implemented aiming to exploit the utility of each relay, ensuring that RelaySpot is able to increase the utility of the overall network, and not only of specific links.
- Multiple relay selection: Each node can be self-elected as relay for more than one session at the same time. Moreover, one communication session can be helped by one or more relays in each moment in time, leading to a system with wireless diversity of one or higher. If the destination selects more than one relay for a single session (diversity >1) and gets corrupted frames from all relays, in a worse case scenario, it can perform frame combination in order to create a good frame. In this paper we consider a diversity of one during the experimental evaluation.

In the following, the components of RelaySpot are described followed by the formats used for frames.

3.1 Components

In the following, RelaySpot's building blocks are explained. First the relays are selected opportunistically, and then the destination schedules the potential relays for the following transmissions. If there are other better relays, then the relays can be switched.

3.1.1 Opportunistic Relay Selection

Relay selection is a challenging task, since it greatly affects the design and performance of a cooperative network. However, relay selection may introduce extra overhead and complexity, and may never be able to find the best relay in dynamic scenarios. Hence, the major goal of RelaySpot is to minimize overhead introduced by cooperation, with no performance degradation, by defining an opportunistic relay selection process able to take advantage of the most suitable self-elected relay.

This section describes the functionality proposed to allow self-elected relays to avoid high interference and to guarantee high data-rates to a destination, while preventing waste of network resources. Relay selection is performed in three steps: First, each node checks if it is eligible to be a relay by verifying two conditions, overheard a good frame sent by the source and be positioned within the so called

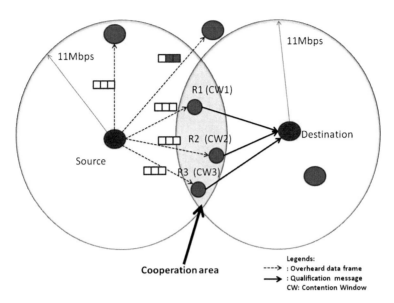

Fig. 2 Opportunistic relay selection

cooperation area; Second, eligible nodes start the self-election process by computing their selection factor; Third, self-elected relays set their CW based on their selection factor, and send a qualification message towards the destination after the contention window expires, as shown in Fig. 2.

The relay selection process is only executed by nodes that are able to successfully decode frames sent by a source. These relays start by verifying if they are inside the cooperation area by computing their Cooperation Factor (CF) as given in Eq. 1, where R_{sr} is the rate of the source-relay channel, and R_{rd} of the relay-destination channel rate. The rate of the source-relay and relay-destination channel is computed by overhearing RTS and CTS frames exchanged between source and destination. The CF ensures that potential relays are closely bounded with the source while having good channel towards the destination: an eligible relay must have a CF that ensures a higher data-rate than over the direct link from source to destination.

$$CF = (R_{sr} * R_{rd}) / (R_{sr} + R_{rd}), \quad CF \in [0, \infty[\tag{1}$$

The qualification of a node (that is able to decode the source frame and is within the cooperation area) as a relay depends solely upon local information related to interference (node degree plus load), mobility and history of successful transmissions towards the specified destination.

Node degree, estimated by overhearing the shared wireless medium, gives an indication about the probability of having successful relay transmissions: having information about the number of neighbors allows the minimization of collision and blockage of resources. However, it is possible that nodes with low node degree are overloaded due to: (i) local processing demands of applications (direct interference); (ii) concurrent transmissions among neighbor nodes (indirect interference). Hence, RelaySpot relies upon node degree and traffic load generated and/or terminated by the potential relay itself, to compute the overall interference level that each potential relay is subjected to.

Equation 2 estimates the interference level that a potential relay is subjected to as a function of node degree and load. Let N be the number of neighbors of a potential relay, T_d and T_i the propagation time of direct and indirect transmissions associated to the potential relay, respectively, and N_i and N_d the number of nodes involved in such indirect and direct transmissions. Adding to this, T_p is the time required for a potential relay to process the result of a direct transmission. The interference factor (I) affecting a potential relay has a minimum value of zero corresponding to the absence of direct or indirect transmissions.

$$ I = \sum_{j=1}^{N_d} \left(T_{dj} + T_{pj} \right) + \sum_{k=1}^{N_i} T_{ik}, \quad I \in [0, \infty] \tag{2} $$

Figure 3 shows a scenario where a node R is selected as a potential relay for nodes S and D. Node $N1$ is the direct neighbor of node R, while there are several other indirect neighbors ($N2, N3, N4, X$). Apart from R, node X also seems to be a relay candidate due to its low interference level. But it may be difficult to select R or X due to the similar interference levels: while R has a short transmission from a neighbor and a long transmission from the source, X is involved in an inverse situation. The selection of R or X as a relay can be done based on two other metrics of the RelaySpot framework: history of successful transmissions towards destination; stability of potential relays.

The goal is to select as relay a node that has low interference factor, which means few neighbors (ensuring low blockage probability), and fast indirect and direct transmissions (ensuring low delays for data relaying).

By using the interference level together with the history and mobility factors, the probability of selecting a node as a relay for a given destination is given by Eq. 3: the Selection Factor (S) is proportional to the history of successful transmissions that a node has towards the destination and its average pause time, and inversely proportional to its interference level.

$$ S = \frac{H * M}{1 + I}, \quad S \in [0, 1] \tag{3} $$

The History Factor (H) is the ratio between the number of successful transmissions and the total number of transmissions towards destination, as given by Eq. 4:

$$ H = \frac{N_{successful}}{N_{total}}, \quad H \in [0, 1] \tag{4} $$

Fig. 3 Opportunistic relay
selection: example scenario

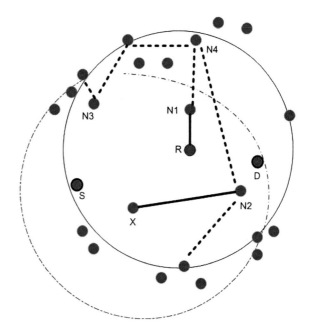

If self-elected to operate as a relay, the node computes its CW, as shown in Eq. 5. The CW plays an important role in scheduling relay opportunities. The goal is to increase the probability of successful transmissions from relays to the destination by giving more priority to relays that are more closely bounded to the destination, have less interference and have higher pause times.

$$CW = CW_{min} + (1 - S)(CW_{max} - CW_{min}) \tag{5}$$

From a group of nodes that present good channel conditions with the source, the opportunistic relay selection mechanism gives preference to nodes that have low degree, low load, good history of previous communication with the destination, as well as low mobility. In scenarios with highly mobile nodes, opportunistic relay selection is expected to behave better than source-based relay selection (e.g., CoopMAC), since with the latter communications can be disrupted with a probability proportional to the mobility of potential relays, and relays may not be available anymore after being selected by the source.

3.1.2 Cooperative Relay Scheduling

As illustrated in Fig. 4 the selection mechanism may lead to the qualification of more than one relay (R1, R2, R3), each one with different values of S, leading to different sizes of CW (e.g., R3 transmits first). Selected relays will forward data towards the destination based on a cooperative relay scheduling mechanism.

Fig. 4 Opportunistic relay
election

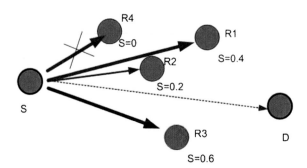

Fig. 5 Relay scheduling:
example scenario

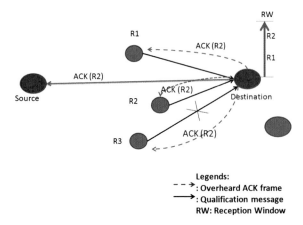

Based on the CF (i.e., R_{sr} and R_{rd}) of the qualification messages received from all
self-elected relays, the destination estimates which of the involved relays are more
suitable to help in further transmissions. To get multiple qualification messages the
destination only processes the received qualification messages after a predefined time
window, i.e., Reception Window (RW). As shown is Fig. 5 the size of the reception
window is of major importance, since it will have an impact on the number of
qualification messages that will be considered by the destination.

After the expiration of the reception window the destination processes all the
received qualification messages based on their received signal strength (R_{rd}) and R_{sr}
(R_{sr} is carried by QM). Depending on the configured diversity level the destination
will select one or more relays to help the current transmission. If diversity is set
to one, the destination sends an ACK frame to the source including the ID (i.e.,
MAC address) of the selected relay, which will continue sending the frames to the
destination.

When the destination is configured to operate with a diversity higher than one,
the destination sends an ACK frame to the source including the MAC addresses
of the selected relays. During data transfer, the destination sends the received data
to the application as soon as a correct frame arrives from any of the selected relays.

Fig. 6 Frame combining at destination by scheduler

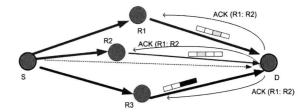

If the selected relays start sending corrupted frames (e.g., because they moved to a faraway position) the destination waits until a good frame is received, until it received data from all selected relays, or until a predefined timeout occurs. If the destination only got corrupted frames it will try to combine them to create a good frame. If such process is possible, the destination will send an ACK to the source including the MAC addresses of the relays that sent the frames which combined produced a good frame. As shown in Fig. 6, where the primary relay R3 fails to relay the frame.

In this paper we consider a diversity of one during the experimental evaluation, which means that the transmission is diverted by the source to a unique relay selected by the destination.

RelaySpot solution (in proactive mode) is destination based, because the destination chooses the best set of relays via scheduler. In reactive mode (i.e., reaction to a failed link), the scheduler is not used, because the relay only forwards the failed data frame on behalf of source. Therefore, RelaySpot in reactive mode is relay-based, because decision about cooperation initiation and selection is taken on relays.

3.1.3 Relay Switching

Since relays are selected opportunistically, based on local information, there is the possibility that the best relay will not be able to compute a small contention window, losing the opportunity to relay the frame. In order to overcome this situation, as well as to support the failure of selected relays, RelaySpot includes a relay switching operation.

All potential relays are able to compute their own CF, as well as the CF of the selected relay. The former is possible by overhear the ongoing RTS/CTS, which are used to compute the cooperation factor from the signal quality. The CF of the currently selected relay can be computed based on its source-relay and relay-destination data-rate, that any other potential relay can collect by overhearing data and ACK frames.

If a potential relay is not selected in the relay selection procedure, it compares its CF with the CF of the selected relay. If its CF is better, which means that it can provide better gain, it sends a Switching Message (SM) to the destination, by means of a dummy data frame, informing it about its own CF. This way the current relay can be switched to the newly selected relay, since: (i) by overhearing the frame sent by the new relay, the source will send the next data frame towards that relay: (ii) by receiving the frame sent by the new relay, the destination knows that the next data frame will be sent by it.

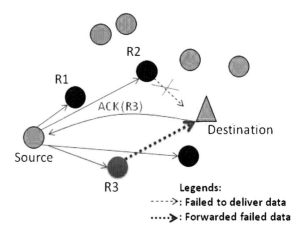

Fig. 7 Relay switching due to failure of current selected relay

Relay switching is suitable for dynamic scenarios where a previously selected relay may not be efficient at some stage, due to mobility, fading, or obstacles, for instance. Hence, unlike prior-art, relay switching can overcome such variations in network conditions making the deployment of cooperative relaying possible for dynamic networks.

While the use of relay switching can be used to improve the performance of a communication, by replacing a good relayed transmission by a better one, relays can be switched implicitly when a potential relay detects a missing ACK for an already relayed communication. In this situation, the relays try to forward the overheard data frame on behalf of the relay that failed the transmission. If successful, the destination notifies the source about the MAC ID of the new relay within and ACK frame.

Figure 7 shows an example of implicit relay switching, where a previously selected relay, R2, fails to relay data to the destination. In this case, instead of retransmitting the failed data and re-selecting the relays, another potential relay (in this case R3) that has better cooperation factor than R2 reserves the channel for sending the failed data frame to the destination. If it is successful, the destination sends an ACK frame with indication of R3 as relay. This way the source switches from R2 to R3 starts from the next data frame.

3.2 Frames

After associating with an AP, nodes start by sending RTS/CTS frames to gain access to the shared medium, and all nodes start listening for control and data frames sent out by others on the shared channel: the overhearing process is required by 802.11's DCF mechanism as all nodes in the network need to correctly update their Network Allocation Vector (NAV). In RelaySpot, potential relays are self-elected based on the state of data transmission from nodes to AP (destination). In the remaining of this section, we explain how RTS, CTS, DATA and ACK frames are used by RelaySpot.

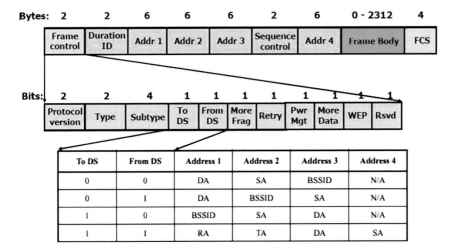

Fig. 8 802.11 frame structure used by RelaySpot

Table 1 Control frame bits to code rate information

Bit 11	Bit 12	Bit 13	Code (in Mb)
0	0	0	No rate information
0	0	1	Link is 1
0	1	0	Link is 2
0	1	1	Link is 5.5
1	0	0	Link is 11

As mentioned before, RelaySpot does not require any new frame, making use of the RTS, CTS, DATA and ACK frames already specified by the 802.11 standard. To ensure standard compatibility, the generic 802.11 frame structure is considered by RelaySpot (c.f. Fig. 8).

During the operation of RelaySpot, a node may need to inform other nodes about the data reception rate. To exchange this information bits 11–13 of the frame control field are used to code the data-rate of the transmission link. Table 1 illustrates the codes used by RelaySpot to identify the data-rate in the 802.11b nodes (the number of codes is enough to identify also the data-rates in the 802.11g and 802.11n). The usage of these bits does not jeopardize the operation of nodes that do not execute RelaySpot, since it only use bits that are not used within control frames.

3.2.1 RTS Frame

The usage of RTS frame is different during the cooperative transmission phase and when reacting to failed transmissions. In each of these cases the RTS frame is used as follows:

- For cooperative transmissions: Duration field is set to accommodate two transmissions (source-relay and relay-destination), while address 4 accommodates the address of the relay. More-frag bit is set to 0 indicating cooperation phase.
- Reaction to failed transmissions (or implicit switching): Address 4 is set with the source address, and the More-frag bit is set to 1 indicating the retransmission of failed transmission from the specified source.

3.2.2 CTS Frame

Since RelaySpot is triggered by the relays themselves, these need to gather as much information as possible about the surrounding transmissions, in order to detect nodes that need to be helped. CTS frames allow overhearing nodes to get information about the data-rate of direct links. To provide this information bits 11–13 in the CTS frame control are used to code information about the data-rate of the direct link, as illustrated in Table 1.

3.2.3 ACK Frame

RelaySpot uses the address 4 in the ACK frame (marked as unused by the 802.11 standard) to allow the destination to inform the source about the address of the relay for the subsequent data frames. While bits 11–13 indicate the data-rate of the relay-destination channel, according to Table 1, as the source needs it to reserve the channel. If address 4 and/or relay-destination rate is not used, the source keeps using the direct transmission.

If the ACK indicates the reception of failed data via a relay, the destination sets the More-frag bit to 1.

3.2.4 Data Frame

DATA frames can be sent on the direct link, prior to relay selection, or via the relay node, after relay selection. Data frames sent over the direct link have a ToDS/FromDS code of (1:0). In this case the address code is defined as follows: Address 2 indicates the source address; Address 3 the destination address. Relayed data frames have a ToDS/FromDS code of (1:1) and are sent over the source-relay link and over the relay-destination link. In each of these two cases the address code is defined as follows:

- Over the source-relay link: Address 1 indicates the relay address; Address 2 the source address; Address 3 the destination address.
- Over the relay-destination link: Address 2 indicates the relay address; Address 3 the destination address; Address 4 the source address. The relay also sends the rate of the source-relay channel encoded in bits 11–13 of the frame control according to Table 1.

3.2.5 Qualification Message

If a node is able to elect itself as a potential relay, the self-elected relay uses a frame of 112 bits (of type CTS) to inform the destination about its willingness to operate as relay for that source-destination transmission. The QM contains information about the rate of the source-relay channel encoded in bits 11–13 of the frame control according to Table 1. Although relays compute their CF to participate in the selection phase, the destination needs to know those values to identify the most suitable relay. Since the destination already knows R_{rd}, it only needs information about R_{sr} to estimate the CF of a specific relay.

3.2.6 Switching Message

If a node is able to successfully decode a data frame sent by a relay (with ToDS/FromDS bits set to (1:1)), it can elect itself as a potential replacement of that relay if: (i) detects a missing ACK from the destination; (ii) detects that its source-relay and relay-destination links provide better data-rates than the current source-relay-destination path.

The node self-elected to replace the current relay uses an empty data frame to inform the destination about its willingness to relay data frames related to that source-destination transmission. This data frame has ToDS/FromDS bits set to (1:1) where address 2 indicates the relay address, address 3 the destination address and address 4 the source address. Moreover, the relay sends its CF (the rate information about source-relay and relay-destination channels) in bits 11–13 of the frame control. The more-frag bit in frame control is set to 1 to indicate that it is cooperation switching frame.

4 Example Illustrations

In this section we illustrate the operation of RelaySpot with examples to describe: (i) Proactive operation, by selecting relays followed by a cooperative transmission to improve the direct transmission; (ii) Reactive operation, by retransmitting on behalf of source; and (iii) Sequence chart to show explicit and implicit switching.

4.1 Proactive: Relay Selection and Cooperative Transmission

Figure 9, illustrates the RelaySpot operation in a scenario where we have a poor link between the source and the destination. If direct link exists, it means that RelaySpot can work in proactive mode. Within proactive mode, the slow direct link can be improved by being replaced by fast relayed links. The decision to initiate cooperation

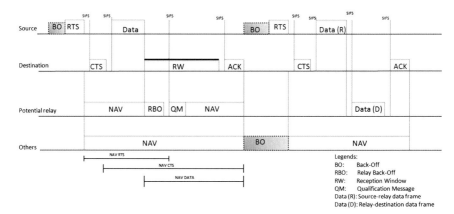

Fig. 9 An illustration of RelaySpot's proactive mode

is taken by the destination, after detecting a poor direct link. In the 802.11b standard poor link means either 1 or 2 Mbps, which can be helped by relays. If the direct link has 5.5 Mbps, then it can not be helped, because the condition for cooperation is false ($CF > R_{sd}$). The destination implicitly indicates the need for cooperation within CTS frame. Destination also increases the duration within CTS frame by a RW amount, to allow relays to send their QMs. By overhearing CTS, all nodes update their NAV accordingly.

In this example, first the relay selection takes place. After overhearing a data frame from the source, the relays contend according to Eq. 5 and try to send QMs to the source. It is assumed that the destination received at least one non colliding QM from a potential relay within RW, and the data frame from source was also received correctly. Hence, destination sends ACK with potential relay ID, after RW expires.

The cooperative transmission takes place after the relay has been selected. After the usual handshake, the data frame is forwarded to the selected relay, which relays the data frame to the destination without contention. The destination confirms the reception of data via the relay within ACK. This cooperative transmission continues until the last frame, or if the direct link gets better.

4.2 Reactive: Relay Selection and Retransmission

The reactive mode is triggered when the direct link fails. In short, RelaySpot can also operate in a reactive mode, aiming to replace a failed direct link, if potential relays detect a failed transmission (by missing ACK). In this case potential relays try to send the failed data frame to the destination on behalf of the source, based on their CW. If a potential relay gets the channel, it starts sending data by sending an RTS frame with address 4 set with the source ID (i.e., MAC address) and More-frag set to 1. By overhearing this RTS frame the source stops the retransmission process. In this case no scheduler is used at the destination.

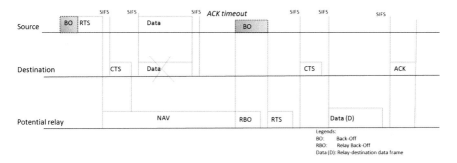

Fig. 10 An illustration of RelaySpot's reactive mode

In case of reactive relaying, the role of destination is not vital as in case of proactive relaying. Reactive relaying is initiated by relays themselves, followed by opportunistic relay selection. The example illustrated in Fig. 10, shows that the destination did not receive the data frame from the source. As a result there is no ACK sent to the source. We assume that there is a potential relay that was successful accessing the medium before the source. In this case the relay forwards the frame on behalf of the source, avoiding retransmissions.

Another example of reactive mode is when relays react to failed relayed transmissions, i.e., implicit switching, which is discussed in next section.

4.3 Sequence Chart

Figure 11 illustrates the message exchange used by RelaySpot in a scenario with three potential relays (R1, R2 and R3), and one source-destination pair. In this scenario we assume a low data-rate direct link.

The destination uses CTS frames to piggyback the source-destination data-rate, since this is a low data-rate link that may need to be helped. Eligible relays compute their cooperation factor based on the rate information collected by overhearing RTS/CTS frames; if an eligible relay is qualified to help the direct channel, it sends a qualification message (QM) by setting its contention window based on the selection factor computed after overhearing a data frame from the source. After receiving a data frame from the source, the destination does not send an ACK immediately. Rather, it starts the RW in order to give opportunity to qualified relays to send QMs. After the RW expires, the destination selects the relay based on the received QM, and confirms the selected relay in an ACK message, which includes the relay-destination data-rate (R_{rd}).

After relay selection and for the next data frames, the source sends an RTS message to the destination to reserve the channel for accommodating source-relay and relay-destination transmissions. After receiving the CTS frame sent by the destination, the source forwards a data frame to the selected relay, which sends the data frame to the

Fig. 11 RelaySpot sequence chart

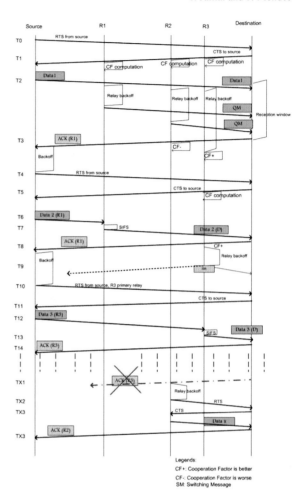

destination after a Short Interframe Space (SIFS) amount of time. After successful reception of the relayed data frame, the destination sends an ACK message to the source.

In this example we assume that, due to mobility, R3 becomes a better relay than R1 in order to illustrate the proactive behavior (in term of explicit switching) of RelaySpot. In this case R3 sets its contention window according to Eq. 5 after overhearing the ACK frame, and sends an SM to the destination carrying its CF. Upon overhearing this frame, the source start using R3 as a relay for the next data frames.

Now we assume that R3 fails to relay data at some instant in time to illustrate the reactive behavior (in term of implicit switching). If potential relays detect missing ACKs, they try to retransmit overheard data frames on behalf of R3. In this example, R2 tries to set its CW according to Eq. 5. If R2 is successful to forward the failed data frame, the destination sends an ACK frame to the source with R2 ID and

Table 2 Simulation parameters

Parameter	Values
Playground size	$200 \times 200\,\mathrm{m}^2$
Path loss coefficient	4
Carrier frequency	2.412e9 Hz
Max transmission power	100 mW
Signal attenuation threshold	-120 dBm
MAC header length	272 bits
MAC queue length	14 frames
Basic bitrate	1 Mbps
Rts-Cts threshold	400 bytes
Thermal noise	-110 dBm
MAC neighborhood max age	100 s
Speed	1 m/s
Reception window size	1504 us
Payload size	1024 Bytes

R2-destination data-rate (R_{rd}), while the source learns about R_{sr} from overhearing transmissions from R2. Then, the source switches to R2 for next data frames. This way implicit relay switching takes place.

5 Performance Evaluation

In this section, we evaluate the performance of RelaySpot by means of simulations, and compare it with the IEEE 802.11 standard, as well as two versions of RelaySpot: one that is not aware of mobility and another that does not use relay switching. The reason for not comparing RelaySpot with other cooperative MAC protocols is that RelaySpot is an hybrid protocol (reactive and proactive), that combines opportunistic and cooperative behavior, while the other proposals belong to different categories.

5.1 Simulation Environment

Evaluation is based on simulations run on the MiXiM framework of the OMNeT++ 4.1 simulator using 2D linear mobility model. Table 2 lists the simulation parameters. Each simulation has a duration of 300 s and is run ten different times in order to provide results with a 95 % confidence interval.

Simulations consider a scenario based on a wireless local network with one static AP and up to 25 mobile nodes. Each mobile node is a source of data towards the AP, and can be a potential relay of the transmissions started by other mobile nodes. Each simulation starts by randomly placing the group of mobile nodes (1 to 25) in

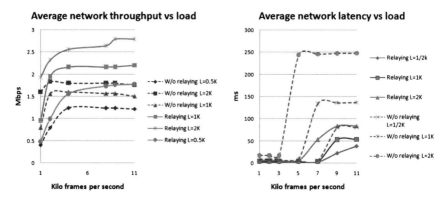

Fig. 12 Impact of frame size

a square of $200 \times 200\,\mathrm{m}^2$, having the AP at its center. Each node is equipped with only one half-duplex transceiver and has a unique MAC address. All the nodes in the network transmit control frames and data frames with the same power, and the network load is uniformly distributed among all nodes.

Wireless communications are done over one unique channel shared by all nodes. The used wireless channel supports four different data rates (1, 2, 5.5 and 11 Mbps) determined by the distance of the node towards the AP, while control frames are transmitted at a basic rate set to 1 Mbps [5].

To setup the mobile scenario with the most suitable configuration, we start by performing some experiments in a static scenario (1 AP and 25 nodes) to get the most suitable value for the frame size and the size of the reception window at the AP. The latter has an impact on the number of QMs that the AP can get from potential relays, which influences the selection of the most suitable relay. The frame size can have an impact on the quality of the number of successful transmissions.

In what concerns the frame size, simulation results (c.f. Fig. 12) shown that the performance gets better as the frame size increases, since more bits are transmitted in each transmission opportunity. Since RelaySpot leads to a reduction of frame retransmissions, the potential bad impact of handling large frames is diminished.

In what concerns the size of the reception window, results (c.f. Fig. 13) show that it is better to have a big reception window to allow the AP to grab a larger number of QMs, allowing it to select the best relay with high probability.

A very small reception window allows the AP to receive only one QM, which means that the destination has only one relay to select from. Such relay is with high probability a node closer to the source, since such nodes overhear good copies of source frames first. Moreover, in case of collision of QMs, the destination is not able to select a relay, leading to low throughput and high latency. Our findings show that a reception window of size 1504 us provides an overall network gain of 46 % in terms of throughput and 154 % in terms of latency, under a network load of 11 K frames per second, and payload size of 1 Kb. Contrary to what could be expected,

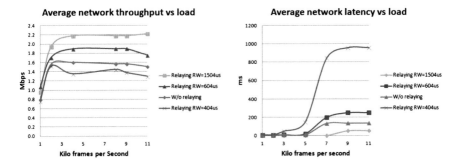

Fig. 13 Impact of reception window

our findings show that latency decreases with a large reception window. The reason is that, although the reception window introduces a delay in the response of the destination, this only occurs during relay selection and not during the process of data relaying. A reception window of size 1504 us also provides highest gain under varying network density [6].

5.2 Network Capacity

In this section we analyze the performance of RelaySpot based on its impact on the overall transmission capacity of a wireless local network. This is done by measuring the overall network throughput and latency when all 25 mobile nodes transmit to the AP, while moving with random pause time between 10 to 100 s. The goal is to understand if RelaySpot can increase the transmission capacity of the network by increasing the overall throughput and decreasing the overall latency in the presence of nodes with different levels of mobility. Moreover, we also measure the tradeoff between the number of successful helped transmissions and the number of transmissions whose performance was degraded due to the action of a relay.

In this set of simulations the network load is uniformly distributed among all 25 nodes. Figure 14 compares the average network throughput achieved by RelaySpot (which is aware of mobility by means of factor M in Eq. 3), with a version of RelaySpot without mobility-awareness and with the 802.11 standard, all under a series of different traffic load (frame size equal to 1 Kb). Simulation results show that RelaySpot can achieve higher throughput than the 802.11 standard and the mobility unaware RelaySpot even with high load, mainly because it is able to select stable relays (with low mobility), which are more likely to help for longer time. RelaySpot achieves an average throughput gain of 42 % in relation to 802.11. RelaySpot without mobility-awareness can still achieve an average throughput gain of 17.6 % in relation to 802.11, due to the scheduler at the destination, which is able to select a relay with a pair of channels (source-relay; relay-destination) with better throughput than the direct link.

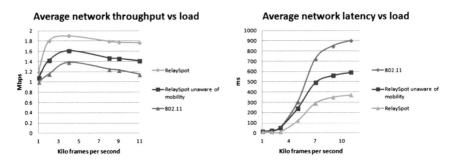

Fig. 14 Analysis of network capacity

In what concerns the overall network latency, Fig. 14 shows that RelaySpot achieves an average gain of 152 % in relation to a direct 802.11 transmission, while the mobility unaware RelaySpot achieves an average gain of 17.8 % in relation to the direct transmission. The main reason for the gain that RelaySpot has in relation to 802.11 is the fact that with RelaySpot the selected relay does not contend, thus reducing the delay. What differentiates RelaySpot from its mobility unaware version is the fact that by selecting relays with high pause time RelaySpot reduces the overall communication delay, by avoiding re-selection of relays during the communication session. Furthermore, since RelaySpot allows low data rate nodes to release the wireless medium faster, other nodes can access the medium more frequently, leading to less overall network latency, even in scenarios with high mobility.

As is well known, the network load has a great impact on the performance of any MAC protocol. Figure 14 observes this effect for both RelaySpot and 802.11 protocols: the performance gets better as the load increases, since more bits are transmitted at each transmission opportunity. However, when the network is overloaded (4 kilo frames per second) then the margin gain is reduced mainly due to collisions. Since RelaySpot operation leads to a reduction of frame retransmissions, the potential bad impact of retransmissions in a heavy loaded network is diminished.

In comparison to static scenarios (Fig. 12), it is clear that with RelaySpot the overall network capacity does not decrease significantly in the presence of mobility. Even the mobility unaware RelaySpot has gains over 802.11. The reason is that the overhead of relay failure due to mobility is smaller than the benefit achieved from helping poor communication sessions.

5.3 Impact of Relay Switching

The aim of this experiment is to analyze how much can relay switching contribute to a good network capacity, by rectifying the impact of relay failures. In this set of simulations we consider a scenario with one AP and 25 mobile sources, each one generating a traffic load of 10 Kb/s. Several simulations are run with different

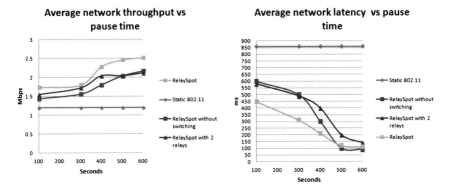

Fig. 15 Analysis of impact of switching

levels of mobility, from simulations where all nodes have 100 s of pause time to simulations where nodes pause time is of 600 s (static nodes, since the pause time equals the simulation time).

Figure 15 shows the benefits of RelaySpot over 802.11 and highlights the benefits of relay switching. The advantage of switching between relays during the lifetime of a communication session is analyzed by comparing RelaySpot with two benchmark versions of itself: a version without relay switching; a version where switching is done between two relays only (if one fails the other relay forwards the data).

Results illustrated in Fig. 15 clearly show that RelaySpot has always better performance than 802.11. This performance gain is still clearer even when RelaySpot switches between two predefined relays only, or when RelaySpot does not switch at all.

In what concerns the analysis of relay switching, our first finding shows that the performance of RelaySpot increases with its capability to switch between any qualified relay: relay switching gives RelaySpot an average throughput gain of 20 % in relation to the RelaySpot version that does not use switching at all. Moreover, it is also clear that the flexibility of being able to switch between any qualified relay brings additional performance to RelaySpot: Fig. 15 shows that RelaySpot has an average throughput 13.5 % higher than the RelaySpot version that uses only two predefined relays in the switching process.

In static scenarios (pause time of 600 s) the throughput gain of RelaySpot increases 16 and 19 % in relation to RelaySpot versions without switching and with 2 relays only, respectively. These results show the advantage of switching even in scenarios without mobility: in these scenarios switching is mainly useful to overcome the impact of interference over relay operations: a relay can be subjected to different interference levels depending upon the number of neighbor nodes transmitting, and the amount of data generated and consumed by the relay itself.

In static scenarios, switching traffic between just two relays does not bring any major gain (c.f. Fig. 15). The reason is that the usage of two relays brings some extra overhead that does not compensate the small throughput gain that comes from switching a communication session between two relays that may be under similar interference conditions with high probability. Such probability is lower when we increase the number of relays involved in the switching process, as happens with RelaySpot (which is evident from the results illustrated in Fig. 15). Moreover, when compared with RelaySpot, the probability of non optimal relay selection is higher when we consider only two relays.

The problem of using only a small number of predefined relays to switch upon (two relays in this experiment) is also evident in terms of latency, as shown in Fig. 15. The overall latency of the RelaySpot version with two relays is higher than the version not using switching for pause times higher than 300 s. These experiments show that for the majority of the scenarios, switching among two relays does not bring any advantage, due to the high probability of having the two relays under the same interference conditions.

The advantage of switching starts to be more evident when we use all the potential relays, as RelaySpot does. In relation to the version that does not use switching at all, RelaySpot brings better performance in terms of latency as soon as mobility increases (for pause time lower than 500 s). The reason is that by exploiting a significant number of potential relays (all qualified nodes) RelaySpot increases the probability of finding a node with low interference at a certain moment in time.

For more static scenarios the advantage of switching is not significant in this experiment (RelaySpot as a latency 16 % lower than the RelaySpot version without switching) since all nodes have the same set of neighbor during the simulation and all nodes have the same traffic load.

6 Conclusions

In this paper, we present RelaySpot, the first cooperative MAC protocol that is able to increase the performance of dynamic wireless networks, by being aware of the level of mobility, interference and transmission success rate of potential relays. RelaySpot comprises three building blocks: opportunistic relay selection, cooperative relay scheduling, and cooperative relay switching. In order to operate in dynamic scenarios, and unlike previous work, RelaySpot does not require the maintenance of CSI tables, avoiding periodic updates and consequent broadcasts.

RelaySpot can effectively increase the transmission capacity of wireless local area networks, even in the presence of wireless interference and mobile nodes. Experimental results show that RelaySpot brings an overall average throughput and latency gains of 42 and 152 %, respectively, in relation to 802.11, and of 17.6 and 17.8 % in relation to a version of RelaySpot that is unaware of node mobility [9, 11].

In very dynamic scenarios, where a selected relay may not the best choice for the entire duration of a communication session, our experiments show that the relay switching capability of RelaySpot brings an overall average throughput and latency gains of 20 and 21 %, respectively, in relation to a version of RelaySpot that does not perform relay switching. This shows RelaySpot capability to improve the utilization of spatial diversity by switching in real time to the relay that offers the best throughput and latency conditions within the cooperation area.

As future work, we plan to extend the operation of RelaySpot with the inclusion of a chain relaying capability, in which the operation of a poor relay is compensated by a second relay, located closer to the destination. We plan to compare the performance of this new functionality with the relay switching approach described in this paper.

References

1. Ahmad R, Zheng F, Drieberg M, Olafsson S (2008) An enhanced relay-enabled medium access control protocol for wireless Ad Hoc networks. In: Proceedings of IEEE vehicular technology conference (VTC), Singapore, May 2008
2. Elmenreich W, Marchenko N, Adam H, Hofbauer C, Brandner G, Bettstetter C, Huemer M Building blocks of cooperative relaying in wireless systems. Electrical and Computer Engineering, Springer
3. Feeney LM, Cetin B, Hollos D, Kubisch M, Mengesha S, Kar H (2007) Multi-rate relaying for performance improvement in IEEE 802.11 WLANs. In: Proceedings of WWIC, Coimbra, Portugal, May 2007
4. Jamal T, Mendes P (2010) Relay selection approaches for wireless cooperative networks. In: Proceedings of IEEE WiMob, Niagara Falls, Canada, Oct 2010
5. Jamal T, Mendes P (2013) 802.11 medium access control in MiXiM. Technical Report. SITILabs-TR-13-02 University Lusófona, March 2013
6. Jamal T, Mendes P (2013) Analysis of hybrid relaying in cooperative WLAN. InL Proceedings of IFIP wirelessdays, Valencia, Spain, Nov 2013
7. Jamal T, Mendes P (2013) Cooperative relaying for dynamci networks. EU Patent EP13182366.8, Aug 2013
8. Jamal T, Mendes P (2014) Cooperative relaying for wirelesss local area networks. In: Wireless networks moving objects, Springer LNCS WiNeMo, March 2014
9. Jamal T, Mendes P, Zúquete A (2011) Interference-aware opportunistic relay selection. In: Proceedings of ACM CoNEXT student workshop, Tokyo, Japan, Dec 2011
10. Jamal T, Mendes P, Zúquete A (2011) RelaySpot: a framework for opportunistic cooperative relaying. In: Proceedings of IARIA ACCESS, Luxembourg, June 2011
11. Jamal T, Mendes P, Zúquete A (2012) Opportunistic relay selection for wireless cooperative network. In: Proceedings of IEEE IFIP NTMS, Istanbul, Turkey, May 2012
12. Jamal T, Mendes P, Zúquete A (2012) Wireless cooperative relaying based on opportunistic relay selection. Int J Adv Netw Serv 05(2):116–127
13. Liu P, Tao Z, Narayanan S, Korakis T, Panwar S (2007) CoopMAC: a cooperative MAC for wireless LANs. IEEE J Sel Areas Commun 25(2):340–354
14. Lu M-H, Steenkiste P, Chen T (2009) Design, implementation and evaluation of an efficient opportunistic retransmission protocol. In: Proceedings of IEEE MobiCom, Beijing, China, April 2009
15. Tan K, Wan Z, Zhu H, Andrian J (2007) CODE: cooperative medium access for multirate wireless Ad Hoc network. In: Proceedings of IEEE SECON, California, USA, June 2007
16. Zhu H, Cao G (2006) rDCF: a relay-enabled medium access control protocol for wireless Ad Hoc networks. In: IEEE Transactions on mobile computing, 5 March 2006

Resource Allocation in User-Centric Wireless Networks

Huseyin Haci, Huiling Zhu and Jiangzhou Wang

Abstract This paper presents a behavioral fairness-based resource allocation (RA) algorithm for the uplink transmission of a multiuser orthogonal frequency division multiplexing system. The aim of fairness-based RA is to boost users' willingness to start and keep cooperating with each other and behave goodly to keep their trust and reputation high. An interesting notion of differentiated streams is also included in the proposed RA, which can fine-tune RA result to transfer excess bandwidth and power from real-time streams to non-real-time streams, i.e. increase useful throughput. In order to reduce the complexity of RA, a set of contiguous subchannels are grouped into a chunk to allocate resource chunk by chunk. The effects of users' cooperation value and the number of subchannels grouped in a chunk on the average throughput is analyzed and shown by numerical simulations.

Keywords Behavioral fairness based allocation · Chunk-power-bit allocation · Cooperative communications

1 Introduction

The fundamental paradigm of User Centric Networks (UCNETs), which aim to share subscribed Internet access and services among users, is started with peer-to-peer (P2P) Internet service model, where users become active service providers instead of being just service consumers. In UCNETs, end-users share their Internet access freely and transparently among themselves in a way that is legally and technically independent from infrastructure and access providers. Existing commercial examples

H. Haci · H. Zhu (✉) · J. Wang
School of Engineering and Digital Arts, University of Kent, Canterbury, UK
e-mail: h.Zhu@kent.ac.uk

H. Haci
e-mail: hh219@kent.ac.uk

A. Aldini and A. Bogliolo (eds.), *User-Centric Networking,*
Lecture Notes in Social Networks, DOI: 10.1007/978-3-319-05218-2_10,
© Springer International Publishing Switzerland 2014

of UCNETs are FON, OpenSpark, Whisher, etc [1]. These commercial networks spread by means of the end-user willingness to cooperate and share connectivity.

To provide good and robust connectivity and service in such networking environment, UCNET models should incorporate some essential properties. These properties, as declared in [1] includes: connectivity sharing, cooperation, trust and self-organization. Our main objective in this paper is to motivate users to "cooperate" with others and keep their trust and reputation at high level through proper fairness based resource allocation. The term "cooperate" does not only mean relaying others' traffic, but also includes providing other services, such as multimedia content sharing, printer facility sharing, etc. The importance of user cooperation in UCNETs can be easily understood as such a wireless system will be very limited in coverage and not reliable with only few services locally available in case of low or even medium number of users cooperatively sharing their connectivity.

Besides setting one of the goals to boost user-cooperation and connectivity sharing, the next fundamental issue to consider is to provide high speed transmission over broadband wireless channels to cope with the requirement from the growth of wireless multimedia and data applications. The effect of severe frequency-selective fading in broadband channels is one of the most important obstructions on achieving today's satisfactory transmission rates. Orthogonal frequency division multiplexing (OFDM) is a promising technique to eliminate this obstruction by dividing the broadband channel into a number of narrowband subchannels. Accordingly, the channel may be transformed to a sequence of frequency non-selective fading channels. This feature makes OFDM-based orthogonal frequency division multiple access (OFDMA) an attractive multiple access technique for high speed wireless communications systems and adopted in UCNETs.

In OFDMA, different subchannels are dynamically allocated to different users with the aim of optimizing a system performance, such as maximizing throughput or minimizing energy, while satisfying system constraints, such as bit-error-rate (BER), data rate or fairness. Multiuser diversity [2] is exploited to improve and achieve maximum possible performance in multiuser wireless systems by properly allocating resources to users depending on instantaneous conditions. It is provided in [3, 4] that by allocating a single subchannel to the user which has the best channel condition, the best performance of the system can be achieved. However, when the number of subcarriers is large, subchannel-based allocation may have large overhead and be very complicated.

To mitigate the complexity and overhead of subchannel-based allocation schemes, the correlation between neighboring subchannels in OFDM systems can be exploited. A simplified RA, chunk-based RA, can be adopted by properly grouping set of adjacent subchannels into a chunk and do relevant RA per chunk [4]. The performance of chunk-based-allocation can approach to the performance of subchannel-based allocation. This is because, in OFDM systems most likely when a subchannel is good (or bad), its adjacent subchannels are also good (or bad) with quite similar channel quality [4].

The RA algorithm imposed in this paper constitutes chunk, power and bit allocation approaches. To maximize users' willingness for contributing to the network and behave goodly, the achieved data rate proportions among users and resource allocation priorities are related to cooperation, trust and reputation (composite value) of users. Therefore, a user that wants to enjoy high speed communications, high quality multimedia stream, non-intermittent connectivity, and so on, should become and stay as a highly cooperative and good behaving user. The RA algorithm also maintains the BER constraint of various applications and users' transmit power threshold.

In the proposed algorithm, non-greedy real-time multimedia streams and greedy non-real-time data streams for a user are considered and differentiated. In order to be more precise at managing resource allocations for users' applications. In most work on RA, it is usually assumed that there is a single greedy application for a user, where there is no upper limit on usable data rate [3]. This is not very realistic for real-time streams which has maximum encoding rate. The unlimited data rate assumption for real-time applications can result in providing excess (unusable) resources to applications, which can lead to inefficiently using resources. Because of differentiating streams, an upper bound can be given to the data rate of real-time streams. In this paper, according to this upper bound, resources allocated to users are fine-tuned to reduce waste of resources. Note that when increasing the number of subchannels per chunk in the chunk-based RA, the maximum bits per symbol per chunk increases. When the number of subchannels per chunk is large, its achievable maximum bits per symbol per chunk may be larger than the upper bound data rate of the real-time application. Therefore, an important system design issue, the number of subchannels per chunk, in chunk-based systems can be related to this upper bound as well. The tradeoff here is between the reduction in computational complexity of RA algorithm (as number of subchannels in a chunk increases) and the increase in difference between total achievable data rate per chunk and real-time stream's upper bound, which will be analyzed in this paper.

2 System Model

The OFDMA system is assumed to have M subcarriers and K active users. Each user is assumed to have two data streams—one real-time multimedia stream and one non-real-time data stream. Accordingly, there are $2K$ data streams in the system. In order to reduce the complexity of RA every M' contiguous subchannels are grouped in a chunk [4]. Hence the system has $N = M/M'$ chunks and each of these chunks is allocated to one stream of a user. RA period is taken to be τ ms.

All subchannels in the system are considered to be Rayleigh fading channels that introduce an additive white Guassian noise (AWGN) with a double-side power spectral density of $N_0/2$. The frequency response, $h_{k,m}$, of user k in channel m is complex Guassian distributed and its magnitude, called fading factor, $\alpha_{k,m}$, is Rayleigh distributed with unitary mean square, $E\{\alpha_{k,m}^2\} = 1$ for all k and m. $h_{k,m}$ is further assumed independent for all users.

For the adaptive modulation scheme, denoting L as the level of modulation, $L - ary$ QAM is employed in the system. Accordingly, the set from which L can take values is $L = \{0, 2^2, \ldots, 2^b, \ldots, 2^B\} = \{0, 4, \ldots, L_b, \ldots, L_B\}$ where b represents the total number of bits per symbol in the QAM and

$$
L_b = \begin{cases} 0, & \text{if } b = 0 \\ 2^b, & \text{if } b \text{ is even and } 0 < b \le B. \end{cases}
$$

If the mth subchannel is allocated to the kth user and the allocated power is $\varepsilon_{k,m}$, the instantaneous bit-error-rate $\beta_{k,m}$, can be approximated in a closed form as [5]

$$
\beta_{k,m} \approx 0.2 \exp\left(-\frac{1.6 \cdot SNR_{k,m}}{l_{k,m} - 1}\right) = 0.2 \exp\left(\frac{c \cdot \varepsilon_{k,m} \cdot \alpha_{k,m}^2}{l_{k,m} - 1}\right) \tag{1}
$$

where the received signal to noise ratio, $SNR_{k,m}$, is given by

$$
SNR_{k,m} = \frac{\varepsilon_{k,m} \cdot T_s \cdot \alpha_{k,m}^2}{N_0}, \tag{2}
$$

and $c = -1.6T_s/N_0$.

From (1), the achievable bits/symbol, $r_{k,m}$, of the kth user on the mth subchannel can be derived as

$$
r_{k,m} = b_{k,m} = \log_2 l_{k,m} = \log_2\left(1 + \frac{c \cdot \varepsilon_{k,m} \cdot \alpha_{k,m}^2}{\ln(5\beta_{k,m})}\right). \tag{3}
$$

It can be seen that the modulation level, $l_{k,m}$, and the achievable bits/symbol, $r_{k,m}$, are proportional to fading factor (channel quality).

In the case of chunk-based resource allocation, where a number of contiguous subchannels are grouped into a chunk, the modulation level and transmit power per subchannel is the same for all subchannels within a chunk. By defining $\tilde{l}_{k,n}$ and $\tilde{\varepsilon}_{k,n}$ as the modulation level and the assigned power per subchannel in the nth chunk for the kth user, we can obtain $l_{k,m} = \tilde{l}_{k,n}$ and $\varepsilon_{k,m} = \tilde{\varepsilon}_{k,n}$, where $m = nM, \ldots, (n + 1)M' - 1$.

Therefore, the average BER, $\beta_{k,n}$, of the kth user on nth chunk can be given as

$$
\beta_{k,n} = \frac{1}{M'} \sum_{m=nM'}^{(n+1)M'-1} \beta_{k,m} = \frac{1}{M'} \sum_{m-nM'}^{(n+1)M'-1} 0.2 \exp\left(\frac{c \cdot \varepsilon_{k,m} \cdot \alpha_{k,m}^2}{l_{k,m} - 1}\right). \tag{4}
$$

According to [4] $\beta_{k,n}$ can be approximated as

$$
\beta_{k,n} \approx 0.2 \exp\left(\frac{c \cdot \tilde{\varepsilon}_{k,n} \cdot \tilde{\alpha}_{k,n}^2}{\tilde{l}_{k,n} - 1}\right) \tag{5}
$$

where

$$\tilde{\alpha}_{k,n}^2 = \frac{1}{M'} \sum_{m=nM'}^{(n+1)M'-1} \alpha_{k,m}^2 . \tag{6}$$

Furthermore, considering the achievable bits/symbol, $r_{k,m}$, since modulation levels of all subchannels within a chunk are the same, $r_{k,m}$ on all subchannels will also be the same. Hence, the achieved bits/symbol/subchannel on the nth chunk for the kth user can be given as

$$\tilde{r}_{k,n} = \tilde{b}_{k,n} = \log_2 \tilde{l}_{k,n} = r_{k,nM'} \tag{7}$$

where $r_{k,nM'}$ is given by (3). And the total bits/symbol/chunk, rate achieved by a chunk, $r_{total}(k, n)$, is calculated by $M'\tilde{r}_{k,n}$.

3 Chunk, Power and Bit Allocation

In UCNET, enjoying high-speed data and high-quality multimedia services rather than slow data and low-quality multimedia services, in a fair way depending on users' composite value, is an adequate incentive for network users to share their resources and provide cooperation. Therefore, in order to boost users' willingness of sharing resources and cooperation, a dynamic behavioral fairness-based resource allocation scheme is proposed in this section to provide better service to better behaving users, by adopting a composite value c_k to represent the kth user's cooperation, trust and reputation in the network. That is, more cooperative, trustful and reputed users will get more resources than average and/or bad behaving users. To achieve the target of the proposed RA, the aim of the dynamic resource allocation approach is to maximize the uplink throughput meanwhile to balance the tradeoff between capacity and fairness, which is related with credit of users.

An objective function is set up to maximize the uplink throughput under relevant constraints, one of which is to ensure proportional fairness with respect to users' credits. Also the BER constraint of applications and the transmit power constraint of users are taken into account.

As mentioned in Sect. 1, to be able to differentiate between streams of a user, an additional parameter q is introduced in the formulation to represent the index of streams ($q = 0$ for non-real-time, $q = 1$ for real-time). Hence, for the credit-based RA, the optimization problem is formulated as

$$R_{total} = \max_{\tilde{b}_{k,n,q}, \rho_{k,n,q}} \sum_{k=0}^{K-1} \sum_{n=0}^{N-1} \sum_{q=0}^{1} \rho_{k,n,q} M' \tilde{b}_{k,n,q} \tag{8}$$

subject to

$$\rho_{k,n,1} = \{0, 1\} \text{ and } \sum_{k=0}^{K-1}\sum_{q=0}^{1} \rho_{k,n,q} \leq 1 \; \forall n, \tag{9}$$

$$\sum_{n=0}^{N-1}\sum_{q=0}^{1} \rho_{k,n,q} M' \tilde{\varepsilon}_{k,n,q} \leq \varepsilon_0 \; \forall k, \tag{10}$$

$$\sum_{n=0}^{N-1} \rho_{k,n,1} M' \tilde{b}_{k,n,1} \leq \bar{R}_{\text{max,Realtime}} \; \forall k, \tag{11}$$

$$R_{0,0} : R_{0,1} : \cdots : R_{K-1,q=0} : R_{K-1,q=1} = \gamma_{0,0} : \gamma_{0,1} : \cdots : \gamma_{K-1,q=0} : \gamma_{K-1,q=1} \tag{12}$$

where $\rho_{k,n,q}$ indicates whether nth chunk is allocated to the qth stream of kth user (i.e $\rho_{k,n,q} = 1$) or not (i.e $\rho_{k,n,q} = 0$). Constraint (9) ensures that a chunk can be allocated only to one data-stream of a user. Constraint (10) makes sure that the total aggregated transmit power of a user does not exceed the preset transmit power threshold ε_0. The maximum achievable encoded data rate limit for real-time streams is provided by constraint (11). $\bar{R}_{\text{max,Realtime}}$ is the threshold for maximum data rate that real-time streams can be enabled. This upper limit is useful to prevent waste of allocating excess data rate (and transmit power) since real-time applications cannot benefit from greater data rate than what they actually need. The last constraint (12), is used to ensure reception of proportional capacity between users based on their composite credit. $\left\{R_{k,q}\right\}_{k=0,q=0}^{K-1,1}$ is the total data rate reached by the qth application of kth user and it is defined as

$$R_{k,q} = M' \sum_{\Omega_{k,q}} \log_2\left(1 + \frac{c \cdot \tilde{\varepsilon}_{k,n,q} \cdot \tilde{\alpha}_{k,n}^2}{\ln(5\bar{\beta})}\right) \; \forall k, q \tag{13}$$

where $\Omega_{k,q}$ is the set of chunks allocated for qth application of kth user. $\left\{\gamma_{k,q}\right\}_{k=0,q=0}^{K-1,1}$ in (12) is the proportion value of the qth application of the kth user to ensure proportional fairness based on users' credit, and it is defined as

$$\gamma_{k,0} = \gamma_{k,1} = c_k/(c_0 + c_1 + \cdots + c_{K-1}), \; \forall k. \tag{14}$$

Optimization problem in (8) is difficult to solve as it contains both continuous variables, such as $\tilde{\varepsilon}_{k,n,q}$, and binary variables, such as $\rho_{k,n,q}$. Further the nonlinear constraints of the problem increase the difficulty in finding the optimum solution. Therefore, a suboptimal approach dividing the resource allocation process into two steps is proposed in this section to simplify the RA problem. The first step is chunk allocation that is based on users' proportional fairness requirements and their channel quality. The second step is the joint power and bit allocation, which is based on the users' total transmit power and applications BER constraint.

Chunk Allocation The basic idea is to allocate chunks with best channel quality to users as much as possible. At each chunk allocation iteration if qth stream of the kth user has lowest proportional capacity $R_{k,q}/\gamma_{k,q}$, for this user, the chunk with best channel quality is detected among set of available chunks A. Initially, A includes all the chunks. Once the chunk is allocated to the qth stream, this chunk is removed from A and the stream's value of achieved proportional capacity is updated. At chunk allocation, equal power allocation is assumed across all chunks. The results in a suboptimal allocation but it should be done to avoid prohibitive computational burden, in case of joint allocation of chunks and power. The algorithm for suboptimal chunk allocation is described as follows

---

Algorithm 1 Chunk Allocation

---

1. Initialization

    (a) Set $R_{k,q} = 0$, $\Omega_{k,q} = \emptyset$ $\forall k, q$ and $A = \{0, 1, \ldots, N-1\}$ where $\emptyset$ stands for an empty set.

2. Chunk Allocation

    (a) Find k and q satisfying
        $\frac{R_{k,q}}{\gamma_{k,q}} \leq \frac{R_{i,q}}{\gamma_{i,q}}$ among all $i = \{0, 1, \ldots, K-1\}$ and $q = \{0, 1\}$
    (b) Find n satisfying $\tilde{\alpha}_{k,n}^2 \geq \tilde{\alpha}_{k,j}^2$ $\forall j \in A$
    (c) Let $\Omega_{k,q} = \Omega_{k,q} \cup \{n\}$, $A = A - \{n\}$ and update $R_{k,q}$ according to (13)

After chunk allocation, the set of chunks allocated to the qth stream of the kth user is, $\Omega_{k,q}$, known for all users and streams and $\Omega_{k,q}$ are disjoint for all k and q and $\Omega_{0,0} \cup \Omega_{0,1} \cup \cdots \Omega_{K-1,q=0} \cup \Omega_{K-1,q=1} \in \{0, 1, \ldots, N-1\}$. Now our goal of maximizing sum capacity, while satisfying constraints and maintaining proportional fairness, becomes a computationally feasible problem.

Power and Bit Allocation For a certain determined chunk allocation, the optimization problem can be formulated as

$$R_{\text{total}} = \max_{\tilde{b}_{k,n,q}} \sum_{k=0}^{K-1} \sum_{q=0}^{1} \sum_{n \in \Omega_{k,q}} M' \tilde{b}_{k,n,q} \qquad (15)$$

subject to

$$\sum_{q=0}^{1} \sum_{n \in \Omega_{k,q}} M' \tilde{\varepsilon}_{k,n,q} \leq \varepsilon_0 \ \forall k \tag{16}$$

$$R_{k,1} = \sum_{n \in \Omega_{k,1}} M' \tilde{b}_{k,n,1} \leq \bar{R}_{\max,\text{Realtime}} \ \forall k \tag{17}$$

$$R_{0,0} : R_{0,1} : \cdots : R_{K-1,q=0} : R_{K-1,q=1} = \gamma_{0,0} : \gamma_{0,1} : \cdots : \gamma_{K-1,q=0} : \gamma_{K-1,q=1} \ \forall k, q. \tag{18}$$

The optimization problem in (15) can be solved by using Lagrangian method, as shown in [6]. The solution of problem is equivalent to finding the maximum of following function

$$L = M' \sum_{k=0}^{K-1} \sum_{q=0}^{1} \sum_{n \in \Omega_{k,q}} \tilde{b}_{k,n,q} + \sum_{k=0}^{K-1} \lambda_k \left(\sum_{q=0}^{1} \sum_{n \in \Omega_{k,q}} M' \tilde{\varepsilon}_{k,n,q} - \varepsilon_0 \right)$$

$$+ \sum_{k=0}^{K-1} \mu_k \left(\sum_{n \in \Omega_{k,1}} M' \tilde{b}_{k,n,1} - \bar{R}_{\max,\text{Realtime}} \right)$$

$$+ \sum_{k=1}^{K-1} \sum_{q=0}^{1} \eta_{k,q} \left(R_{0,0} - \frac{\gamma_{0,0}}{\gamma_{k,q}} R_{k,q} \right) + \eta_{0,1} (R_{0,0} - R_{0,1}) \tag{19}$$

Following the Lagrangian method to find the optimum power allocation for user k, we differentiate (19) with respect to $\tilde{\varepsilon}_{k,n,q}$ and set each derivative to 0 to obtain

$$\left. \frac{\partial L}{\partial \tilde{\varepsilon}_{k,n,q}} \right|_{k \geq 1} = M' \frac{\partial \tilde{b}_{k,n,q}}{\partial \tilde{\varepsilon}_{k,n,q}} + \lambda_k M' + \mu_k M' \frac{\partial \tilde{b}_{k,n,1}}{\partial \tilde{\varepsilon}_{k,n,1}} + \eta_{k,q} \left(-\frac{\gamma_{0,0}}{\gamma_{k,q}} \frac{\partial R_{k,q}}{\partial \tilde{\varepsilon}_{k,n,q}} \right) \tag{20}$$

$$\left. \frac{\partial L}{\partial \tilde{\varepsilon}_{0,n,0}} \right|_{k=0,q=0} = M' \frac{\partial \tilde{b}_{0,n,0}}{\partial \tilde{\varepsilon}_{0,n,0}} + \lambda_0 M' + \sum_{k=0}^{K-1} \sum_{q=0}^{1} \eta_{k,q} \left(\frac{\partial R_{0,0}}{\partial \tilde{\varepsilon}_{0,n,0}} \right) \tag{21}$$

$$\left. \frac{\partial L}{\partial \tilde{\varepsilon}_{0,n,1}} \right|_{k=0,q=1} = M' \frac{\partial \tilde{b}_{0,n,1}}{\partial \tilde{\varepsilon}_{0,n,1}} + \lambda_0 M' + \mu_0 M' \frac{\partial \tilde{b}_{0,n,1}}{\partial \tilde{\varepsilon}_{0,n,1}} + \eta_{0,1} \left(\frac{\partial R_{0,1}}{\partial \tilde{\varepsilon}_{0,n,1}} \right). \tag{22}$$

By assuming that chunks n and m are allocated to qth application of kth user: i.e. m and $n \in \Omega_{k,q}$, optimal power distribution between chunks can be derived from (20) to (22) as follows. From (20), one obtains

$$\frac{\partial \tilde{b}_{k,n,q}}{\partial \tilde{\varepsilon}_{k,n,q}} = \frac{\partial \tilde{b}_{k,m,q}}{\partial \tilde{\varepsilon}_{k,m,q}}. \tag{23}$$

Since $\tilde{b}_{k,n,q}$ is given by (7),

$$\frac{\partial \tilde{b}_{k,n,q}}{\partial \tilde{\varepsilon}_{k,n,q}} = \frac{c \cdot \tilde{\alpha}_{k,n}^2}{\ln(5\bar{\beta}) + c \cdot \tilde{\alpha}_{k,n}^2 \cdot \tilde{\varepsilon}_{k,n,q}} \cdot \frac{1}{\ln 2}. \tag{24}$$

Replacing derivatives in (23) by (24) and letting $H_{k,n} = (c \cdot \tilde{\alpha}_{k,n}^2)/\ln(5\bar{\beta})$, one obtains

$$\frac{H_{k,n}}{1 + \tilde{\varepsilon}_{k,n,q} H_{k,n}} = \frac{H_{k,m}}{1 + \tilde{\varepsilon}_{k,m,q} H_{k,m}}. \tag{25}$$

Without loss of generality, it is assumed that $H_{k,0} \leq H_{k,1} \leq \cdots \leq H_{k,N_{k,q}-1}$ for all k and define $N_{k,q}$ as the number of chunks in $\Omega_{k,q}$, n_i^q as the ith chunk in the ascending list of chunks allocated to qth stream, where $i = \{0, 1, \ldots, N_{k,q} - 1\}$. Then, (25) can be rewritten as

$$\tilde{\varepsilon}_{k,n_i^q,q} = \tilde{\varepsilon}_{k,n_0^q,q} + \frac{H_{k,n_i^q} - H_{k,n_0^q}}{H_{k,n_i^q} H_{k,n_0^q}}, \tag{26}$$

for $i = 0, 1, \ldots, N_{k,q} - 1, k = 0, 1, \ldots, K - 1$ and $q = \{0, 1\}$. Equation (26) shows the power distribution on chunk i for the qth application of user k. It is shown that more power will be put on the chunks with higher channel quality. This matches the water-filling concept in resource allocation [7].

Defining $\varepsilon_{k,total}$ as the total power allocated for user k and using (26), $\varepsilon_{k,total}$ can be expressed as

$$\varepsilon_{k,total} = \sum_{q=0}^{1} \sum_{i=0}^{N_{k,q}-1} \tilde{\varepsilon}_{k,n_i^q,q} = N_{k,q=0}\tilde{\varepsilon}_{k,n_0^{q=0},q=0} + N_{k,q=1}\tilde{\varepsilon}_{k,n_0^{q=1},q=1}$$

$$+ \sum_{q=0}^{1} \sum_{i=1}^{N_{k,q}-1} \frac{H_{k,n_i^q} - H_{k,n_0^q}}{H_{k,n_i^q} H_{k,n_0^q}} = \varepsilon_0 \tag{27}$$

for $k = 0, 1, \ldots, K - 1, i = 0, 1, \ldots, N_{k,q} - 1$ and $q = \{0, 1\}$.

Further considering $\gamma_{k,0} = \gamma_{k,1}$ from (14) we can argue that the data rate achieved by both applications should be the same

$$\sum_{i=0}^{N_{k,q=0}} \log_2 \frac{\left(\ln(5\bar{\beta}) + \left(\tilde{\varepsilon}_{k,n_0^{q=0},q=0} + \frac{H_{k,n_i^{q=0}} - H_{k,n_0^{q=0}}}{H_{k,n_i^{q=0}} H_{k,n_0^{q=0}}} \right) \cdot H_{k,n_i^{q=0}} \right)}{\ln(5\bar{\beta})}$$

$$= \sum_{i=0}^{N_{k,q=1}} \log_2 \frac{\left(\ln(5\bar{\beta}) + \left(\tilde{\varepsilon}_{k,n_0^{q=1},q=1} + \frac{H_{k,n_i^{q=1}} - H_{k,n_0^{q=1}}}{H_{k,n_i^{q=1}} H_{k,n_0^{q=1}}} \right) \cdot H_{k,n_i^{q=1}} \right)}{\ln(5\bar{\beta})} \tag{28}$$

The optimal power allocation $\tilde{\varepsilon}_{k,n_i^q,q}$ for all k, q and i can be determined by applying numerical algorithms, such as Newton's root-finding method [6], to find the zero of (27) and (28). Then, achievable bits/symbol $\tilde{b}_{k,n,q}$ is obtained from (3). Now as fine-tuning step, excess bandwidth and energy from real-time stream of a user is transferred (recycled) to its non-real-time stream depending on the real-time stream's upper bound, b_{th}, and achieved rate.

4 Numerical Results

In this section users' average throughput as a function of credit value c_k and the average system throughput with respect to the number of subchannels in a chunk (chunk size) is analyzed. Analysis is done to show the effectiveness of credits in RA and differentiated streams on system design. Unless noted otherwise, the system parameters are assumed as follows: number of subchannels $M = 48$ and chunk size $M' = 2$. The number of active users in the system $K = 8$ and the set of user credit values $c_k = \{1, 2, 3, 4\}$. Further, the BER of streams $\bar{\beta} = 10^{-3}$. The average signal to the noise ratio on each subchannel is 20 dB.

Figure 1 shows the users' average throughput as a function of credit value c_k. The results are presented as *bars* in the figure, where $bar(k, q)$ represents $E(\tilde{b}_{k,n,q})$, the average bits/symbol of kth user's qth stream, e.g. $bar(2, 0)$ represents average bits/symbol achieved by second user's real-time stream. With increased credit value, the rising trend of achieved bits/symbol for both streams can be clearly seen from the figure. This boosts cooperation incentives of users to aim for more credits. Further, for greedy non-real time streams there is no limit for increase in $E(\tilde{b}_{k,n,q})$, see $bar(k, 1)$ at Fig. 1. However, due to upper limit (i.e. 12 bits/symbol in our experiments) the growth in $E(\tilde{b}_{k,n,q})$ is limited for real-time streams, $bar(k, 0)$. The effect of upper limit is obvious in Fig. 1 when credit $c_k = 4$. In RA fine-tuning, excess resources are recycled to non-real time streams to achieve higher useful throughput (see $bar(7, 1)$ and $bar(8, 1)$ at Fig. 1).

Average system throughput as a function of *SNR* for various chunk sizes M' is shown in Fig. 2. Significant increase in average system throughput with respect to increasing *SNR* shows the benefit of adaptive modulation. More important aspect at Fig. 2 is the system performance regarding chunk sizes M'. It can be seen that the

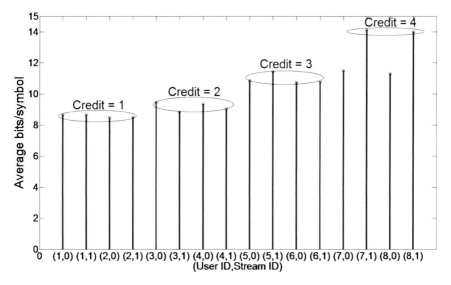

Fig. 1 Average bits/symbol versus (User ID, Stream ID) as function of credit

Fig. 2 Average system throughput versus *SNR*

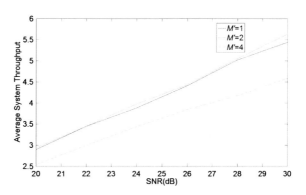

performance is similar when $M' = 1$ and 2, one can achieve similar performance with subchannel-based RA while reducing complexity. However, when $M' = 4$ there is a significant degradation in system throughput. This is because, with more subchannels in a chunk, the achieved bits/symbol per chunk will be high and b_{th} will be violated frequently. Hence more recycling (re-allocation) of bandwidth and energy will cause larger deviation from the optimal chunk and power allocation. As result user's average throughput and system's throughput will reduce.

5 Conclusion

A credit-based chunk, power and bit allocation algorithm is presented. At the proposal, achievable rate proportions and RA priorities are related to composite credit of users to maximize user willingness to contribute and behave well. The notion of differentiated streams in RA is another important aspect considered by the paper, as it enables resource recycling to achieve higher useful throughput/reduce waste of resources.

References

1. Sofia R, Mendes P (2008) User-provided networks: consumer as provider. IEEE Commun Mag 46(12):86–91 NULL
2. Knopp R, Humblet PA (1995) Information capacity and power control in single-cell multiuser communications. In: Proceedings of IEEE international conference on communications, pp 331–335
3. Wong CY, Cheng RS, Lataief KB, Murch RD (1999) Multiuser OFDM with adaptive subcarrier, bit, and power allocation. IEEE J Sel Areas Commun 17(10):1747–1758
4. Zhu H, Wang J (2009) Chunk-based resource allocation in OFDMA systems—part I: chunk allocation. IEEE Trans Commun 57(9):2734–2744
5. Chung ST, Goldsmith A (2001) Degrees of freedom in adaptive modulation: a unified view. IEEE Trans Commun 49(9):1561–1571
6. Boyd S, Vandenberghe L (2004) Convex optimization. Cambridge University Press, New York
7. Cover TM, Thomas JA (1991) Elements of information theory. Wiley, New York

Autonomous Resource Allocation and Strategy Optimization in Wireless Networks

Mürsel Yildiz and Manzoor Ahmed Khan

Abstract In this paper, we propose an autonomous control framework, where the focus remains on autonomous resource utilization and strategy optimization. The control framework aims at an immensely dynamic, time varying, and less predictable network environment. The first main motivation of the proposal is to eliminate redundant resource utilization for specific network objectives while reducing the network response time. To this end, the decision making engines of the control framework characterize the network environment in terms of the correlation with specific network objectives and improve the resource utilization accordingly. The second and the last motivation is to match network strategies with different environmental conditions in terms of impact rates on the environment. We make use of Partially Observable Markov Decision Processes (POMDP) based control loop to improve the network strategies against alarm conditions for different environmental conditions. The control framework is implemented on WLAN APs and the proposed approach is validated by means of real implementation and simulation of different network scenarios.

Keywords Autonomous network control · Strategy optimization · Resource allocation · POMDP · Expectation maximization

1 Introduction

The comprehensive and large-scale organizational attempts on the conceptual and standardization level for the autonomic network architectures (ANA) have recently been placed in the limelight for the future Internet (FI) literature. Although a clean slate or evolutionary approaches are the moot point between recent proposals, researchers agree on increasing the autonomic capability of the network with

M. Yildiz (✉) · M. A. Khan
DAI-Labor/Technische Universität Berlin, Berlin, Germany
e-mail: muersel.yildiz@dai-labor.de

A. Aldini and A. Bogliolo (eds.), *User-Centric Networking*,
Lecture Notes in Social Networks, DOI: 10.1007/978-3-319-05218-2_11,
© Springer International Publishing Switzerland 2014

learning and the decision making engines (DMEs). To this end, different disciplines in the literature of artificial intelligence techniques are introduced to the network entities in order to increase the intelligence level of the DMEs in the network.

The IBM's control loop [1] has been immensely studied in the FI literature as a technique for the autonomous behavior in terms of the analysis and aggregation of raw data and hence making intelligent decisions. Researchers mainly focus on the periodic execution of control loops by means of gathering sensor data, the analysis and the aggregation of collected raw data, the reasoning and decision making, and finally the execution. However, for the realistic scenarios, this process might increase the complexity for network operation or result in redundant data gathering and analysis under normal conditions. In other words, an optimization mechanism for the activation instances or the duration of autonomous functionality might improve the resource utilization in the network. Additionally, a standardization for different ANA proposals has been recently started [2, 3], where a conflict free architecture is proposed for the harmonization of different solution approaches on the similar problematic cases in the network. This allows the network hold various strategies in an algorithm repository. We believe an evaluation process on the impact rates of the executed algorithms based on different environmental conditions might help for the network strategy optimization.

In this paper, we propose an autonomous control framework focusing on resource utilization and strategy optimization in dynamic and time varying network environments. The very first main motivation of the proposal is to eliminate redundant resource utilization for specific network objectives while reducing the network response time. To this end, the decision making engines of the proposed control framework characterize the network environment in terms of the correlations with specific network objectives and optimize the resource utilization accordingly. The second and the last motivation is to match network strategies with different environmental conditions in terms of impact rate. With a POMDP modified control loop, as an optimization policy, the proposed solution improve the network strategies against alarm conditions depending on the environmental characteristics and action success rate.

Next section provides the literature review mostly on different large-scale FI proposals. In Sect. 3, we respectively detail the environment description and system settings at which this work mainly aims, the motivation and problem statement and finally the solution statement. In Sect. 4 an overview on the mathematical models and approaches is provided, Sect. 5 introduces the proposed model in details. Section 6 details the technical implementation followed by the evaluation and tests section. Finally, we comment on the proposed solution and provide a summary of the proposal.

2 Related Work

An ongoing research for the standardization of autonomic future Internet architecture is conducted by ETSI [3, 4]. The main motivation of the standardization effort is the harmonization of proposals by means of providing a roadmap for the corresponding studies and preventing conflicts among them. The GANA (Generic Autonomic Network Architecture) [4, 5] is proposed as a fundamental NA (Network Architecture), which mainly investigates the conflicts raised due to the different objective related decisions among the network entities. To this end, three different types of inter-relations are defined, namely, hierarchical, peering and sibling based on four hierarchical levels of abstractions. A similar generic NA approach, the Autonomic Network Architecture (ANA) [6], intends on the dynamic adaptation and self-organization depending on end user preferences.

In FOCALE [7, 8] architecture, the network is composed of managed elements (ME) and autonomic management elements (AME), which are responsible for monitoring the current states of MEs, comparing with the desired states, and taking actions respectively. The AME units ensure the network policy by continuously executing FOCALE control loop based on IBM's control loop [1] in order to capture and analyze the current status of the network. To this end, nested control loops are implemented throughout the network in a hierarchical way. The control loops are mainly classified as top (maintenance) loop for detecting anomalies and bottom (adjustment) loop for reconfiguration due to anomalies or the business goals. Depending on the context, policies and semantics of the interpreted data, FOCALE architecture proposes changing functionalities of control loops.

In the CONCERN project [9, 10] distributed cognitive cycles are proposed for system and network management (DC-SNM) in a hierarchical manner, aiming to facilitate the promotion of distributed management. The management and control functions are distributed among the network elements in order to achieve faster decision and execution with continuous local optimization and knowledge building capabilities. Hints, recommendations or requests are disseminated among the hierarchical levels in order to indicate the change in the network administration decisions, i.e. policies. Two different types of agents are defined, namely, Network Element Cognitive Manager (NECM) as the local agents on the network entities and Network Domain Cognitive Manager as the domain agent orchestrating the local agents.

In SOCRATES (Self-optimization and self-organization in wireless networks) [11], a bottom to top vision is presented with the key objectives of operational expenditure reduction, enhancement of network coverage, resource utilization and conflict management due to resource allocation or service quality optimization. Conflict cases are simulated providing reasoning models for the control engines. Additionally, a subset of use cases are covered [12] and is implemented on the 3GPP E-UTRAN access technology. The main focus of the SOCRATES approach is to achieve O/CAPEX (operational and capital expenditures) for a SON (Self-organizing networks) application by reducing human intervention to the network. Detailed information on the chosen architecture types with respect to the use-cases can be found in [13, 14].

In addition to the organizational and large-scale projects, there exist many significant small-scale contributions for the autonomic network control and management. Costa et al. [15], focus on alarm correlation and root cause analysis, by means of a rule-management module, which is based on data mining for rule generation and reinforcement learning. Basically, historical alarm data is stored inside local databases and correlation rules are extracted periodically by alarm management system. Famaey et al. [16], propose a generic NA in which the distributed autonomic components are structured in a hierarchical manner in order to decrease the complexity of component interactions. The network overhead associated with management and control is reduced by grouping autonomic management components into a hierarchy. To this end, the dissemination of network context and policies are efficiently orchestrated. Lim et al. [17], suggests a decentralized self-organizing network management and control system based on concept of navigation patterns. Graph traversal algorithms are the core mechanism that control the execution of management operations. Moreover, graph traversal algorithms are used to control and coordinate the dissemination, analysis and aggregation of management information among the network.

A randomization-based control and management approach is studied in [18]. Brunner et al. studies probabilistic management paradigm in order to cope with the overhead of redundant information gathering and processing in dynamic and unpredictable environments. Basically, dynamic behavior inside the network results in a lack of real time view of system and prevents coordinative functionality especially when the complexity of the network environment is high. In this study, it is claimed that the cooperation based decentralized control and management functionalities might fail in real dynamic environments while trying to assure the synchronization among the control entities. Due to these reasonings, a random activation of management and control functionalities is proposed.

The multi-agent systems (MAS) are widely issued in the autonomous network literature. In [19], Jones et al. suggest a MAS, which bears features of scalability, interoperability, survivability, concurrent diagnosis and anomaly detection capabilities focused on accurate alarm rate. The semi-autonomous agents have their own world model consisting of actions, history of past actions for further enhancement and thumbprints blocks in order to dynamically change world-detection model for adaptation. In [20], Timon et al. propose mobile agents for efficient distributed network control and management. The mobile agents, which barry management or control tasks, are delegated to remote hosts to execute the task and carry the measurements back to the sender hosts. A similar agent-delegation approach is studied in [21].

3 Problem Statement

This section focuses on the description of environment, the problem statement and finally the proposed solution.

3.1 Environment Description and System Settings

This work mainly aims at an immensely dynamic, time varying, and less predictable network environment. The motivation for this consideration comes from the fact that the proposed framework targets the wireless networks, which exhibit dynamics caused by the wireless medium, clients' behavior, traffic demands, stakeholders' preferences, and telecommunication infrastructure deployment, etc.

The behavior of focused network context is unsteady due to the alternating interference degree causing different packet loss rates, shortcomings of network protocols in wireless medium, contingent hardware or software related failure of network elements or decision making entities, unpredictable traffic behavior of clients even during daytime, alternating demand on different network objectives and etc. Additionally, in a realistic network scenario; (i) the perceptual aliasing (same observations might show up in different states) (ii) noisy/faulty sensors or (iii) a combination of both situations result in partial and unreliable information for reasoning [22]. Finally, the dynamics of the system might not be captured in real time owing to the problem of recency or the reliability of information and hence the decisions might rely on partial information.

3.2 Motivation and Problem Statement

For each network objective, an extensive range of strategies are proposed in the literature based on a set of assumptions describing different network settings. An autonomous network approach requires an integration of these strategies supported by evidence and the execution in accordance with a match of strategy-observation pairs. In other words, an autonomous control system should in a way optimize the strategy by choosing algorithms, which have more impact on specific alarm conditions.

We believe that autonomous network agility in strategy optimization might come from tracking impact rate of executed algorithms based on triggering factors. We envision triggering factors as an indicator of instantaneous environmental conditions. Due to unreliable partial information, however, false triggers might not necessarily define an alarm condition in the network or indicate a less impact of an executed strategy.

Owing to the fact that decision instances are sensitive to execution time dictates that timely decision(s) should be made i.e. waiting long for all the decisions' required inputs should be avoided and take the decision on partial/inferred observations. Intuitively, such decisions may not be graded as optimal ones, however, literature encamp a number of well known approaches (e.g. partially observable MDP, Bayes' model and or various learning approaches), which have been used extensively in the research literature in different settings. We envision the network objectives as characteristics of demand and supply. Hence, defining instantaneous environment characteristic in

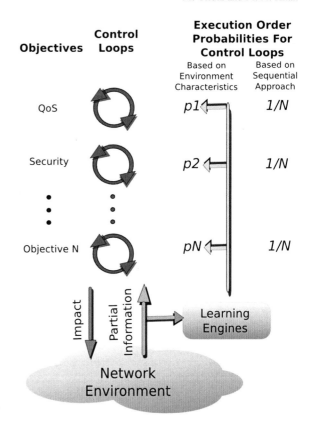

Fig. 1 Resource allocation based on the environmental conditions

terms of demand for each specific network objective and dynamically prioritizing various objectives may help optimizing the network resource utilization and network response time.

3.3 Solution Statement

We propose a two-level learning mechanism targeting the aforementioned problems. As indicated in Fig. 1, the lower level mechanism characterizes the network environment in terms of demand and assigns relative priorities to each network objective. Hereafter, this process is referred as *network objective prioritization*.

The higher level mechanism; (i) associate different algorithms specific to each objective, (ii) helps in optimization of objective specific network strategy by tracking impact rate of each algorithm on a specific environmental condition and finally (iii) handles false triggers. Hereafter, we refer this higher level learning mechanism as *strategy optimization*.

4 Background

In this section, we briefly comment on the mathematical models and approaches used in the proposed framework.

4.1 Partially Observable Markov Decision Processes

In most of the real world environments, decision making engines (DME) might not necessarily have a complete perception of their environment [23]. Observation-action mapping may rarely be executed as expected due to the partial observability of the operated stochastic environment [24, 25]. In the real-world applications, introducing randomness to DMEs provides with taking different actions for the same environmental conditions, which may increase the success rate [25].

POMDP framework is a good fit to model partially observable real-world sequential decision processes, which introduces the randomization in decision making instances. In a POMDP model, the controlled world has pre-defined environmental states and actions of DME might cause a transitions between world-states. The instant world states or the transitions between states are not completely vivid, but only partially observable, through imperfect observations. Therefore the DME takes periodically observations and keeps a Bayesian estimate of the likelihood of each state. The solution model includes (i) transition probability distributions based on impacts of DME actions on the instant belief state (ii) observation probability distribution function indicating how likely an observation might be taken for the state and a specific action, and finally, (iii) reward or cost functions associated to each state, action couples. Consequently, the POMDP framework is a six tuple of states set, action set, state transition function, reward set, observation sets and observations probabilities. The details of mathematical formulation of POMDP can be found in [24].

4.2 Markov Arrival Processes

In continuous time Markov processes (CTMP), the state transitions of an irreducible continuous time Markov Chain (CTMC) are associated with special events having Poisson distribution with different parameters. In addition to the analysis of spontaneous transitions of states, MAPs define *arrivals* of special characters that are associated with the phases and result additional transitions from associated state to another. The *sojourn* time of the Markov Chain at a specific state depends on the spontaneous transitions of CTMC and the arrival instances. The versatility of modeling various stochastic systems and the maneuverability makes MAPs popular in different scientific branches [26]. Detailed information on specific types of MAPs and mathematical formulations can be found in [26–30].

It should be noted that for the proposed framework, multiple parameters need to be estimated at different time instances. For the parameter estimation of distribution for a random variable becomes more complicated in case of uncompleted data set due to hidden variables [31]. EM algorithm is an iterative optimization algorithm proposed for estimating unknown parameters based on a measurement data [32]. The EM algorithm relates the missing measurement data with a lower-bound posterior distribution and optimizes the bound with a maximization step. More on the mathematical formulation of EM algorithm can be found in [31–33].

5 Proposed Model

In this section the proposed solution is detailed. As a formulation of network policy, we suggest to use the ontology, as a knowledge platform for reasoning, relationships and restrictions [34, 35] in the network. The details are provided in the following sections.

5.1 Ontological Characterization of Network Policy

The ontology graph of the network policy is constructed in a gradual manner defining network objectives. The graduation is performed based on relations between network objectives, their subbranches, corresponding triggers (i.e. observation), and finally the relevant actions (proposed algorithms as solutions). Based on the entity type and deployment in the network, the related assignments are restricted to certain branches of the general network policy specific for each network entity. In order to provide a better understanding of the proposed approach, and as a sample case, we focus on an access point (AP) of which responsibility branches are defined as illustrated in Fig. 2. However, it should be noted that the proposed control framework is generalized in the proceeding sections.

5.2 Network Objective Prioritization

Marked Markov Arrival Processes are utilized in order to construct and detect state transitions of the CTMC for each objective and observation. The critical observations in specific repositories are mapped over special events of *arrivals* in the MMAP models. The arrival rates are estimated using expectation maximization (EM) algorithm and MMAP model is updated respectively. Prioritization is performed based on the steady-state probability distributions of environment states. The higher steady-state probability of critical environmental states for each objective implies a demand from the environment on the prioritization of the corresponding objective or the observation.

Fig. 2 Sample network ontology

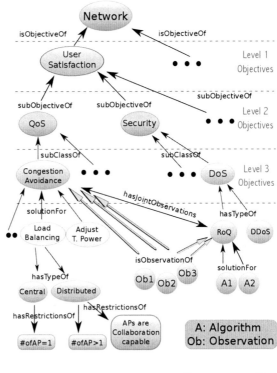

Fig. 3 State transition diagram of the environment

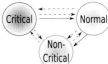

In the following section, we provide technical details and mathematically formulate the network objective prioritization.

5.2.1 MMAP Construction

We introduce three system states, these states are characterized by system objective functions. On an abstract level, we assume that all the objective functions may be encamped in any of the proposed three states. The common three-state CTMC for level-2 objective and observations in corresponding level-3 observation repositories is depicted in Fig. 3. Each critical observation instance is a different type of *arrival* event and is analyzed independently in the observation repository Ω_{o_1,o_2,o_3}. Here Ω_{o_1,o_2,o_3} is defined as the observation repository for level-3 objective o_3, which belongs to level-2 objective o_2 and level-1 objective o_1. It should be noted that we assume that observations coming from different levels are independent. Each *arrival* results only in a transition from less critical state to a more critical one. Finally an *imaginary* event is defined causing transitions between any state. This is due to false trigger like problems and the possibility of inadequacy for the defined observations in covering all the network dynamics for the related objective.

The D_0 Matrix, in Eq. 1, defines the parameters for each state transitions without an arrival and rates of each transition from state i to j is defined as λ_{ij}.

$$D_0 = \begin{pmatrix} -\lambda_1 & \lambda_{12} & \lambda_{13} \\ \lambda_{21} & -\lambda_2 & \lambda_{23} \\ \lambda_{31} & \lambda_{32} & -\lambda_3 \end{pmatrix} \tag{1}$$

Due to transition characteristic of imaginary event (*see the MMAP Construction Section*), the state transition rates are equal: $\lambda_{12} = \lambda_{13} = \lambda_{21} = \lambda_{23} = \lambda_{31} = \lambda_{32} = \lambda$.

For a single arrival Ob_x (representing the xth observation in $\Omega_{o1,o2,o3}$), the D_{Ob_x} is defined in the form of:

$$D_{Ob_x} = \begin{pmatrix} 0 & \gamma_{Ob_i} & 0 \\ 0 & 0 & \gamma_{Ob_i} \\ 0 & 0 & 0 \end{pmatrix} \tag{2}$$

where γ_{Ob_i} defines rate of the *arrival*. As it is not possible for an arrival to cause a transition from lower critical state to a more one, corresponding parameters are set to 0. The final generator matrix D is in the form:

$$D = D_0 + \sum_{Ob_i \in \Omega_{o1,o2,o3}} D_{Ob_i} \tag{3}$$

Using Eq. 3 and for $i \in \Omega_{o1,o2,o3}$ the final minimal generator matrix is:

$$D = \begin{pmatrix} -2\lambda - \sum_i \gamma_{Ob_i} & \lambda + \sum_i \gamma_{Ob_i} & \lambda \\ \lambda & -2\lambda - \sum_i \gamma_{Ob_i} & \lambda + \sum_i \gamma_{Ob_i} \\ \lambda & \lambda & -2\lambda \end{pmatrix} \tag{4}$$

Under the constraints for the steady-state probability distributions $\pi \cdot D = 0$ and $\pi \cdot \left(1\ 1\ 1\right)^T = 1$, the steady state probability distribution of CTMC are given as:

$$P\{Non\text{-}Critical\} = \frac{\lambda}{3\lambda + \sum_{i \in \chi} \gamma_{Ob_i}} \tag{5}$$

$$P\{Normal\} = \frac{3\lambda^2 + 2\lambda \sum_i \gamma_{Ob_i}}{(3\lambda + \sum_i \gamma_{Ob_i})^2} \tag{6}$$

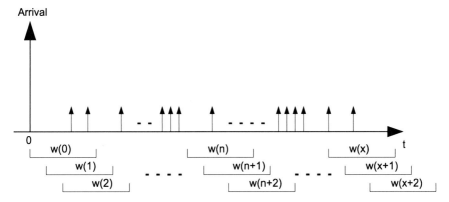

Fig. 4 An example of arrival rate versus time

$$P\{Critical\} = \frac{3\lambda^2 + 3\lambda \sum_i \gamma_{Ob_i} + (\sum_i \gamma_{Ob_i})^2}{(3\lambda + \sum_i \gamma_{Ob_i})^2} \tag{7}$$

When it comes to the observation prioritization, we use similar methodology where a single arrival exists. The interpretation of these results are:

- Setting larger λ value would set a static priority for each level-2 objective or observations as the steady-state probabilities becomes homogenous.
- Smaller λ parameters would leave the prioritization of objectives or observations totally to the environmental behavior.
- Different λ values on different concepts results in a predetermined prioritization of specific concepts against others. In other words the network administration might guarantee specific services independent from the environment.
- The increase in the number of observations in a specific repository would provide the control framework with stronger prediction capability for the environment and network conditions.

5.2.2 Arrival Rate Estimation

The *arrival* events are counted periodically as shown in Fig. 4, where $w(n)$ represents the nth window.

The arrival rate estimation is carried out using EM algorithm. In accordance with the three-state formulation of CTMC for each level-2 objective and corresponding observations, and due to the fact that Poisson mixtures fit the observations in a much better way in case of multiple routines [36], the EM algorithm is applied on a mixture of three Poisson distributions.

The probability of counting k number of the same observation in a fixed window, given the set $\Theta = \left(\gamma_{Ob_i}^{(nC)}, \tau_{Ob_i}^{(nC)}, \gamma_{Ob_i}^{(N)}, \tau_{Ob_i}^{(N)}, \gamma_{Ob_i}^{(C)}, \tau_{Ob_i}^{(C)} \right)$, where γ's are defined as the arrival rates and τ's represent the instant probability distribution of states;

$$p(k|\Theta) = \sum_{s \in \{nC,N,C\}} \tau_{Ob_i}^{(s)} \frac{e^{-\gamma_{Ob_i}^{(s)}} \left(\gamma_{Ob_i}^{(s)}\right)^k}{k!} \tag{8}$$

For readability of the following equations, we refer γ_i instead of γ_{Ob_i} and τ_i instead of τ_{Ob_i}. We define the environmental states as the hidden variables of EM model with state space $Z = (z_{nC}, z_N, z_C)$. The probability distribution of hidden variables are $\tau_{nC}, \tau_N, \tau_C$ respectively.

Having the counter vector $K = < k_1, k_2 \ldots k_n >$ as the number of occurrence, i.e. case of critical observation, and given $z = z_j, z_j \in Z$,

$$Pr\{K, z_j \,|\Theta\} = \prod_{i=1}^{n} \left[\tau_j \left(\frac{e^{-\lambda_j} \lambda_j^{n_i}}{n_i!} \right) \right] \tag{9}$$

$$log\{Pr(K, z_j \,|\Theta)\} = \sum_{i=1}^{n} [log\{\tau_j\} - \lambda_j + n_i log\{\lambda_j\} - log\{n_i!\}] \tag{10}$$

using Bayes theorem:

$$Pr\{z_j \,|k_i, \Theta\} = \frac{\tau_j e^{-\lambda_j} \lambda_j^{k_i}}{\tau_{nC} e^{-\lambda_{nC}} \lambda_{nC}^{k_i} + \tau_N e^{-\lambda_N} \lambda_N^{k_i} + \tau_C e^{-\lambda_C} \lambda_C^{k_i}} \tag{11}$$

referring to the results Eqs. 8–11, the **E-Step** is:

$$E_{Z|K;\Theta} = \sum_{i=1}^{n} \sum_{z_j \in Z} Pr\{z_j \,|k_i, \Theta\} log\{Pr(K, z_j \,|\Theta)\} \tag{12}$$

under the constraint $\tau_{nC} + \tau_N + \tau_C = 1$ and utilizing nth estimated set of $\Theta_{\lambda(n)} = \left(\lambda_{nC}^{(n)}, \lambda_N^{(n)}, \lambda_C^{(n)}\right)$, the $(n+1)$th estimation of probability distribution over the environmental sets as the start point of **M-Step** is given:

$$\tau_L^{(n+1)} = \arg\max_{(\tau_L)} E_{Z|K;\Theta_{\lambda(n)}^n}(\tau_L) \tag{13}$$

using the above equation and $(n+1)$th arrival rates of corresponding environmental states are:

$$\gamma_L^{(n+1)} = \arg\max_{(\gamma_L)} E_{Z|K;\Theta}(\gamma_L) \tag{14}$$

$\forall L \in nc, N, C$.

Fig. 5 Positions of the proposed mechanisms in the generic control loop

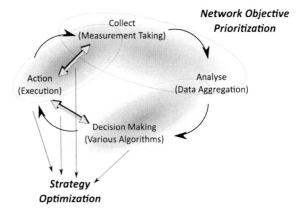

In this model, we interpret the case of an increase in the probability distribution of a specific environment state as a transition to the corresponding state. Hence the associated arrival rate is forwarded to the MMAP formulation process.

5.3 Strategy Optimization

It is the higher level mechanism, in which the control loop for each objective are executed in case of a trigger from lower level objective prioritization process. The control loops are modified based on POMDP. With the modified control loop different strategies are associated with various environmental conditions by monitoring the success rates. In the proceeding sections, we briefly discuss the construction of proposed control loops.

5.3.1 Control Loop

The construction and conceptual background of the IBM generic control loop is discussed in [37]. Based on this, we illustrate in Fig. 5 the aggregation of the proposed mechanisms into the generic control loop. The first stage mechanism, i.e. the network objective prioritization, runs at the measurement collection and the analysis steps. Depending on the objective observation repository, the analysis is performed and in case of a critical instance, corresponding triggers are forwarded to the higher mechanism, i.e. the strategy optimization. The proposed strategy optimization mechanism has an impact on the decision making and the execution steps, by means of the selection of corresponding network actions. Moreover, we consider on potential supplementary cycles between decision-making, execution and the measurement taking for the further analysis on the executed actions.

Fig. 6 The proposed general
policy for the control loop

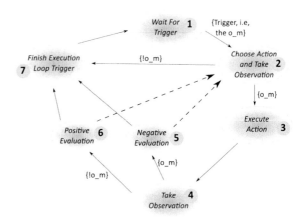

We propose mainly an elimination mechanism on the network action repository by means of selecting the actions having more impact on the specific network objective tackle. To this end, we propose monitoring the success rates of the executed actions in order to determine potentially effective actions for an existing problem in the network. Having these objectives we illustrate the Fig. 6 as the basic methodology loop for the analysis of the executed actions, which helps for the strategy optimization. The continuos lines stands for the straightforward transitions between iterative steps to be taken for the control loop. We comment on the dashed lines representing the additional transitions in proceeding parts of this section.

The description of the states represented in Fig. 6 is as follows:

- Wait for trigger state: In case of a critical observation the lower level objective prioritization phase forwards an execution trigger together with the information model of level-2 objective and corresponding observation.
- Choose action and take observation: Based on the observation-algorithm match, a more successful algorithm is selected for execution. More details on the selection of algorithms is provided later in this section.
- Execute state: The agent executes chosen algorithm.
- Take observation state: After the execution state, a pre-defined time period is waited for the stabilization of network under the impact of executed algorithm and an additional observation is taken in order to observe judge on success rate of the executed algorithm.
- Positive evaluation state: The success rate of executed algorithm is increased.
- Negative evaluation state: The success rate of executed algorithm is decreased. It should be highlighted that this state is a normalization state for the elimination of algorithms at state S_4. In other words, the relative probability over the executed action for being selected at state S_4 is increased for remaining available actions.
- Finish execution loop trigger state: The control loop hands the token to the lower level objective prioritization mechanism.

Table 1 Summary of parameter set for the control loop dynamics

Parameter	Description
$A^{(ij)}$	Action set for third and corresponding second level objective i, j respectively, with a size of K
$Ob^{(ij)}$	Observation set for third and corresponding second level objective i, j respectively, with a size of M
$a_k^{(ij)}$	The kth action in $A^{(ij)}$ $\left(a_k^{(ij)} \in A^{(ij)}\right)$
$o_m^{(ij)}$	The mth observation in $Ob^{(ij)}$ $\left(o_m^{(ij)} \in Ob^{(ij)}\right)$
$\Theta_m^{(ij)}$	$\left\{\left(\gamma_{Ob_m}^{(nC)}, \tau_{Ob_m}^{(nC)}, \gamma_{Ob_m}^{(N)}, \tau_{Ob_m}^{(N)}, \gamma_{Ob_m}^{(C)}, \tau_{Ob_m}^{(C)}\right)\right\}$ set for $o_m^{(ij)}$, determined by the network objective prioritization stage
$tr_m^{(ij)}$	The mth observation trustworthiness
$\rho_k^{(ij)}$	Success rate for the action $a_k^{(ij)}$
$\gamma_m^{(ij)}$	Expected rate parameter of observation $o_m^{(ij)}$
$t_{\mathfrak{J}\mathfrak{J}}^{(ij)}$	The state transition probability from $S_I^{(ij)}$ to $S_J^{(ij)}$, where $S_{\mathfrak{J}}^{(ij)}, S_{\mathfrak{J}}^{(ij)} \in S^{(ij)} = \left\{S_1^{(ij)}, \ldots, S_7^{(ij)}\right\}$
W	Window size for arrival observations
$o\_m^{(ij)}$	Critical instance of $o_m^{(ij)}$ observation
$!o\_m^{(ij)}$	Non-critical instance of $o_m^{(ij)}$ observation
$a\_k^{(ij)}$	The success instance of $a_k^{(ij)}$

It should be highlighted that this improved control loop represents a basic approach for the network environments, where the complete knowledge on the world states, i.e. the state indicators or the network observations, is available. Nevertheless, the proposed model attempts for very dynamic and stochastic environments, where only partial information on the corresponding network states and the transitions may be available. This is due to the previously mentioned reasonings such as false alarms, recency complications, unreliable or incomplete information, sensitivity for the execution times and network stabilization durations, etc. Thus further analysis is required for the proposed control loop by means of additional optimization attempts for an optimal policy decision under uncertainty.

A summary of the parameter set for the control loop dynamics is provided in Table 1, which have a great impact on the transitions between control loop states. Considering a specific loop for a specific objective (third and corresponding second level objective i, j), a critical instance for any observation $\left(o_m^{(ij)} \in Ob^{(ij)}\right)$ generates a trigger for the control loop, which is represented as o_m in Fig. 6. Thus state transition rate from the first state to the second one depends on the total critical instance rates for corresponding objective observations:

$$t_{12}^{(ij)} = P\{Trig.\} = \sum_{\left(o_m^{(ij)} \in Ob^{(ij)}\right)} P\left\{o_m^{(ij)} \text{ is selected, } o\_m^{(ij)}\right\}$$

$$= \sum_{\left(o_m^{(ij)} \in Ob^{(ij)}\right)} P\left\{o\_m^{(ij)}|o_m^{(ij)}\right\} P\left\{o_m^{(ij)}\right\} \tag{15}$$

$$= \sum_{m=1}^{M} \left(\frac{\gamma_m^{(ij)}}{\sum_{m=1}^{M} \gamma_m^{(ij)}}\right) \frac{\gamma_m^{(ij)}}{W}$$

where,

$$\gamma_m^{(ij)} = \gamma_{Ob_m}^{(nC)} \tau_{Ob_m}^{(nC)} + \gamma_{Ob_m}^{(N)} \tau_{Ob_m}^{(N)} + \gamma_{Ob_m}^{(C)} \tau_{Ob_m}^{(C)} \tag{16}$$

As an additional observation is taken at $S_2^{(ij)}$, the $t_{23}^{(ij)} = t_{12}^{(ij)}$ and $t_{27}^{(ij)} = 1 - t_{23}^{(ij)}$ as no other state transition is allowed. The $t_{34}^{(ij)} = 1$. By separating the $S_3^{(ij)}$ and $S_4^{(ij)}$ we represent the required time duration for the network stabilization depending on the sequential form of the selected action and observation tuples. One may expect a high dependency on the selected action at $S_2^{(ij)}$, e.g. $a_k^{(ij)}$, and its success indicator, i.e. the $\rho_k^{(ij)}$, for the transition character from $S_4^{(ij)}$ to either $S_5^{(ij)}$ or $S_6^{(ij)}$. In other words:

$$t_{46}^{(ij)} = \sum_{\left(a_k^{(ij)} \in A^{(ij)}\right)} P\left\{a_k^{(ij)} \text{ is selected, } a_k^{(ij)} \text{ is successful}\right\}$$

$$= \sum_{\left(a_k^{(ij)} \in A^{(ij)}\right)} P\left\{a\_k^{(ij)}|a_k^{(ij)}\right\} P\{a_k^{(ij)}\} \tag{17}$$

$$= \sum_{k=1}^{K} \left(\frac{\rho_k^{(ij)}}{\sum_{k=1}^{K} \rho_k^{(ij)}}\right) \rho_k^{(ij)}$$

$$t_{45}^{(ij)} = 1 - t_{46}^{(ij)} \tag{18}$$

The positive/negative evaluation states are the most significant states for the proposed control loop, where two factors are of critical importance. These factors are (i) the impact of the selected action and (ii) the dynamic and stochastic behavior of the environment with an incomplete feedback. *At the first glance*, there is a great uncertainty on the additional observation at state $S_4^{(ij)}$ due to the ambiguity for the trustiness of the observation, e.g. whether a critical observation instance stems from the less impact of the executed action or from the dynamics of the network environment. *Nevertheless*, this complexity is the focused item of the improved control loop in this paper. For a better understanding of the remarked ambiguity, we focus on a network scenario related with resource allocation.

Considering a congestion case in the network due to the high client arrival rates, who have high bandwidth requirements; we assume that a load balancing activity is executed by the control loop on an access point of the network, which is based on forcing client migrations. Due to higher bandwidth utilizations and arrivals, the additional observation at $S_4^{(ij)}$ becomes critical despite the executed load balancing action. This situation resembles the complexity which is described with the afore-mentioned ambiguity. On the other hand, one may argue that the proposed algorithm is not effective for the predefined environment, which is a cut for the admission control based approaches rather than forcing client migrations. Thus the proposed control loop eliminates a migration based load balancing action not because of the ineffectiveness but because of the instant environment dynamics.

The very simple additional parameter, which is left as a design parameter, is the trustworthiness of the observations, which describe the transitions between states $S_5^{(ij)}$ to $S_2^{(ij)}$, $S_5^{(ij)}$ to $S_7^{(ij)}$, $S_6^{(ij)}$ to $S_2^{(ij)}$ and finally $S_6^{(ij)}$ to $S_7^{(ij)}$. With the trustworthiness, we define a subjective term for a specific observation class, which defines (i) how this specific observation is capable of handling recency problems (ii) how certainly can it be utilized as a reasoning in describing the network environment (iii) how frequently the observation may create false triggers, etc. A useful discussion on trustworthiness of an observation can be found in [38]. At this point, It is worth mentioning that although subjective probabilities may be assigned in an intuitive way for the state transition probabilities in the construction of the system, the validation of the assigned probabilities is as discussion point as they may be descriptively inaccurate especially when the complexity of the problem is high [39]. We comment more on this parameter in Sect. 5.4.

$$t_{62}^{(ij)} = t_{52}^{(ij)} = 1 - \frac{1}{M} \sum_{m=1}^{M} tr_m^{(ij)}$$
$$t_{67}^{(ij)} = t_{57}^{(ij)} = \frac{1}{M} \sum_{m=1}^{M} tr_m^{(ij)}$$
(19)

Clearly, the transition parameters dynamically changes based on the environmental conditions. Additionally these parameters are dependent on the objective and more concretely the corresponding action/observations. Nevertheless, considering on the instant shots of the proposed control loop, the policy state transitions hold memoryless property as discussed before, i.e. the future state transitions depend only on the current state of the loop. In other words the proposed policy loop forms a Markov Chain with the specifications such as (i) the world state set $S^{(ij)} = \left\{ S_1^{(ij)}, \ldots, S_7^{(ij)} \right\}$, (ii) the previously discussed instant state transition probabilities $T^{(ij)} = \left\{ t_{\Im\Im}^{(ij)} | \Im \neq \Im, 1 \leq \Im, \Im \leq 7 \right\}$, (iii) the start state is $S_1^{(ij)}$.

5.3.2 POMDP Formulation

The main motivation behind using the POMDP rather than Markov Decision Process (MDP) is the previously mentioned trustworthiness of the observation set. An MDP framework is represented with the form of $<S^{(ij)}, A^{(ij)}, T^{(ij)}(s, a, s'), R^{(ij)}(s, a)>$, where in addition to the previously discussed parameters, the $R^{(ij)}$ is defined as the reward for the action $a_k^{(ij)}$ given that the state is s. The reward concept is the utility definition that the agent may except. In an MDP formulation, although there is a great uncertainty for the transitions of the states based on the executed action, the control agents are provided with a complete information on the current world states. In this study, however, due to the trustworthiness of the taken observations, additional complexity arises in determining the real world state. Rather than having a complete world state information, there is the *belief state* concept, where the agent tracks a set of believes set Z by means of the additionally taken observations. To this end, the observation function $O(s', a, o)$ is defined, which is the likelihood of making the observation o, given that the next state is s' and the action a is executed. In our model, the $O\left(s_l^{\prime(ij)}, a_k^{(ij)}, o_m^{(ij)}\right)$ takes either $\frac{1}{M}\sum_{m=1}^{M} tr_m^{(ij)}$ or $1 - \frac{1}{M}\sum_{m=1}^{M} tr_m^{(ij)}$ depending on the expectation of either critical instance or a non-critical instance. The POMDP model includes the *state-estimator* as an additional dynamic in order to provide a belief for the current state of the agent world. For a specific objective, the next state estimation $b_{l+1}^{(ij)}$ at time step l is a function of the current belief state $b_l^{(ij)}$, the taken action $a_l^{(ij)}$ and the observation $o_l^{(ij)}$.

$$b^{(ij)}\left(s_{l+1}^{(ij)}\right) = P\left\{s_{l+1}^{(ij)} | o_k^{(ij)}, a_k^{(ij)}, b^{(ij)}\left(s_l^{(ij)}\right)\right\}$$
$$\sim O\left(s_{l+1}^{(ij)}, a_l^{(ij)}, o_l^{(ij)}\right) \sum_{s^{(ij)}\in S^{(ij)}} T^{(ij)}\left(s_l^{(ij)}, a_l^{(ij)}, s_{l+1}^{(ij)}\right) b^{(ij)}\left(s_l^{(ij)}\right)$$

$$(20)$$

Now considering an l-step non stationary policy A and a k-step move of the agent, we are interested in calculating the expected total sum of the rewards gained as the value function. As the first reward is simple the reward associated to the known state:

$$V_k(s) = R(s, A(s))$$
$$+ \gamma \sum_{s'\in S} T(s, a_k, s') \sum_{o_k\in O} O(s', A(s'), o_k) V_{o_k}(s')$$

$$(21)$$

Remembering that the state information is not provided for the agent, the expected optimal policy is calculated over the belief-states as the belief state value functions are given as $V_k(s) = \sum_{s\in S} b(s) V_k(s)$. Kaelbling et al. [24] proposes a more convenient method in representing the value functions, where the $\alpha_k = <V_k(s_1), \ldots, V_k(s_n)>$

vector represents the value function distribution among possible states and $V_{k+1}(b) =$ $\text{argmax}_{k \in p} b . \alpha_k$.

The iterative calculation of the complex models may require a long computation time and hence computation power. In this study, we re-estimate the POMDP based optimal policy for the previously discussed control loop in case certain model parameters, i.e. the state transition parameters, differs over a previously determined threshold ϵ_t.

5.4 Framework Restrictions and Requirements

A certain amount of framework parameters are required for the control operation. These parameters are:

- λ, which is discussed in Sect. 5.2.1
- a network ontology model for the graded network objectives
- information models for the action and observations including definitions for corresponding network objectives
- the ϵ_t as a threshold for the sensibility of framework to the changing control loop parameters to construct new POMDP based control policies
- $tr_m^{(ij)}$, the trustworthiness for each observation, discussed in Sect. 5.3.1.

6 Software Architecture

This section is devoted to the details of the software implementation. The proposed control framework is implemented using Java programming language, and we make widely use of dynamic class loading in Java programming [40].

6.1 Block Diagram

The block diagram of a cognitive network entity is provided in Fig. 7. As we selected an access point (AP) in our technical network scenarios, additional AP related blocks are added into the illustration.

As can be inferred from Fig. 7, the proposed control framework is implemented as the network controller unit having a set of observations, actions and the proposed network objective prioritization unit as the decision making engine (DME).

Additionally, we implemented a dynamic action/observation loader mechanism from a repository in order to provide the cognitive entities with the ability of gaining more efficient strategies for each objective defined in the network ontology. The network administration can dynamically change, update or upgrade observation/action

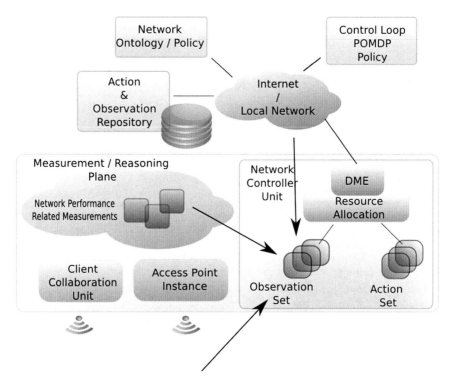

Fig. 7 Access point block diagram

sets for the corresponding cognitive entities. The similar mechanism is implemented also for the POMDP based control loop policies. The cognitive entity polls periodically the related repositories for the updates on the network policy, the POMDP modified control loop and the action/observation repositories. This is to provide the researchers using proposed autonomous control framework with the ability of dynamically changing the framework related parameters, which are described in Sect. 5.4.

The observation sets are in conjunction with the measurement/reasoning plane in order to ease the implementation. Additionally various observations can be added to the system in order to populate various observations at different points in the network, e.g. from the backbone network or the last hop. Finally, an additional client collaboration unit is added for the technical scenarios, which are explained later in Sect. 7.

6.2 Software Flow Charts

The main DME software consists of two main blocks. The first block is responsible for the organizational issues and the flow chart of this block is given in Fig. 8.

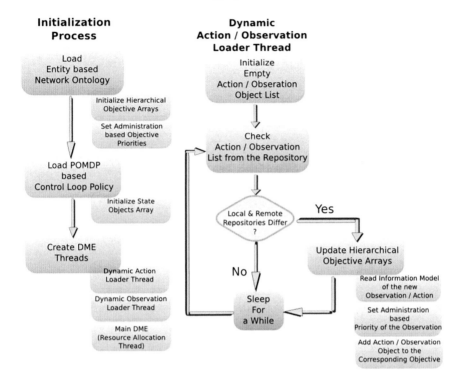

Fig. 8 Organizational threads' flow charts

The initialization process starts with loading and analyzing the network ontology based on the network entity type. Hierarchical objective arrays are generated during analysis phase and corresponding administration based object-priorities are assigned. A similar action is done for the information model of POMDP based control loop policy. The initialization process ends up after creating DME threads, namely, dynamic action loader, dynamic observation loader and the main resource allocation/control loop executor thread. The second organizational thread type, namely, the dynamic action or observation loader thread, is responsible for updating hierarchical objective arrays by attaching new observations or action objects in case of an update.

In this model, we make widely use of information models (IM) for the boot up processes and also dynamically handling newly added action or observation sets or changes in the framework parameters. We illustrate in Fig. 9 examples for the information model of a specific observation and action.

The second type of threads is assigned as the main DME thread and is responsible for prioritizing the network objectives and executing the corresponding POMDP based control loops. The flow chart of this thread is given in Fig. 10. A probability distribution function (pdf) is created based on the priorities of second level objectives and a random second level objective is chosen according to the created pdf. A similar randomization approach is utilized for the observation set of the corresponding second level object and the POMDP based control loop is executed in case the

Fig. 9 Example information
models (IM) for an observation and action

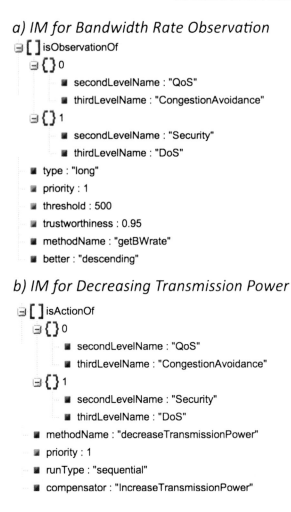

Fig. 9 Example information models (IM) for an observation and action

chosen observation is critical. Finally depending on the taken observation, the priorities for corresponding second level objective and the observation is updated using
the proposed network objective prioritization algorithm.

7 Evaluation and Tests

In this section we describe the evaluation methodology for the proposed control
framework. The behavior and decisions of the control framework is tracked through
simulation and real experimentations. Following subsections give the details of experiment environments, test scenarios and the corresponding results.

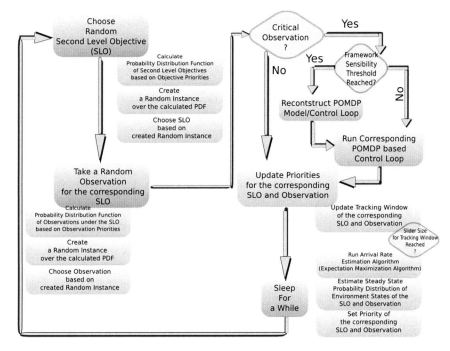

Fig. 10 Dynamic resource allocation thread flow charts

7.1 Simulation Results

7.1.1 Scenario Description

To illustrate and evaluate the performance of the proposed control framework, this section presents a sample simulation test scenario. A cognitive network entity has the sample ontology as the network policy, which is represented in Fig. 2. Referring to this Figure, the action and the observation sets for both the Congestion Avoidance and Reduction of Quality (RoQ) mitigation objectives consist of three functionalities. We prepare a Java based simulation software in order to realize this network scenario. This software basically resembles the network environment, where critical instances are generated based on the provided probability distribution models. The generated critical conditions may be handled if an additional probability distribution model is provided to the software. These conditions are maintained if no input is provided. Such situations are assumed to be handled once the generated random numbers, i.e. one as the critical case and one as a corresponding solution, are compatible with each other. Our main objective is to make use of this property in testing (i) the prioritization of network objectives, which are resembled by the probability distributions with a higher expected occurrence rate and (ii) the elimination of the corresponding actions, which are resembled with the probability distributions generating compatible random numbers, i.e. as impact.

We name the actions in the action set of Congestion Avoidance as *Action* 1, *Action* 2 and *Action* 3. Similarly the observations for the corresponding objective are named as *Event* 1, *Event* 2 and *Event* 3. We model the occurrence of events and successful impacts of the actions with Poisson distribution. In the beginning of the simulation the occurrence rates of the events have the following relation:

$$\upsilon_{event\,3} > \upsilon_{event\,2} > \upsilon_{event\,4} > \upsilon_{event\,1} > \upsilon_{event\,5} > \upsilon_{event\,6} \tag{22}$$

In other words, we perform the simulation in a mostly congested network environment. We change the rates during the simulation from time to time. The relation between successful impact rates of the corresponding actions are:

$$\varphi_{Action\,3} > \varphi_{Action\,2} > \varphi_{Action\,1} \tag{23}$$

and

$$\varphi_{Action\,6} > \varphi_{Action\,5} > \varphi_{Action\,4} \tag{24}$$

We run a 3 h simulation and observe the behavior and the decisions of the control framework under these conditions.

7.1.2 Results and Evaluation

As can be inferred from Fig. 11 the instant probability distributions of the belief states for specific objectives are dependent on the observation rates of special events in the corresponding observation set. Hence, the priorities of network objectives are changed during the simulation and the network resources are allocated accordingly and the network response time is decreased for the prioritized objectives. Moreover, the observation of events with higher rates of occurrences are prioritized under the same objective observation set. The proposed control framework takes 946 times a measurement in order to track *Event* 3, 324 times for *Event* 2 and 275 times for *Event* 1. Additionally, the network adapts to the changing conditions in terms of resource allocation. It is observed from Fig. 11 that the priority of Security objective is increased once the rates of occurrences for security related events are increased.

In case of an increase in the success rate of an action, the proposed control framework has a tendency for performing the same action for the critical situations, which are defined under same objective observation set. As illustrated in Fig. 12, *Action* 3 has the most effective impact on the environment in case of a congestion and is taken 99 times, while the opponent algorithms *Action* 2 is taken 50 times and *Action* 1 is taken only 19 times.

Fig. 11 Object prioritization based on event arrivals

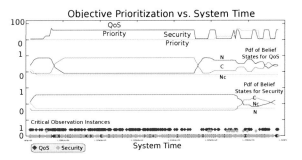

Fig. 12 Success rates and instances of network actions versus system time

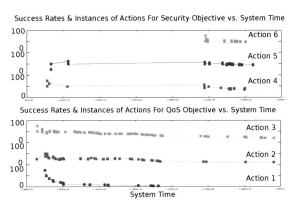

7.2 Testbed Experiments

7.2.1 Scenario Description

Two wireless access points (AP) have the aforementioned sample ontology, which is represented in Fig. 2. Based on different combinations of observation sets and the different environmental conditions, we define two test scenarios. For the first scenario, the action set for the Congestion Avoidance objective consists of three functionalities, namely, *increasing the transmission power*, *decreasing the transmission power* and finally *migrating the client with worst performance feedback to the second AP*. For this last action, we make use of a similar methodology with the [41] in order to off-load a specific client to another available AP. We define the *instantaneous bandwidth utilization of AP* as the single observation for this objective. *Decreasing the transmission power* is the single action for the RoQ mitigation objective. Similarly the single observation of this objective is *threshold delay time for the network performance observation from each client perspective*. With this configuration we perform the experiment where the network environment is highly congested from time to time.

For the second scenario, we keep the action and observation sets for the Congestion Avoidance same as it is in the first scenario. In the beginning of experiment, no

Fig. 13 Test bed

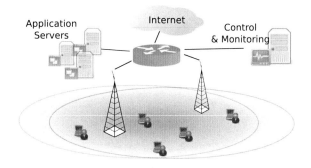

functionality exists in the action repository for the RoQ mitigation objective. We add the *instantaneous bandwidth utilization of AP* as the second observation for this objective. In other words, both objectives have the common observation set. The network environment includes malicious clients and is rarely congested during the experiment. We add later an *attack mitigation* action, which is a similar methodology proposed in the [41], to the repository during the experiment.

7.2.2 Test Bed

For the aforementioned test scenarios, we partially make use of the user centric wireless testbed, details of which can be found in [42]. A simplified illustration for the utilized testbed portion is provided in Fig. 13. The test bed is composed of two wireless access points, a central router, application servers, a control and monitoring computer and finally a server as the repository for network ontology, POMDP control loop policy and action/observation functionality classes.

Two Alix boards of PC Engines [43] are configured as the wireless access points, running Voyage Linux [44] as the operating system. Linux PCs are used as well-aimed clients and as traffic generators. We utilize Iperf tool [45] for the traffic generation during the scenario in order to congest the APs.

A client network manager (CNM) software is written additionally, which provides the client devices the collaboration capability with the APs. The CNM is composed of a messaging mechanism with APs, AP connection builder and network performance tracking blocks, which are illustrated in Fig. 14. Distributed internet traffic generator (DITG) [46] is utilized for the network performance tests. The client devices are capable of performing a handover to the other APs when they receive the corresponding message from the camped AP.

7.2.3 Results and Evaluation

For the scenario 1, we illustrate the behavior of control framework on one of the APs and the network performance tests of three clients in terms of average delay time

Fig. 14 Client network
manager block diagram

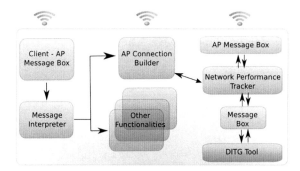

Fig. 15 Object prioritization
based on event arrivals
and average delay time
measurements on client
devices

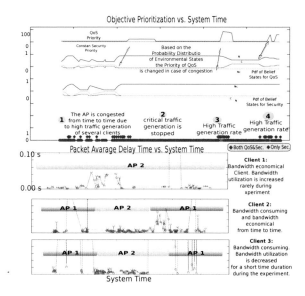

Fig. 16 Success rates and
instances of quality of service
(QoS) objective actions versus
system time

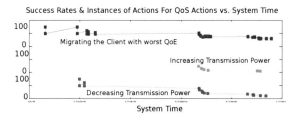

in Figs. 15 and 16. Based on the observation arrivals, the AP successfully organized objective priorities and the network response time in critical situations is improved. With an additional mechanism, the AP detects the clients with higher bandwidth utilization and balances the load in a fair way. Moreover, it is clearly observed that the AP is capable of choosing better actions against congestion.

Fig. 17 Object prioritization based on event arrivals and average delay time measurements on client devices

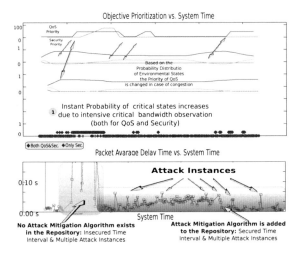

Fig. 18 Success rates and instances of network actions versus system time

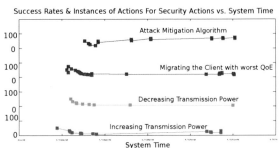

The Figs. 17 and 18 illustrate the behavior of the AP in the aforementioned scenario 2. The control framework successfully detects the new attack mitigation algorithm in the attack/observation repository and takes the action against RoQ attacks. We observe that the performance tests of the clients are improved even in intensive attack instances.

8 Discussion and Summary

In this study, the action and observation repositories are limited to the executables, which are restricted to a finite run-time period and are composed of a sequential order of function calls. In other words, the repository elements that are dynamically loaded and executed, are guaranteed to be terminated in a short time period. This aspect may be strengthen by splitting algorithms of loop characteristics into various function sets, which may in turn be compiled as separate executables and added to the corresponding repositories. Nevertheless, we define this limitation as vulnerable sector, which we will address as an extension of this study.

Another attention catching dimension in this study is the natural question, "how do we define the false triggers?" In this work, we limited the definition of false triggers to a single perspective, i.e. a case of critical observation might not define an alarm condition in the network or indicate a less successful impact on the network environment for an executed action. This factor is represented with the design parameter tr, i.e. a grade representing the trustworthiness of the observation. However, an additional perspective on the false triggers, i.e. the missing data, where no trigger interrupt the control framework despite of an anomaly situation in the network is left as a future work for this study.

When it comes to shedding the light on the designer inference factor, i.e. the λ parameters in D_0 Matrix (ref. to Sect. 5), which are defined as constants and are shown to be the normalizing factors in the control framework's believes and effecting the priorities of network objectives. λ values are set by the network administration at the boot up time. With this single tuning parameter, the impact of the network administration can be adjusted in the decision instances of the proposed control framework. This parameter provides flexibility for the designers to interfere on the operation of the network as a human to machine mediator. On the other hand, for very dynamic environments boot up assignments of these parameters may not be appropriate and sufficient as presumed by the designer. We believe that the optimization on the control of λ parameters may potentially be helpful for the stability and reliability of the proposed control framework. The certainty level of the environment behavior, i.e. the measure for the environment routines, can easily be adapted for the solutions provided. With the certainty, we mean the variances in a generated probability distribution function, i.e. how stable the drawn data is stable. Though one may argue that the design complexity would increase depending on the administration preferences as the proposed framework targets a more comprehensive perspective. Thus excluding the administration preferences may not be suitable for various network scenarios and architectures. Hence we keep the mediator parameter in the proposed framework without including further machine inference on the administration preferences.

References

1. IBM (2004) Understand the autonomic manager concept. Available http://www.ibm.com/developerworks/library/ac-amconcept/. Accessed 21 Oct 2004
2. Chaparadza R, Ciavaglia L, Wodczak M, Chen C-C, Lee B, Liakopoulos A, Zafeiropoulos A, Mancini E, Mulligan U, Davy A, Quinn K, Radier B, Alonistioti N, Kousaridas A, Demestichas P, Tsagkaris K, Vigoureux M, Vreck L, Wilson M, Ladid L (2009) Etsi industry specification group on autonomic network engineering for the self-managing future internet (etsi isg afi). In: Vossen G, Long D, Yu J (eds) Web information systems engineering—WISE 2009, ser LNCS, vol 5802. Springer, New York, pp 61–62
3. Uzsak F, Simon C (2011) Monitoring system for EFIPSANS networks. In: 34th International Conference on IEEE Telecommunications and Signal Processing (TSP 2011)

4. ETSI (2011) Etsi gs afi 001: autonomic network engineering for the selfmanaging future internet (afi): scenarios, use cases, and requirements for autonomic/self-managing future internet. This ETSI Specification was published by ETSI

5. Chaparadza R (2008) Requirements for a generic autonomic network architecture (gana), suitable for standardizable autonomic behaviour specifications of decision-making-elements (dmes) for diverse networking environments. In: Annual review of communications, vol 61. Springer, Berlin

6. ANA (2006–2009) European union information society technologies framework programme 6 project ana 2006–2009. Available http://www.ana-project.org/

7. Raymer D, van der Meer S, Strassner J (2008) From autonomic computing to autonomic networking: an architectural perspective. In: Fifth IEEE workshop on engineering of autonomic and autonomous systems, EASE 2008, pp 174–183

8. Strassner J (2007) The role of autonomic networking in cognitive networks. Wiley, New York, pp 23–52. Available http://dx.doi.org/10.1002/9780470515143.ch2

9. Mihailovic A (2009) First report on system capabilities, functional entities definition and system architecture for cognitive management of future internet elements, deliverable d1.2. Available https://www.ict-selfnet.eu

10. Consern home page (2008–2010) Available https://www.ict-selfnet.eu

11. van den Berg J, Litjens R, Eisenblatter A et al (2008) Socrates: self-optimisation and self-configuration in wireless networks. COST 2100 TD(08)422, Wroclaw, 6–8 Feb 2008

12. Scully N et al (2008) Use cases for self-organising networks, deliverable d2.1. Available http://www.fp7-socrates.org/

13. Scully N et al (2009) Review of use cases and framework, deliverable d2.5. Available http://www.fp7-socrates.org/

14. Scully N et al (2009) Review of use cases and framework ii, deliverable d2.6. Available http://www.fp7-socrates.org/

15. Costa R, Cachulo N, Cortez P (2009) An intelligent alarm management system for large-scale telecommunication companies. In: Lopes L, Lau N, Mariano P, Rocha L (eds) Progress in artificial intelligence, vol 5816. ser LNCS, Springer, New York, pp 386–399

16. Famaey J, Latrea S, Strassner J, De Turck F (2010) A hierarchical approach to autonomic network management. In: Network operations and management symposium workshops (NOMS Wksps), 2010 IEEE/IFIP, April 2010, pp 225–232

17. Lim K-S, Adam C, Stadler R (2004) Decentralizing network management. IEEE Electron Trans Netw Serv Manage (eTNSM) 1.2

18. Brunner M, Dudkowski D, Mingardi C, Nunzi G (2009) Probabilistic decentralized network management. In: IFIP/IEEE international symposium on integrated network management, 2009. IM '09. June 2009, pp 25–32

19. Jones LW, Carpenter TL (2002) Towards decentralized network management and reliability. In: Proceedings of the 2002 international conference on security and management

20. Du TC, Li EY, Chang A-P (2003) Mobile agents in distributed network management. Commun ACM 46:127–132. Available http://doi.acm.org/10.1145/792704.792710

21. Goldszmidt G, Yemini Y (1998) Delegated agents for network management. IEEE Commun Mag 36(3):66–70

22. The pomdp page. Available http://www.pomdp.org/

23. Russel SJ, Norvig P (2010) Uncertain knowledge and reasoning. In: Artificial intelligence: a modern approach. Prentice Hall, New Jersey, pp 480–684

24. Kaelbling LP, Littman ML, Cassandra AR (1998) Planning and acting in partially observable stochastic domains. Artif Intell 101:99–134

25. Littman ML (1994) Memoryless policies: theoretical limitations and practical results. The MIT Press, Cambridge, pp 238–245

26. He Q-M (2010) Construction of continuous time markovian arrival processes. J Syst Sci Syst Eng 19:351–366. http://dx.doi.org/10.1007/s11518-010-5139-5. doi:10.1007/s11518-010-5139-5

27. Fischer W, Meier-Hellstern K (1993) The markov-modulated poisson process (mmpp) cookbook. Perform Eval 18(2):149–171. Available http://www.sciencedirect.com/science/article/pii/016653169390035S

28. Cordeiro JD, Kharoufeh JP (2011) Batch markovian arrival processes (bmap). Wiley Encyclopedia of Operations Research and Management Science

29. Klemm A, Lindemann C, Lohmann M (2003) Modeling ip traffic using the batch markovian arrival process. Perform Eval 54(2):149–173. Modelling techniques and tools for computer performance evaluation. Available http://www.sciencedirect.com/science/article/pii/S0166531603000671

30. He Q-M, Neuts MF (1998) Markov chains with marked transitions. Stoch Process Appl 74(1):37–52. Available http://www.sciencedirect.com/science/article/pii/S0304414997001099

31. Moon T (1996) The expectation-maximization algorithm. Sig Process Mag IEEE 13(6):47–60

32. Dellaert F (2002) The expectation maximization algorithm. Available http://hdl.handle.net/1853/3281

33. Borman S (2004) The expectation maximization algorithm: a short tutorial. Unpublished paper available at http://www.seanborman.com/publications. Technical report

34. Huhns M, Singh M (1997) Ontologies for agents. IEEE Internet Comput 1(6):81–83

35. Nejdl W, Olmedilla D, Winslett M, Zhang C (2005) Ontology-based policy specification and management. In: Gomez-Perez A, Euzenat J (eds) The semantic web: research and applications, ser LNCS, vol. 3532. Springer, New York, pp 133–144. Available http://dx.doi.org/10.1007/11431053_20. doi:10.1007/11431053_20

36. Church KW, Gale WA (1995) Poisson mixtures. Nat Lang Eng 1:163–190

37. Jacob B, Lanyon-Hogg R, Nadgir DK, Yassin AF (2004) A practical guide to the ibm autonomic computing toolkit. IBM, International Technical Support Organization, Durham

38. Cassandra T (2003–2013) Pomdps: who needs them? Available http://www.pomdp.org/pomdp/talks/who-needs-pomdps/index.shtml

39. Hey JD, Lotito G, Maffioletti A (2010) The descriptive and predictive adequacy of theories of decision making under uncertainty/ambiguity. J Risk Uncertain 41(2):81–111

40. Liang S, Bracha G (1998) Dynamic class loading in the java virtual machine. In: ACM SIGPLAN Notices, vol 33, no 10. ACM, pp 36–44

41. Yildiz M, Toker AC, Sivrikaya F, Camtepe SA, Albayrak S (2012) User facilitated congestion and attack mitigation. In: Mobile networks and management. Springer, Berlin, pp 358–371

42. Yildiz M, Toker AC, Sivrikaya F, Albayrak S (2012) User centric wireless testbed. In: Testbeds and research infrastructure. Development of networks and communities. Springer, Berlin, pp 75–87

43. Pc engines home page. Available http://www.pcengines.ch/

44. Voyage linux home page. Available http://linux.voyage.hk/

45. Iperf. Available http://iperf.sourceforge.net/

46. Botta A, Dainotti A, Pescapé A (2012) A tool for the generation of realistic network workload for emerging networking scenarios. Comput Netw 56:3531–3547

An Autonomous Load Balancing Framework for UCN

Mürsel Yildiz

Abstract In this chapter, we present an autonomous load balancing framework for the user centric networks (UCN), where the end-user participates to the wireless community either with (i) the provider role or (ii) the consumer role. The introduced framework is an improved application of a more comprehensive control and strategy optimization scheme [1]. Our main objective is to provide the UCN access points (AP) with the ability of executing load balancing algorithms and actions with higher success percentages by means of adapting to the changing environmental conditions. The proposed mechanism is a two layer decision-making framework. Firstly the probability distributions are generated over certain indicators for the environmental conditions. We call this step as the characterization stage for the operation environment, which constructs the belief states of the framework. To this end, we make use of Expectation Maximization (EM) algorithm. We introduce a new approach for the utilization of EM algorithm, where we neatly sidestep the necessity of prior assumptions over the distribution for a random parameter. In the second step, the success rate of available load balancing algorithms and actions are monitored based on the percentage of healing critical overloaded conditions. Hence more effective algorithm and environment pairs are constructed. For this objective, we introduce our Partially Observable Markov Decision Process (POMDP) based control loop. We discuss on the necessity of POMDP utilization for the generic control loop in order to meet our goal of matching load balancing algorithms and different environments. This framework is fully implemented using the Java & C programming languages. The solution is validated with real testbed experimentations in order to verify the feasibility of the proposed model.

Keywords Load balancing in wireless LAN · UCN · POMDP · Fourier series · EM algorithm

M. Yildiz (✉)
Technische Universität Berlin, Germany, Berlin, Germany
e-mail: muersel.yildiz@dai-labor.de

A. Aldini and A. Bogliolo (eds.), *User-Centric Networking*,
Lecture Notes in Social Networks, DOI: 10.1007/978-3-319-05218-2_12,
© Springer International Publishing Switzerland 2014

1 Introduction

The last decade has been a true success story, where rapid innovations in hardware and software technologies raise the tendency for a specific novelty, i.e. all-in-one attitude. This approach, in turn, opens the door for the smart device era. The elegant features of smart devices, such as portability, computation power and the high performant network capabilities enables the participation of these devices to the collaboration based applications, which entail the long term availability and low cost interoperability. This fact draws network providers' attention to the utilization of end-user device capacity for network control and optimization related activities. These advances in telecommunications literature prepare a convenient ambiance for the concept of end-user centricity, which brings about the opportunity for the end-user to play a key role in the network.

The evolution of the Internet towards multidisciplinary services and applications improves the significance of end-user, i.e. the client, by means of various business models. In parallel to this progress and in an autogenous manner, the user centricity becomes a comprehensive notion for various networking scenarios and research fields. Alongside the role of the end-user in control or management related decision-making, many proposals focus on interpreting the well-equipped end user device as a key component of the Network. For instance, empowering the end-user as a new network entity with a provider role, is a promising concept in extending the coverage of low-cost Internet access.

The Internet providers and end-users are attracted by the concept of user centric networking (UCN), where the end-user, as an Internet stakeholder, plays the micro network provider role aside from being the consumer [2]. The worldwide availability, efficiency and low-cost for broadband access technologies are the main attractive features of the IEEE Wireless LAN (WLAN) standard, which renders this novelty feasible. The main motivation of participating the end-user to the network infrastructure, not only as a consumer but also as a producer of the content, is to expand the coverage for Internet access by means of wireless fidelity (Wi-Fi) [3]. This social community like fresh approach necessitates improvements in core enabling functionalities of the wireless networks. For instance, it is crucial to achieve a balanced load in the UCN in order to prevent potential congestion regions, which can demotivate members for further cooperations. Load balancing, on the other hand, helps for a homogenous distribution of quality of services (QoS). With the homogenous QoS distribution, we mean the effort of UCN providers to keep a certain performance indicators over an acceptable threshold throughout the UCN. This is due to the motivation of fairness in incentive mechanisms [4].

The load balancing problem is broadly discussed in the network literature as one of the core factor for improving resource allocation. The term, a balanced load, bears various meanings depending on the network dynamics and on the type of resources where the load balancing activity takes place. Hence, the term attains different levels of attraction in different layers or mechanisms. Various studies are conducted accordingly, based on prior knowledge over the operation environment and network resources. The autonomous formation of UCN networks, however, raises the difficulty for the load balancing as the characteristics of the operation environment may not be know beforehand. Additionally the highly stochastic behavior of members and cases for limited resources forms different environments from the load balancing perspective. Therefore, an adaptation capability for the UCN providers is strictly required in order to capture the environmental characteristics and match with suitable load balancing strategies. In this chapter, we present an autonomous load balancing framework for the the ability of executing load balancing algorithms and actions with higher success percentages by means of adapting to the changing environmental conditions. The organization of this chapter is as follows: In Sect. 2 we firstly provide a literature overview on the load balancing in WLAN. This section is a summary for various load balancing strategies based on different environments. The Sect. 3 briefly introduces the UCN and provides relevant issues from the load balancing perspective. The proceeding section, namely the Sect. 4, introduces the proposed autonomous framework and its application to the load balancing problem. The Sect. 5 is devoted for the implementation, experiments and evaluation. Finally we provide a summary of the study and comment on proposed framework in Sect. 6.

2 Related Work

In [5], Wenxiao discusses a network architecture for the load balancing among heterogenous wireless networks. This study proposes grid architecture for the heterogeneous networks and a semi-centralized control unit, which is responsible for load balancing among heterogeneous networks. Fischer et al. [6] propose a distributed load balancing algorithm by means of adaptive channel allocation strategies. They propose usage of simulated annealing, graph coloring, neural networks as the implementation techniques for the channel allocation problem of cellular networks. Mishra, however [7], discusses a client driven channel allocation method for a balanced load among wireless channels. The proposed channel assignment problem is based on the conflict set coloring spectrum. The load balancing aspect in this proposal is related with the client-AP associations and refers to the [8] for the client-AP association model once the channel assignment is met.

One another precaution against unbalanced load is the transmission power control based approaches, which is known as the cell-breathing techniques for the operator networks. In one of the these approaches, the [9] focuses on a technique, which differs from the ordinary coverage shaping in the sense that only the transmission power for the AP beacon messages are dynamically changed. In [10], the solution consists of two steps, firstly the most congested AP is detected and then after in the second phase the transmission power is decreased in discrete steps in order to find out the optimal users assignment. In a similar approach [11], Wang et al. focus on the AP coverage shape. The study [12] points out another load balancing approach using cell breathing techniques and focuses on the traffic types in order to heal congestion prevention by means of QoS mechanism provided in IEEE802.11e standards. Soudani et al. [13]

discusses the signal strength effect for the QoS of the network and criticizes such mechanisms as a self-congesting network by means of retransmissions due to packet losses.

Frame drop rate of real-time sessions in an access point's transmission queues are proposed as the load measurements in [14]. Nevertheless, the backward compatibility of this method is criticized due to the implementation complexity [15]. Similarly, delay time between scheduled and actual transmission time of periodic beacon frames can be a good measure for the load of an AP as proposed by [16]. Having a measure of the load on the AP, it is possible to balance the load among APs in the network through both wireless station (WS)-based solutions or network-based solutions. A possible remedy for potential ping-pong problem is to assign random waiting times and number of measurement instances for each WS before executing the handover [17].

In an association control method based load balancing approaches, Bejerano et al. [8] propose a min–max load-balancing algorithm where the clients take measurements for the channel quality and provide the measurement parameters to a central controller deciding the client-AP associations. Pawelczak et al. [18] focuses on the quantization of the relationship between the random channel access length, system channel utilization and random user blocking probability. In [19] two-phase control strategy is adopted for the load balancing schemes. In the first phase the statics on the QoS guarantee are provided to the stations, whereas in the second phase, dynamic vertical handoffs are performed among the stations to prevent fluctuations in terms of performance. Ning et al. [20] discusses an algorithm, which is claimed to balance the load based on the simulations by means of instructions for the new client arrivals to hand off to the suitable under-loaded network. Finally, in another association control mechanism [21], a joint group call admission control (JGCAC) algorithm is proposed for heterogeneous networks.

AP selection problem is extensively studied in the literature for a balanced load. The [22] focuses on an alpha-optimal load balancing approach, where alpha refers to rate, throughput, delay and load equalizing terms. A distributed user association policy based on the optimal load vector mapping is iteratively applied, which globally adapts the spatial load. Miyata et al. [23], endorse the common AP selection algorithm based on RSSI by arguing *performance anomaly* [24] problem. They propose an algorithm, which imposes the collaborative behavior of the new comers to the WLAN in terms of willingness to move to a network guided appropriate AP. On the contrary Nicholson et al. [25], claim that the AP selection based on RSSI results in not a better throughput and bit-rate in the WLAN in comparison with the selection of APs in a randomized way.

In [26], the delay in the actual beacon frames are assumed to be an indicator for the load of the AP and the contention on the wireless medium, which in turn gives the client the ability to estimate available bandwidth and affiliate for the APs accordingly. Similarly, [27] proposes an AP selection method based on the available bandwidth, which is estimated by passively monitoring the transmission channels and calculating the channel utilization total time in a period. A similar real-time measurement based AP selection technique [28], two different algorithms are proposed for the estimation

Fig. 1 End user
collaborations in UCN

of traffic loads among the APs during camp on time. In this study the load of APs are estimated by observing the delays between probe request and the probe response frames.

3 An Overview of the UCN and Related Issues

As illustrated in Fig. 1, the UCN is a wireless social community, which is formed by means of dynamically changing stakeholders' roles. From time to time a member can become either an AP or a client. The main motivation is to form a dynamic network for bandwidth sharing, i.e. expanding the coverage for Internet access, based on social community dynamics. Various members, with a provider role, acts as a mediator between UNC-clients and the Internet in order to fulfill this motivation. The APs are freely chosen by the UCN-clients in the coverage area. Due to the incentive mechanisms and trust management in a UCN environment, the stakeholders go through a negotiation phase once a client camps on an AP. These negotiations are based on the utilization of network resources, their amount, unit resource costs, etc. The fundamental exchange units as a cost or a revenue may either be based on virtual currencies or trust/reputation metrics. More details on the UCN dynamics can be found in [2–4, 29].

The UCN is a highly stochastic and dynamic telecommunication landscape due to fully uncertain coalescences of various stakeholders. The main focus of the realization of UCN remains in the autonomic deployment at local-loops by means of feasible business models, adequate incentives and various network core functionalities' evolution. The user centric wireless local loop (ULOOP) [3, 30–32] are example large scale projects in the realization of these concepts.

3.1 Challenges and Relevant Issues from the Load Balancing Perspective

Alongside the core enabling functionalities, the system complexity drastically increases in the UCN environment due to the diversity of end-user devices, subscribed to UCN community, in terms of software or hardware capabilities (ref. to Fig. 1). The dynamic environment and diversity of user entities (UE) joining to the

UCN as micro-providers or as clients, are certain factors forming several challenges for UCN. For instance, the frequency and the duration of congestion instances play a great role for the persistency of the instant and future UCN communities. This is due to the fact that the stakeholders would start leaving the community in case of unacceptable service quality. The uncoordinated operation of UCN-APs, combined with growing traffic demands would hurt the users' quality of experience (QoE). Thus a tolerable certain level of service quality should be guaranteed throughout the UCN environment.

A generic definition of congestion or an unbalanced load among the UCN-APs may not be convenient for the UCN. This sophistication stems mainly from the entity Wi-Fi performances and the personal preferences of the stakeholders. Besides, the highly dynamic characteristics raise the complexity in the decision of solution approaches for the load balancing. The very brief certain determinants, which characterize the complexity in load balancing can be listed as:

- Stochastic coalescences of UCN communities in different day-time and the public area, e.g. the load in an Airport may arise from excessive number of client arrivals and departures, whereas in a public social areas the clients may comparably prefer longer session time.
- Different Wi-Fi configuration capabilities among the UCN-APs, e.g., different wireless modules can or can not support various Wi-Fi related configurations, which limits action capabilities for load balancing instances.
- The mobility behavior of UCN stakeholders, i.e. a UCN-AP may operate in different UCN environments with different characteristics.
- Different traffic types depending on the immensely alternating client / community interests or the physical wireless medium characteristics, e.g., the UCN-AP may be operating under heavy interference conditions.
- The fully stochastic deployment/positioning of the UCN-APs in terms of geographical distribution and the number, i.e. a certain number of critically important factors for the signal interference and the load distribution is not preplanned and hence may not be forecasted.

A single strategy solution addressing aforementioned challenges may not be efficient. This is due to the handicap that specific solutions are proposed for pre-assumed environmental conditions. The individual mobility pattern of the UCN-APs doubles the impact of an assumption for highly dynamic environment. Thus an intended tactic may not be sufficient addressing a balanced load in UCN.

4 Proposed Model

In this section, we provide detailed information on our framework. As illustrated in Fig. 2, the main cornerstone of our approach is to maintain distinct control loops [33] for load balancing in different environments. The control loops have the impact over

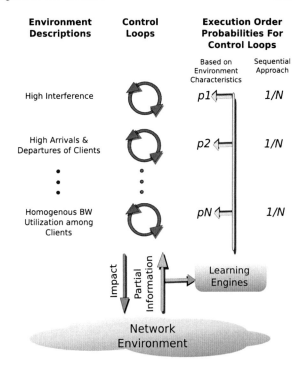

Fig. 2 Prioritization of network objectives based on operation environment

the network environment, i.e. they are the positions where different load balancing algorithms are executed. These control loops monitor the impacts of the executed load balancing strategy and hence evaluates the success rates for future executions. Additionally, they are responsible for collecting partial information from the network environment and forwarding these observations to the learning mechanism proposed in this study. Based on this partial data, the learning engines characterize the environment and assigns believes over the available environment descriptions. Hence in the next iterations, probability distributions over the believes on environment are formed. A random variable is drawn over this probability distribution and the randomly selected control loop is executed. That is to say, the new load balancing strategy is adapted based on the most likelihood on the operation environment and the success rate of the corresponding algorithm.

It should be highlighted that this framework is inspired from a more comprehensive model for autonomous resource allocation and strategy optimization in wireless networks [1]. In this study, however, we improve certain basics of this framework. The details of the proposed model and related framework improvements are provided in the proceeding section.

4.1 A Framework for the Autonomous Resource Allocation and Strategy Optimization (ARASO)

This framework mainly targets immensely dynamic, time varying and less predictable network environments, where the first main motivation is to prevent redundant resource utilization for specific network objectives. By means of preventing the redundant resource utilization, the proposed framework additionally heals the network response time. Secondly, the proposed framework is equipped with additional learning mechanisms in order to match various environmental conditions with different available strategies in terms of strategy impact rates.

In this framework, a gradual composition of objectives are envisioned as independent services to be provided by control entities in the network. An ontology based policy is provided as an input, which describes the entity specific gradual network objectives. These objectives have repositories for various actions and observations. The control entities execute specific control loops for each objective. The definition for a generic control loops in autonomous network literature can be found in [34], which includes the measurement taking, analysis, i.e. reasoning, decision-making and execution steps. In the literature, the very general approach is to maintain these control loops in a sequential manner, where each objective has the same priority. In this framework, on the contrary, a network objective prioritization stage is proposed based on the operation environment characteristics. To this end, each critical instance of a related observation is interpreted as an arrival of service request from the network environment. For each objective a Markov Chain is constructed and the steady-state probability distributions over the chain states are determined by forming objective-specific Marked Markov Arrival Processes (MMAP). Based on determined likelihoods, a common probability distributions for the execution of control loops is constructed. In other words, higher probabilities over critical states in the constructed Markov model is interpreted as an indicator for critical environmental condition and the corresponding objective is hence prioritized.

In this study, however, we improve the object prioritization stage with our new approach for the utilization of EM algorithm. We generate probability distributions over the critical instances of objective observations. As we sidestep the necessity of prior assumptions over these distributions, we generate belief states over the expected number of critical and non-critical observation instances. We briefly comment on the mathematical details of our approach in the following section.

4.1.1 Generation of Believes over the Environmental Conditions

Let the set of observations $\Omega^i = \{ob_1^i, ob_2^i, \ldots, ob_N^i\}$ define the repository of a specific environment $E^i \in E$, where $E = \{E_1, E_1, \ldots, E_M\}$. We are interested in the frequency of critical instances within a specific time window for these observations. In other words, the randomness over these observations are defined as the *number of critical instances counted in a specific window*.

Fig. 3 Example union diagrams of observations. **a** 3 observations **b** 4 observations

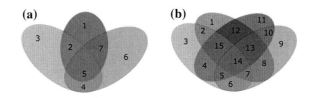

Based on the aforementioned motivation on the random behavior of the observations, the belief on the E^i is clearly an "or" relation among the Ω^i elements. Thus, $P\{E^i\} = P\{ob^i_1 \vee ob^i_2 \vee \cdots \vee ob_N\}$. Without an assumption of independency over these observations, one needs to monitor events of various combinations among specific instances. The exact number of these instances for Ω^i is $2^N - 1$. For a better understanding, we illustrate examples for 3 and 4 observations in Fig. 3.

Considering the independent instances of each observation as an element of an imaginary set, the discovery of these specific instances are simply the *proper subsets* of this imaginary set. Referring to Fig. 3, these independent instances are $\{1, 3, 6\}$ for 3 observations and $\{1, 3, 9, 11\}$ for 4 observations. Therefore we omit providing algorithm flow for finding out specific instances. Obviously, referring these specific instances with a number; the $P\{E^i\} = \sum_{i=1}^{(2^N-1)} P\{(i)\}$. Similarly, the expected critical instances for the event E^i is $\sum_{i=1}^{(2^N-1)} E.v.\{(i)\}$.

Now it is time to shed some light on how we calculate the expected number of critical instances. For a very detailed discussion, the reader is highly recommended to see the reference [35]. Let the random variable $\tilde{i}$ define a specific instance. Let the probability distribution function (pdf) of this random variable has the sample space within $[-L, L]$. We present the pdf of this function with Fourier series expansion as follows:

$$f_{\tilde{i}}(x) = \frac{1}{2L \sum_{i=0}^{\infty}(a_i + b_i)} \sum_{i=0}^{\infty}\left(a_i \cos\left(x\frac{i\pi}{L}\right) + b_i \sin\left(x\frac{i\pi}{L}\right)\right) \quad (1)$$

where the $\{a_i, b_i\}_0^{\infty} \in \mathfrak{R}^+$. We are interested in calculating the expected value, i.e. the $E\{\tilde{i}\}$. The $E\{\tilde{i}\}$:

$$E\{\tilde{i}\} = \int_{-\infty}^{\infty} x f_{\tilde{x}}(x)dx = \int_{-L}^{L} x f_{\tilde{x}}(x)dx$$

$$= \int_{-L}^{L} x \left[\frac{1}{2L \sum_{i=0}^{\infty}(a_i + b_i)} \sum_{i=0}^{\infty}\left(a_i \cos\left(x\frac{i\pi}{L}\right) + b_i \sin\left(x\frac{i\pi}{L}\right)\right)\right] dx \quad (2)$$

$$= \frac{1}{C_L} \sum_{i=0}^{\infty} \int_{-L}^{L} x\left(a_i \cos\left(x\frac{i\pi}{L}\right)\right) dx + \int_{-L}^{L} x b_i \sin\left(x\frac{i\pi}{L}\right) dx$$

where $C_L = 2L \sum_{i=0}^{\infty}(a_i + b_i)$. The Eq. (2) is composed of two integration terms. We make use of the integration by parts.

$$\int_{-L}^{L} x \left(a_i \cos(x \frac{i\pi}{L})\right) dx = \left[uv - \int v du\right] |_{-L}^{L} \tag{3}$$

here the terms u and v are $u = x$, $du = dx$ and $v = \sin(x \frac{i\pi}{L})/(\frac{i\pi}{L})$ respectively. Thus;

$$\int_{-L}^{L} x \left(a_i \cos\left(x \frac{i\pi}{L}\right)\right) dx = \left[\frac{x \sin(x \frac{i\pi}{L})}{(\frac{i\pi}{L})} + \frac{\cos(x \frac{i\pi}{L})}{(\frac{i\pi}{L})}\right] |_{-L}^{L} = 0 \tag{4}$$

With a similar procedure, the second term can be expressed as:

$$\int_{-L}^{L} x \left(b_i \sin(x \frac{i\pi}{L})\right) dx = -L^2 \frac{2\cos(i\pi)}{\pi i} \tag{5}$$

Finally, the expected value is:

$$E\{\tilde{i}\} = \frac{L}{\sum_{i=0}^{\infty}(a_i + b_i)} \left[\sum_{i=0}^{\infty}\left(-b_i \frac{\cos(i\pi)}{\pi i}\right)\right]. \tag{6}$$

Once the expected values for the critical instances of specific environmental indicators are calculated, clearly the belief distribution on the specific environment $B^{E_{(i)}}$ is calculated as follows:

$$B^{E_{(i)}} = \frac{E\{\tilde{i}\}}{\sum_{k=1}^{M} E\{\tilde{k}\}}. \tag{7}$$

Now it is time to find out how we calculate the series coefficients, i.e. the terms a_i's and b_i's. We make use of EM algorithm over our new approach in defining prior pdf form with Fourier series expansion. This approach sidesteps the necessity of an assumption over a specific distribution over the random variable $\tilde{i}$. It should be highlighted that a prior assumption over the distribution characteristic can not hold in partially observable and highly dynamic environments.

The following steps are the EM algorithm steps applied on the previously discussed pdf form. More information on EM details can be found in [35–40]. Given the coefficient set $\Gamma = \{a_i, b_i\}_1^{\infty}$ and the limit L for the sample space, the probability of k critical instances is;

$$P\{\tilde{k} = k | \Gamma\} = \frac{1}{2L \sum_{i=0}^{\infty}(a_i + b_i)} \sum_{i=0}^{\infty} \left(a_i \cos\left(i\pi k/L\right) + b_i \sin\left(i\pi k/L\right)\right) \tag{8}$$

In this model, each sinusoidal component stands for the latent variables and contribute for the future draws. Let these latent variables are represented by the set $Z = \{z_i\}_0^\infty$. For the counter vector $K = < k_1, k_2, \ldots, k_n >$ and given $z_p \in Z$;

$$
\begin{aligned}
& Pr\{z_p \,|\tilde{k} = k_n, \Gamma\} \\
& = \frac{a_p \left(\frac{e^{jp\pi k_n/L} + e^{-jp\pi k_n/L}}{2}\right) + b_p \left(\frac{e^{jp\pi k_n/L} - e^{-jp\pi k_n/L}}{2j}\right)}{\sum_{i=0}^{\infty}(a_i \cos(i\pi k_n/L) + b_n \sin(i\pi k_n/L))} \\
& = T_{p,i}
\end{aligned}
\tag{9}
$$

We make use of Euler's form and rewrite the sine and cosine terms. As an analysis over the cosine and sine terms of each z_p as sub-latent variables as z_p^c and z_p^s respectively;

$$
Pr\{K, z_p^c \,|\Gamma\} = \prod_{i=1}^{n} a_p \left(\frac{e^{jp\pi k_n/L} + e^{-jp\pi k_n/L}}{2}\right)
\tag{10}
$$

Taking the log of Eq. (10);

$$
\log(Pr\{K, z_p \,|\Gamma\}) = \sum_{i=1}^{n} \log \left\{ a_p \left(\frac{e^{jp\pi k_n/L} + e^{-jp\pi k_n/L}}{2}\right) \right\}
\tag{11}
$$

Finally the E-step is:

$$
E_{Z|K;\Gamma} = \sum_{p=0}^{\infty} \sum_{i=1}^{n} T_{p,i} \log \left\{ a_p \left(\frac{e^{jp\pi k_n/L} + e^{-jp\pi k_n/L}}{2}\right) \right\}.
\tag{12}
$$

Using Jensen's Inequality, the higher layer lower boundary for the $E_{Z|K;\Gamma}$ term is constructed as follows, which constitutes the M-step;

$$
\begin{aligned}
E_{Z|K;\Gamma} & \geq \sum_{p=0}^{\infty} \sum_{i=1}^{n} T_{p,i} \log \{a_p\} \left(\frac{e^{jp\pi k_n/L} + e^{-jp\pi k_n/L}}{2}\right) \\
& = \tilde{E}_{Z|K;\Gamma}
\end{aligned}
\tag{13}
$$

The coefficients $\{a_i, b_i\}_1^\infty$ maximizign the lower boundary $\tilde{E}_{Z|K;\Gamma}$ are given in Eq. (13),

Fig. 4 The proposed general
policy for the control loop

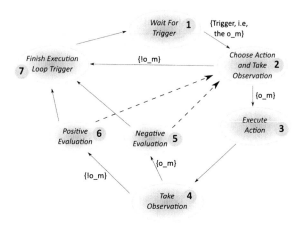

$$\frac{\tilde{\partial}E_{Z|K;\Gamma}}{\partial a_p} = \sum_{i=1}^{n} T_{p,i}\frac{1}{a_p}\cos(p\pi k_i/L)$$

$$\Longrightarrow$$

$$a_p^{(t+1)} = \sum_{i=1}^{n} T_{p,i}\cos(p\pi k_i/L) \tag{14}$$

$$b_p^{(t+1)} = \sum_{i=1}^{n} T_{p,i}\sin(p\pi k_i/L).$$

Next we discuss the POMDP based control policy, as the second stage for healing
load balancing strategies.

4.2 POMDP Based Control Loop

The main motivation of the second stage is to find out more effective algorithms
for different environmental conditions. We provide a general state diagram for this
stage in Fig. 4. As illustrated in this Figure, the second stage waits for a trigger, i.e.
a critical observation instance, to be forwarded from the first stage. A probability
distribution is constructed among the available action/algorithms in network repository, which are associated with the forwarded believes. The constructed probability
distribution is related with the success rate of the corresponding actions/algorithms
and is utilized for the selection of the network strategy. The selected action/algorithm
is executed and additional observations are taken afterwards. Based on the additional
observation, the success rate of the algorithms are updated and the modified control
loop is finished.

Fig. 5 The proposed mechanism position on the IBM's generic control loop [33]

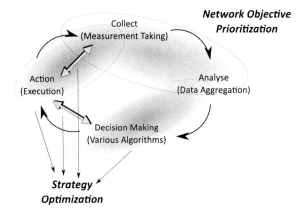

In this model, the generic control loop is slightly modified for the motivation of the proposed framework. The proposed framework works as an aggregation mechanism at different points on the generic control loop [34] as illustrated in Fig. 5. The first functionality, i.e. the prioritization of the network objectives, is an additional step combined with the measurement taking and analysis steps. The strategy optimization stage runs at the decision making and the execution steps. Additionally, one may expect potential supplementary cycles between decision-making, execution and the measurement taking for the further analysis on the executed actions or algorithms.

Next we provide a closer look over the state diagram illustrated in Fig. 4. We shed some light on the dynamic behavior of this control loop and briefly give the motivation for the POMDP formulation.

4.2.1 Motivation for POMDP Formulation

Referring to the Fig. 4 and at the first glance, the detailed activity loop gives the impression of a sequential execution of a control loop. Nevertheless, the transitions from one state to another may not take place with a high certainty as expected. This sophistication may stem from the environmental characteristics, false triggers, complications due to the sensibility to recency for observations or actions, the effectiveness of the selected action, etc. In other words the constructed state transitions appear based on dynamically changing transition probabilities. From state 1 to 2, the transition occurs with a probability of expected appearance rate of the trigger, i.e. critical, observation, which is determined in the first stage, i.e. network objective prioritization step. Similarly, from state 2 to 3 or 2 to 7, the transition probability is directly proportional to the same appearance rate. State 3 and 4 are illustrated separately in order to highlight additional step for observation taking. That is to say, these states may be combined together, i.e. the probability of transition between state 3 and 4 is simply 1. The objective point, i.e. the evaluation of the chosen algorithm, is practiced depending on the transition from state 4 to either 6 or 5. This is bounded up to the impact rate of the algorithm. Therefore the transition probability is directly related with the success rate of the chosen algorithm. Finally, the transitions either from states 6 or 5 to 7 or 2 depends on the trustworthiness of the taken observation. With the trustworthiness we refer a design parameter as an input to the proposed framework. This parameter is related with the degree of capability for the chosen observation in terms of (i) recency related problems, (ii) certainty in describing the network environment, (iii) the frequency of creating potentially false triggers. We strongly recommend reader to have a look at an example discussion of trustworthiness of observations in [41].

The transition parameters change dynamically and are defined with the framework believes on the appearance rates or the success rates of the observations and actions/algorithms respectively. Nevertheless, an instant shots of the constructed model holds memoryless property as as the future state transitions depend only on the current state of the loop. In other words, the proposed policy represents a Markov Chain with certain instant specifications. The process can be modeled as a tuple of (i) states, (ii) transition probabilities, (iii) actions and finally (iv) rewards (which increase the dominance of executions at a specific states in Fig. 4). Thus, one may argue that this process can be modeled as a Markov Decision Process (MDP), where there is a great uncertainty on the transition among model states based on the executed action. In this model, due to the afore-mentioned trustworthiness of the observations as design parameters, a complete knowledge on the current world states may not be available. This feature is the cornerstone for the MDP models. In a partially observable MDP formulation, however, there is additional uncertainty on the current state of the world. In this model, the additional uncertainty is represented with the trustworthiness of the observations. Thus the optimal policy solution for this control loop is determined dynamically with POMDP in this framework.

Next we provide brief information on the POMDP formulation. The detailed formulation can be found in [1].

4.2.2 POMDP Formulation

In this part, we provide details on the POMDP model formulation for the proposed control loop in Fig. 4. The parameter set for the control loop is provided in Table 1. More detailed information is provided in [1].

A critical instance for the observation $(o_m^{(i)} \in Ob^{(i)})$ creates a trigger for the execution of the corresponding control loop, which is referred as o_m. Hence, the state transition from first state to the second clearly depends on expected critical instances of E^i. Thus, $t_{12}^{(ij)} = E \cdot v \cdot \{E^i\} = \sum_{i=1}^{(2^N-1)} E \cdot v \cdot \{(i)\}$. At state $S_2^{(i)}$, the same observation is taken, hence $t_{23}^{(i)} = t_{12}^{(i)}$ and $t_{27}^{(i)} = 1 - t_{23}^{(i)}$. It should be underlined that the $S_3^{(ij)}$ and $S_4^{(ij)}$ are separated in order to highlight the necessary time duration for the network stabilization. This time duration is highly dependent on the executed algorithm. Hence the $t_{34}^{(i)} = 1$. For the algorithm selection at $S_2^{(ij)}$, assuming that

Table 1 The parameter set for the control loop

parameter	Description
$A^{(i)}$	Algorithm set of the environment E^i with a size of K
$Ob^{(i)}$	Observation set for of the environment E^i with a size of M
$a_k^{(i)}$	The kth action in $A^{(i)}$ ($a_k^{(i)} \in A^{(i)}$)
$o_m^{(i)}$	The mth observation in $Ob^{(i)}$ ($o_m^{(i)} \in Ob^{(i)}$)
$tr_m^{(i)}$	The trustworthiness of $o_m^{(i)}$
$\rho_k^{(i)}$	Success rate of the algorithm $a_k^{(ij)}$
$\gamma_m^{(i)}$	Expected rate parameter of observation $o_m^{(i)}$
$S^{(i)}$	Control loop state set (ref. to Fig. 1)
$t_{JJ}^{(i)}$	The state transition probability from $S_I^{(i)}$ to $S_J^{(i)}$
$o\_m^{(i)}$	Critical instance of $o_m^{(i)}$ observation
$!o\_m^{(i)}$	non-Critical instance of $o_m^{(i)}$ observation
$a\_k^{(i)}$	The success instance of $a_k^{(i)}$
$!a\_k^{(i)}$	The failure instance of $a_k^{(i)}$

the algorithm $a_k^{(i)}$ is selected, the success indicator i.e the $\rho_k^{(i)}$ plays a great role for the selection and the future state transitions. This is due to the characteristics of the proposed control loop for the transitions from $S_4^{(i)}$ to either $S_5^{(i)}$ or $S_6^{(i)}$. Therefore;

$$
\begin{aligned}
t_{46}^{(i)} &= \sum_{(a_k^{(i)} \in A^{(i)})} P\left\{a_k^{(i)} \text{ is selected, } a_k^{(i)} \text{ is successful}\right\} \\
&= \sum_{(a_k^{(i)} \in A^{(i)})} P\left\{a\_k^{(i)}|a_k^{(i)}\right\} P\left\{a_k^{(i)}\right\} \\
&= \sum_{k=1}^{K} \left(\frac{\rho_k^{(i)}}{\sum_{k=1}^{K} \rho_k^{(i)}}\right) \rho_k^{(i)} \\
t_{45}^{(i)} &= 1 - t_{46}^{(i)}.
\end{aligned}
\tag{15}
$$

The main motivation of the proposed control loop is to evaluate the executed algorithms is an improved manner. The effectiveness of the selected algorithm and the dynamic and stochastic environmental factors, however, play a great role in the evaluation. This is due to the incomplete information due to the ambiguity for the trustiness of the observation. We highly recommend the reader to review the very useful discussion on trustworthiness in [41]. The trustworthiness is a design parameter in this framework as it highly depends on the selected observation type. The state transitions from $S_5^{(i)}$ to $S_2^{(i)}$, $S_5^{(i)}$ to $S_7^{(i)}$, $S_6^{(i)}$ to $S_2^{(i)}$ and $S_6^{(i)}$ to $S_7^{(ij)}$ is directly proportional to this design parameter. Hence;

$$
\begin{aligned}
t_{62}^{(i)} &= t_{52}^{(i)} = 1 - \frac{1}{M}\sum_{m=1}^{M} tr_m^{(i)} \\
t_{67}^{(i)} &= t_{57}^{(i)} = \frac{1}{M}\sum_{m=1}^{M} tr_m^{(i)}.
\end{aligned}
\tag{16}
$$

Based on the discussion above, the transition parameters are dynamic depending on the design parameters, environmental conditions and the selected observation and algorithm repositories. Focusing on new iteration of the control loop, however, the state transitions hold memoryless property. In other words the future transitions depend only on the current state in the loop. The loop forms a Markov Chain with the previously discussed settings. On another critical parameter, the trustworthiness, is the main tackle leading to the incomplete information on the current state in the loop. Hence, we solve the problem using POMDP rather than MDP. The POMDP model is represented with the form of $< S^{(i)}, A^{(i)}, T^{(i)}(s, a, s'), R^{(i)}(s, a), O(s', a, o) >$. The $R^{(i)}$ stands for the reward definition for each $a_k^{(i)}$ at the belief state s. A POMDP framework is represented with the form of $< S^{(ij)}, A^{(ij)}, T^{(ij)}(s, a, s'), R^{(ij)} (s, a) >$, where in addition to the previously discussed parameters, the $R^{(ij)}$ is defined as the reward for the action $a_k^{(ij)}$ given that the state is s. The reward concept is the utility definition that the agent may except. In an MDP formulation, although there is a great uncertainty for the transitions of the states based on the executed action, the control agents are provided with a complete information on the current world states. In this model, the $O(s_l'^{(i)}, a_k^{(i)}, o_m^{(i)})$ is either $\frac{1}{M} \sum_{m=1}^{M} tr_m^{(i)}$ or $1 - \frac{1}{M} \sum_{m=1}^{M} tr_m^{(i)}$. This is due to the dependency on the criticality of the observation instances. For a specific environment, the next state estimation $b_{l+1}^{(i)}$ at time step l depends on three parameters, namely, the current belief state $b_l^{(i)}$, the taken action $a_l^{(i)}$ and finally observation $o_l^{(i)}$.

$$b^{(i)}(s_{l+1}^{(i)}) = P\left\{s_{l+1}^{(i)} | o_k^{(i)}, a_k^{(i)}, b^{(i)}\left(s_l^{(i)}\right)\right\}$$
$$\sim O\left(s_{l+1}^{(i)}, a_l^{(i)}, o_l^{(i)}\right) \sum_{s^{(i)} \in S^{(i)}} T^{(i)}\left(s_l^{(i)}, a_l^{(i)}, s_{l+1}^{(i)}\right) b^{(i)}\left(s_l^{(i)}\right) \quad (17)$$

Finally considering an l-step non stationary policy A and a k-step move of the agent, the expected total sum of the rewards gained is represented as follows:

$$V_k(s) = R(s, A(s))$$
$$+ \gamma \sum_{s' \in S} T\left(s, a_k, s'\right) \sum_{o_k \in O} O\left(s', A\left(s'\right), o_k\right) V_{o_k}\left(s'\right) \quad (18)$$

The expected optimal policy is calculated over the belief-states, where the value functions are given by $V_k(s) = \sum_{s \in S} b(s) V_k(s)$. Referring to Kaelbling et al. [42], who provides a more convenient method in representing the value functions, the $\alpha_k = < V_k(s_1), ..., V_k(s_n) >$, i.e. the alpha vector, represents the value function distribution among possible states and $V_{k+1}(b) = \text{argmax}_{k \in p} b.\alpha_k$. It should be highlighted that the iterative calculation of the presented POMDP model requires high computation power. Therefore, we relax the updates over POMDP policy by formerly setting threshold values for certain model parameters. The POMDP models are updated once the difference over the model parameters are over the threshold.

4.2.3 Framework Restrictions and Requirements

The following items are a summary of the restrictions and the required inputs for the proposed framework.

- a network ontology model for the graded network objectives
- administration control parameter over the network prioritization, which represents the degree of administration believes on the network operation environment
- information models for the action and observations including definitions for corresponding network objectives
- the threshold parameter for the sensibility of framework to the changing control loop parameters to construct new POMDP based control policies
- the trustworthiness parameter for each observation, discussed

This framework is implemented using JAVA, where dynamic class loading property is broadly utilized [43]. In addition to the sequential type of actions, we include the long-term algorithms into network action repository by introducing an additional parameter for the repetitive activities (loops) of the algorithms. The algorithms are slept for a dynamically alternating duration, which is determined by the positive and the negative evaluation steps of the proposed POMDP based control loop. In this wise, the algorithms and action coalescences are formed with various dominance levels and percentages based on the changing environmental conditions. Next, we describe how this autonomous framework dynamics are adjusted focusing on a single problem, i.e. the load balancing problem in UCN.

4.3 Application of ARASO Framework for the Load Balancing Problem in UCN

The very generic functionalities and components of the introduced framework are replaced with the corresponding load balancing problem conjugates. Referring to the Sect. 4.2, we describe the load balancing policy of the UCN-AP in a graphical and ontological form. We replace the objectives of the network with the environmental setups. Although the framework has no restriction in the amount of environment descriptions, we start with a three definition as the use case. These states are, namely, *high interference*, *higher client arrival and departure rates* and *homogenous bandwidth utilization among clients*. With the *homogenous* term, we only focus on the similarities among network clients in terms of the amount of uploaded and downloaded bytes per time without going into details, i.e. the network activities or the traffic categories. It should be highlighted that these environmental states are not necessarily complementary or independent. In other words, the combinations of these environmental conditions may arise.

As illustrated in Fig. 6, the "Load Balancing" concept is analyzed in three different environmental states. It should be highlighted that the description of environment can be diversified with additional observation and action sets. The observation sets

Fig. 6 Load balancing policy
in ontological form

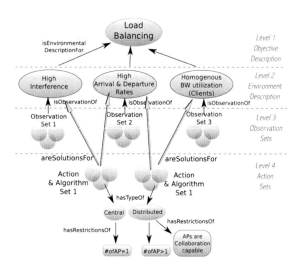

provide information on how likely the environment may be described. Finally, the
action and algorithm sets form the strategy pool as the steps which are likely to be
taken against unbalanced load and which are categorized in the second stage. More
details on the observation and action and algorithm sets are given in the proceeding
sections.

4.3.1 The Observation Set

The observation parameters are kept relatively simple, which can easily be replaced
with additional ones based on extensive researches. The contribution of our study
lies mainly on a skeleton with a generic framework tendency. When it comes to
shedding some light on the the critical parameters as triggers for load balancing;
each algorithm and action are allowed to include also their specific critical parameter
descriptions. In other words, each execution of a specific algorithm or an action for
the load balancing can require information on the specific critical parameters. These
parameters are individual triggers specific to the executed algorithm or the action.
For our use case, (i) the associations and de-association rates of the clients are used in
describing *higher client arrival and departure rates*, (ii) the experienced signal-to-
interference ration (SIR), collected from the network clients are used in describing the
high interference and finally (iii) the bandwidth (BW) utilization of UCN-clients are
monitored for the description of *homogenous bandwidth utilization among clients*.

4.3.2 The Action and Algorithm Set

One of the very first main motivations of the UCN-APs is to guarantee an acceptable
threshold of the QoS for the UCN persistency. Although the load balancing philosophy in UCN necessitates a fair resource sharing among UCN-clients, this may not be

applicable as the distribution and the amount of UCN resources are fully stochastic due to the stakeholders' preferences. In cases of insufficient resources, the system is relaxed with the assumption of potential client off-loads to different telecommunication landscapes or access technologies. Hence the proposed solutions for different dynamics of load balancing problem in UCN may focus on an acceptable QoS degree for the UCN-clients.

In this framework, we make use of three different approaches as action and algorithms. The first approach is a cell-breathing like *transmission power control* technique, where the transmission power of the AP is decreased for a predefined amount of time and set back to its initial value. One may argue for a more enhanced transmission power control approach by means of the introduction of two additional parameters to the control algorithm, namely, (i) the degree to what extend the transmission power is lowered, and (ii) the duration of network operation with a specific transmission power value of the AP. Nevertheless we keep these parameters relatively simple and static as the main focus of this study lies on the autonomous behavior.

The second approach is a *cooperative load balancing* technique based on the UCN dynamics, e.g. incentive mechanisms, reputation parameters, etc. The load is balanced in a collaborative manner among UCN-APs by means of agreements and dynamically migrating resident clients. The proposed algorithm is distributed and hence scalable. For this activity the UCN-APs take two different roles, i.e. either requester or a requestee role. In this model a QoS based crediting mechanism generates sufficient conditions for the motivation of UCN-APs to balance the load in a cooperative way. A utility based cooperative decision is taken among UCN-APs. The utilities of the UCN-APs depend mainly on the unit resource cost, which is calculated based on the performance indicators of the sub-network. The agreement is driven in a pairwise manner over the number of clients to be migrated and the incentives to be exchanged. The mathematical and implementation details of cooperative load balancing algorithm can be found in [4].

The final algorithm is an *admission control* approach having same backbone with the afore-mentioned cooperative load balancing technique. This time the agreement is performed for the newly arriving clients instead of dynamic migrations. We slightly modify the collaboration request parameters and replace the *number of migrated clients* and *incentives* with a single *admission control collaboration* boolean.

4.3.3 The Algorithm Evaluation

It should be highlighted that the algorithms are evaluated based on the previously described load balancing perspective. In other words, the main evaluation parameter is the efficiency in terms of the algorithm impact on keeping the network performance related measurements over an acceptable threshold. One may argue that the UCN dynamics, such as reputation / trust parameters, monetization or the virtual currency are the effective parameters in describing an optimality condition for each network control related activity. Nevertheless, a generic formulation for the real time calculation of the utilities may not be applicable for the UCN due to potential conflicts with proposed algorithms. Moreover, the targeted stabilization time period may not be easily synchronized with a potential generic formulation. Hence we leave the optimality concern to the future approaches, which can easily be adapted to the action and algorithm repository of the model.

5 Evaluation of the Framework

We evaluate the behavior of the proposed framework by means of real implementation and test bed verifications. We partially make use of the user centric wireless testbed, details of which can be found in [44]. The test bed is mainly composed of two wireless access points, a central router, application servers, a control and monitoring computer and finally a server as the repository for network ontology, POMDP control loop policy and action / observation functionality classes. Two Alix boards of PC Engines [45] are configured as the wireless access points, running Voyage Linux [46] as the operating system. Linux PCs are used as well-aimed clients and as traffic generators. Iperf tool [47] is used for the traffic generation during the experiment in order to congest the APs once needed. A client network manager (CNM) software is written additionally, which provides the client devices the collaboration capability with the APs. The CNM is composed of a messaging mechanism with APs, AP connection builder and network performance tracking blocks. Distributed internet traffic generator (DITG) [48] is additionally used for the network performance tests. The client devices are capable of performing a handover to the other APs once they receive the corresponding message from their resident AP.

We patch the Hosatpd [49] in compliance with the admission control based load balancing algorithm and also for the client departures / arrivals observations. For the sake of generating different environmental conditions, we emulate from time to time the client arrivals and departures on different clients by means of additional software block, which simply enables and disables the wireless connectivity. We also connect long-term bandwidth hungry clients in order to emulate the heterogeneity as we believe that the homogenous bandwidth utilization among clients may not hold in a realistic network scenario. Finally we generate heavy traffic on the APs from time to time with the additional network traffic generators in order to create interference on the clients. With this motivation we also make use of traffic shapers on the APs by means of dynamically changing the available bandwidth from time to time. As the impact of each action item depends on the environmental conditions and hence the rate for which the network stabilizes itself, we keep the emulated environmental conditions for longer time periods in order to achieve reliable outcomes.

As illustrated in Fig. 7, the beliefs on, and hence the priorities of, the environmental conditions are changed from time to time. We start with an environment, where the bandwidth utilization among UCN-clients are heterogenous and continue with this setting till the end of the experiment. For the boot-up settings, the arrival and departure rates of the UCN-clients is not critical and we emulate low interferences among the UCN environment, where the clients are not far away from the APs and hence the

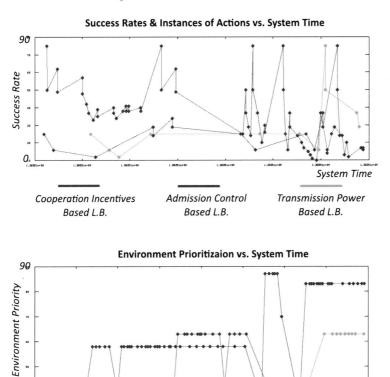

Fig. 7 *Experiment results* Impact rates of various algorithms in a very dynamic UCN environment

RSSI values are well enough. We start congesting the one of the APs with a long-term UCN-client and observe the network actions accordingly. As can be inferred from the Fig. 7, the proposed *cooperation incentives* based load balancing algorithm dominates, receives a high success rate and hence is selected frequently as the strategy by the ARASO software. The loop period of this algorithm is decreased accordingly for frequent cycles. We conclude that the given environmental conditions is cut for the cooperation incentive based load balancing algorithm. For the proceeding phases of the experiment, the environmental conditions are changed stepwise and the arrival / departure rates of the UCN-clients are increased over a critical threshold. The ARASO framework catches the changing environmental conditions and increases the priority of the corresponding environment label as shown with the blue lines. Clearly, the strategies of the UCN-AP are adapted accordingly; the *admission control* and *transmission power control* based algorithms are taken more in comparison to the previous phase, while the success rate of the *cooperation incentives* based load balancing algorithm has less impact rate. This is due to the fact that the second UCN-AP is congested from time to time and becomes non collaborative.

Keeping the second phase environmental conditions for a longer term, we observe that the impact rates of the proposed algorithms get less. Nevertheless, this stems from the characteristics of the environment due to high arrival and departure rates, as the system is congested from time to time and in a very dynamic way, which was due to lack of enough resources, i.e. available UCN-APs. Although it is observed that the proposed framework reacts rapidly to the changing conditions and forces an adaptation to the environment, the availability of resources among UCN environment becomes significantly important. At the final stage of the experiment, where the interference is added to the present settings, this fact plays a great role as various strategies become less effective in extreme environmental conditions. For the higher interference environments, due to additional packet losses and continuous retransmissions, the bandwidth utilization of the AP increased although the throughout and the QoS measurements get worse. It should be highlighted that this increase in the amount of additional bandwidth is slightly perceivable. Nevertheless the system is maintained in a way that the environmental settings include high interferences as we decrease the threshold values. Under these circumstances, the transmission power control based load balancing approach is appraised for low throughout the experiment and clearly results in a degradation of the QoS measurements. We believe that this property is very dependent on the geographical distributions of the clients as well.

6 Discussion and Summary

In this chapter, we introduce a multi-purpose autonomous control scheme for the load balancing problem in UCN. The proposed framework is fully implemented and verified by test-bed experiments. The proposed control software provides a helpful analysis on the focused problem. Interpreting the provided analysis, we conclude the following take home notes for the load balancing problem in UCN:

- A *cooperation incentives* based approach, which is based on the migration of specific clients from the loaded AP to another available one, is an effective solution for the UCN environments, where the clients are quasi static (mobile or static under the coverage of AP pools), the arrival and departure rates are low enough and finally the interference is less. The algorithm may provide significant healing for the unbalanced load among UCN-APs in a fair manner under the assumption that there exists at least two available UCN-APs, i.e. there are enough network resources.
- An *admission control* based approach is highly dependent on the environmental conditions, the traffic behavior of the UCN-clients and the availability of the suitable resources. This approach may play a great role for the load balancing problem

in UCN under the environmental conditions, where the bandwidth utilization of the UCN-clients are homogenous, the arrival and departure rates are high enough.

- The *transmission power control* based approach has less impact rate in comparison to the previously discussed approaches and may even result in critical network performance degradation depending on physical layer characteristics and the geographical distribution of the network clients.
- As a final item, we conclude that the load balancing problem in a highly stochastic and dynamic UCN environment becomes more critical as the decision complexity increases with the behavior of the stakeholders and the lack of pre-organization in UCN. Although various strategies can be adapted rapidly to the changing environmental conditions, this problem may still arise due to various settings such as the shortage of available and collaborative UCN-APs, high interferences and etc.

Acknowledgments The research leading to these results has received funding from the EU IST Seventh Framework Programme ([FP7/2007–2013]) under grant agreement number 257418, project ULOOP User-centric Wireless Local Loop.

References

1. Yildiz M, Khan MA (2014) "Autonomous resource allocation and strategy optimization in wireless networks". In: Aldini A, Bogliolo A (eds) User-centric networking - future perspectives, (LNSN) Springer
2. Sofia R, Mendes P (2008) "User-provided networks: consumer as provider". IEEE Commun Mag 46(12):86–91
3. "European commissions seventh framework programme (fp7) project uloop. grant agreement no. 257418", (2010–2013) [Online]. Available: http://www.uloop.eu/
4. Yildiz M, Khan MA, Sivrikaya F, Albayrak S (2012) "Cooperation incentives based load balancing in UCN: a probabilistic approach", in Globecom 2012 - Next generation networking and internet symposium (GC12 NGNI). Ryerson University, Anaheim
5. Shi W, Li B, Li N, Xia C (2011) "A network architecture for load balancing of heterogeneous wireless networks". J Netw 6(4) [Online]. Available: http://ojs.academypublisher.com/index.php/jnw/article/view/0604623630
6. Fischer S, Petrova M, Mahonen P, Vocking B (2007) "Distributed load balancing algorithm for adaptive channel allocation for cognitive radios". In: Cognitive radio oriented wireless networks and communications. CrownCom 2007. 2nd international conference p 508–513
7. Mishra A, Brik V, Banerjee S, Srinivasan A, Arbaugh W (2006) "A client-driven approach for channel management in wireless lans". In: INFOCOM. 25th IEEE International conference on computer communications. Proceedings INFOCOM, Barcelona p 1–12
8. Bejerano Y, Han SJ, Li EL, (2004) "Fairness and load balancing in wireless lans using association control". In Proceedings of the 10th annual international conference on mobile computing and networking, ser. MobiCom '04. ACM, New York, p 315–329. [Online]. Available: http://doi.acm.org/10.1145/1023720.1023751
9. Bejerano Y, Han S-J (2009) Cell breathing techniques for load balancing in wireless lans. IEEE Trans Mob Comput 8(6):735–749
10. Haidar M, Akl R, Al-Rizzo H, Chan Y, Adada R, (2007) "Optimal load distribution in large scale wlan networks utilizing a power management algorithm". In: Sarnoff symposium, IEEE p 1–5

11. Wang Y, Cuthbert L, Bigham J (2004) "Intelligent radio resource management for ieee 802.11 wlan". In: Wireless communications and networking conference, vol 3. WCNC IEEE, Atlanta, p 1365–1370
12. Brickley O, Rea S, Pesch D, (2008) "Load balancing for qos enhancement in ieee802.11e wlans using cell breathing techniques". [Online]. Available: http://citeseerx.ist.psu.edu/viewdoc/summary?doi=10.1.1.128.4550
13. Soudani A, Divoux T, Tourki R, (2012) "Data traffic load balancing and qos in ieee 802.11 network: experimental study of the signal strength effect". Comput Electr Eng. 38(6):1717–1730 [Online]. Available: http://www.sciencedirect.com/science/article/pii/S0045790612001413
14. Brickley O, Rea S, Pesch D, (2005) "Load balancing for QoS enhancement in IEEE802. In: 11e WLANs using cell breathing techniques". IFIP MWCN, Laxenburg
15. Yen L, Yeh T, Chi K (2009) Load balancing in IEEE 802.11 networks. IEEE Internet Comput. 13(1):56–64
16. Vasudevan S, Papagiannaki K, Diot C, Kurose J, Towsley D (2005) "Facilitating access point selection in IEEE 802.11 wireless networks", In: ACM SIGCOMM, p 26
17. Yen L, Yeh T, (2006) "SNMP-based approach to load distribution in IEEE 802.11 networks". In: IEEE VTC, vol 3 p 1196–1200
18. Pawelczak P, Joshi S, Addepalli S, Villasenor J, Cabric D (2013) Impact of connection admission process on the direct retry load balancing algorithm in cellular networks. IEEE Trans Mob Comput 12(9):1681–1696
19. Song W, Zhuang W, Cheng Y (2007) "Load balancing for cellular/wlan integrated networks". Netw IEEE, vol 21(1):27–33
20. Ning G. Zhu G, Peng L, Lu X (2005) "Load balancing based on traffic selection in heterogeneous overlapping cellular networks". In: The first IEEE and IFIP international conference in central asia on internet, Cambridge University Press, Cambridge
21. Lian R, Tian H, Fei W, Miao J, Wang C (2012) "Qos-aware load balancing algorithm for joint group call admission control in heterogeneous networks". In: IEEE 75th Vehicular technology conference VTC Spring, Yokohama p 1–5
22. Kim H, Veciana G De, Yang X, Venkatachalam M (2012) "Distributed α-optimal user association and cell load balancing in wireless networks". IEEE/ACM Trans Netw. 20(1):177–190 [Online]. Available: http://dl.acm.org/citation.cfm?id=2182751.2182765
23. Miyata S, Murase T, Yamaoka K (2012) "Characteristic analysis of an access-point selection for user throughput and system optimization based on user cooperative moving". In: Consumer communications and networking conference (CCNC) IEEE, p 931–935
24. Heusse M, Rousseau F, Berger-Sabbatel G, Duda A (2003) "Performance anomaly of 802.11b", in INFOCOM. Twenty-second annual joint conference of the IEEE computer and communications. IEEE Soc. 2:836–843
25. Nicholson AJ, Chawathe Y, Chen MY, Noble BD, Wetherall D, (2006) "Improved access point selection". In: Proceedings of the 4th international conference on mobile systems, applications and services, ser. MobiSys '06. ACM, New York p 233–245. [Online]. Available: http://doi.acm.org/10.1145/1134680.1134705
26. Vasudevan S, Papagiannaki K, Diot C, Kurose J, Towsley D (2005) "Facilitating access point selection in ieee 802.11 wireless networks". In: Proceedings of the 5th ACM SIGCOMM conference on internet measurement, ser. IMC '05. Berkeley USENIX association, California p26–26. [Online]. Available: http://dl.acm.org/citation.cfm?id=1251086.1251112
27. Gupta D, Mohapatra P, Chuah CN (2011) "Seeker: A bandwidth-based association control framework for wireless mesh networks". Wirel Netw. 17(5):1287–1304. [Online]. Available: http://dx.doi.org/10.1007/s11276-011-0349-4
28. Chen J-C, Chen T-C, Zhang T, van den Berg E (2006) "Wlc19-4: Effective ap selection and load balancing in ieee 802.11 wireless lans". In: Global telecommunications conference GLOBECOM '06, p 1–6
29. Bogliolo A, Polidori P, Aldini A, Moreira W, Mendes P, Yildiz M, Ballester C, Seigneur J-M (2012) "Virtual currency and reputation-based cooperation incentives in user-centric networks". In: Wireless networking symposium, IWCMC2012-Wireless Nets, Limassol

30. "Openspark community - better than free". [Online]. Available: https://open.sparknet.fi/index. php
31. "Whisher community - building the world's largest wifi network". [Online]. Available: http:// www.whisher.com
32. "Fon community". [Online]. Available: http://www.fon.com
33. IBM, "Understand the autonomic manager concept" (2004). [Online]. Available: http://www. ibm.com/developerworks/library/ac-amconcept/
34. Jacob B, Lanyon-Hogg R, Nadgir DK, Yassin AF (2004) "A practical guide to the ibm autonomic computing toolkit', IBM
35. Yildiz M Khan MA (2013) "Ap selection in ucn: a network performance history based approach", Submitted to IEEE/ACM Transac Netw
36. Moon T (1996) The expectation-maximization algorithm. IEEE Sig Process Mag. 13(6):47–60
37. Bailey T, Elkan C et al (1994) Fitting a mixture model by expectation maximization to discover motifs in bipolymers. Department of Computer Science and Engineering, University of California, San Diego, p 28-36
38. Dellaert F 2002) "The expectation maximization algorithm". [Online]. Available: http://hdl. handle.net/1853/3281
39. Borman S (2004) "The expectation maximization algorithm: a short tutorial. unpublished paper available at http://www.seanborman.com/publications" Technical Report 2004
40. Moon T (Nov 1996) The expectation-maximization algorithm. IEEE Sig Process Mag 13(6):47–60
41. Cassandra T "Pomdps: Who needs them?" [Online]. Available: http://www.pomdp.org/pomdp/ talks/who-needs-pomdps/index.shtml
42. Kaelbling LP, Littman ML, Cassandra AR (1998) Planning and acting in partially observable stochastic domains. Artif Intell. 101:99–134
43. Liang S Bracha G (1998) "Dynamic class loading in the java virtual machine". In: ACM SIGPLAN Notices, vol 33(10) ACM, p 36–44
44. Yildiz M, Toker AC, Sivrikaya F, Albayrak S (2012) "User centric wireless testbed". In: Testbeds and research infrastructure. Development of networks and communities, vol 90 Springer, Dordrecht, p 75–87
45. "Pc engines home page". [Online]. Available: http://www.pcengines.ch/
46. "Voyage linux home page". [Online]. Available: http://linux.voyage.hk/
47. "Iperf". [Online]. Available. http://iperf.sourceforge.net/
48. Botta A, Dainotti A, Pescapè A (2007) Multi-protocol and multi-platform traffic generation and measurement. IEEE INFOCOM, Anchorage p 1–5
49. "Hostapd: Ieee 802.11 ap, ieee 802.1x/wpa/wpa2/eap/radius authenticator". [Online]. Available: http://hostap.epitest.fi/hostapd/

Part IV
Mobility in User-Centric Environments

Mobility Support in User-Centric Networks

Fikret Sivrikaya, Stefano Salsano, Marco Bonola and Marco Trenca

Abstract In this paper, an overview of challenges and requirements for mobility management in user-centric networks is given, and a new distributed and dynamic per-application mobility management solution is presented. After a brief summary of generic mobility management concepts, existing approaches from the distributed and peer-to-peer mobility management literature are introduced, along with their applicability or shortcomings in the UCN environment. Possible approaches to deal with the decentralized and highly dynamic nature of UCNs are also provided with a discussion and an introduction to potential future work.

Keywords User-centric networks · DMM · Distributed mobility management · Dynamic mobility management · Decentralized mobility management · Peer-to-peer

1 Introduction

Today's mobile and wireless infrastructure networks depend on highly reliable network elements connected together with high-quality links to provide global broadband connectivity. Although this architectural approach to building networks has been very successful as manifested by the billions of connected devices, it nevertheless has its drawbacks. CAPEX and OPEX, for example, are rapidly increasing while complexity in operation and management hinder the introduction of novel features. An alternative, unconventional approach to today's mainstream telecommunication standards is to adopt the user-centric networking (UCN) model, which exploits the

F. Sivrikaya (✉)
DAI-Labor, Technische Universität Berlin, Berlin, Germany
e-mail: Fikret.Sivrikaya@dai-labor.de

S. Salsano · M. Bonola · M. Trenca
Dip. Ing. Elettronica, Università di Roma "Tor Vergata", Rome, Italy

A. Aldini and A. Bogliolo (eds.), *User-Centric Networking*,
Lecture Notes in Social Networks, DOI: 10.1007/978-3-319-05218-2_13,
© Springer International Publishing Switzerland 2014

increasing expansion of wireless access networks in order to deploy autonomic and self-organizing wireless community networks.

An empowered Internet end-user lies at the center of UCN and assists in expanding current network operation to the fringes of Internet through several portable and networked devices. UCNs represent a disruption in established Internet communication models in several ways. First, any regular end-user device may behave as a supplier of Internet connectivity and services, and consequently become part of the network. In contrast the "end-to-end" principle, one of the architectural foundation of the Internet, describes a clear splitting between network and end-user systems. Second, UCNs grow spontaneously based on the willingness of users to share subscribed Internet access. Thirdly, connectivity is expected to be intermittent given that UCNs are spontaneously deployed [1].

Mobility support is just beginning to emerge as a topic of research interest in the context of user-centric networking. This paper surveys current literature for architectural designs, protocol elements, and research results that can be employed in user-centric networks. In the process, we follow the evolution from centralized mobility management that depends on a single mobility anchor point to more distributed, and eventually user-provided, mobility support. We present a new solution called UPMT-DAM, which extends an existing host-based per-application mobility management solution and adapts it to more distributed any dynamic environments. We also identify the key research topics for mobility support in user-centric networks and outline main directions for future work in this area.

2 Short Primer on Mobility Management Concepts

The objective of this section is not to present mobility management approaches in detail, but to set the stage for the rest of the paper by introducing the basic concepts and common elements in general mobility management solutions. The reader is referred to the existing literature surveys for a more comprehensive overview on these concepts, e.g., [2–4].

In general, mobility management solutions try to ensure continuity of network services despite physical location changes of the communicating entities, with little or no disruption to the service.

Chan et al. [5] analyzed current mobility management solutions, categorizing them according to the following criteria:

- *Layer*: Application layer, transport layer, network layer, link layer or cross-layer mobility solutions.
- *Controlling Entity*: Network-controlled mobility, mobile-device-controlled mobility, or a combination of both.
- *Architecture*: Centralized, hierarchical and distributed (fully or partially) mobility approaches.

Considering the *Layer* criterion, *Link layer* mobility is often called *Micro-Mobility* and is related to the change of access point within the same subnet or administrative domain. It is responsible for the establishment of a radio link between the Mobile Node (MN) and the Access Point (AP), while the IP address of the MN remains the same. *Network layer* solutions provide mobility features at IP layer and do not make any assumption on the underlying access technologies. They could be provided at the network side, or both at host and network side. *Transport layer* solutions operate at the level of transport protocols, above the network level, therefore they do not require involvement of network nodes. *Higher level* or *Application layer* solutions are usually host side and allow session continuity without help from the network.

Considering the *controlling entity*, the focus is either on the end user equipment or the network side, mainly determining who takes the handover decisions and possibly sets other mobility related parameters. The common approach in cellular networks is to employ a network-controlled approach, where entities in the operator's core network take the handover decision and are in charge of managing the network resources as needed to execute the handover. In any case, the mechanism clearly works with the involvement of mobile device, in particular for reporting local network measurements and other context information. The other approach of granting mobile device the mobility control is employed mostly, but not only, in distributed mobility solutions, as we will cover in more detail in the rest of the text.

In terms of the *Architecture* criterion in the given mobility classification, the first approaches to mobility management followed the centralized hierarchical architecture of cellular networks. In a centralized approach, all mapping information for the fixed session identifier and the changing IP address for a mobile node is kept at a centralized Mobility Anchor (MA), which also intercepts and re-routes packets directed to the MN, as depicted in Fig. 1.

The majority of currently proposed solutions and recommendations in mobility management rely on the separation between host identifier(s) and locator(s) [6]. This separation requires an anchor to maintain an association (often referred to as a binding) between the identifier and the locator, a protocol to update this association, and a data transport method between locators. While the last feature can be provided by normal transport protocols or through tunneling, (mobility) anchor management is a key function, since it provides binding management and is tightly coupled to the update protocol, which affects network performance. Depending on scenario and protocol, the mobility anchor can be a centralized entity in the network [7] or follow a distributed approach, involving the collaboration of several mobility anchors [8].

Mobile IPv6 (MIPv6) has been the focus of several studies [9]. Though the MIPv6 specification was designed to cope with only one binding per Home Address (HoA), extensions for multiple Care-of-Addresses (CoA) [10] allow a node to have multiple addresses per interface as well as having multiple interfaces [11]. In UCNs, users are expected to connect to different networks and communities, and thus the capacity to use different addresses can be advantageous, depending on the connection context, sustaining several bindings.

On the other hand, the centralized nature of the Home Agent (HA) in MIPv6 is a limitation for UCNs. Using a central anchor point reduces the signaling between

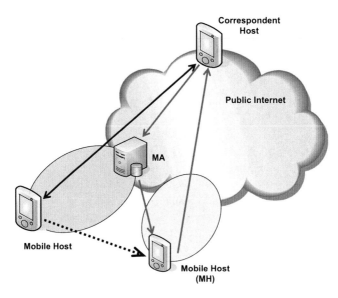

Fig. 1 Typical approach for centralized mobility management

peers, but can introduce overhead when used for bindings and routing. Solutions such as Hierarchical Mobile IPv6 (HMIPv6) [8, 12] and PMIPv6 [13] aim at addressing such performance issues by reducing the required message exchanges during time-critical events, e.g. handovers and bringing the anchor points closer to the node. By introducing the Mobility Anchor Point (MAP), which is a new MIPv6-enabled node located at any level in a hierarchy of routers, the amount of signaling outside the local area is minimized.

Distributed mobility management (DMM) approaches try to address the issues that centralized mobility solutions commonly suffer from. These are identified in [14, 15] as:

- Low scalability due to the need for new mobility anchor deployment with increasing number of mobile nodes and traffic.
- Per node (and not per flow) mobility support, which may unnecessarily increase the congestion on mobility anchors, since applications not requiring mobility support cannot individually bypass the mobility anchor.
- Single point of failure, since the failure of one mobility anchor may affect many mobile nodes.
- Non-optimal routes, resulting in longer delays and unnecessary load in the core network with respect to more distributed mobility solutions.

The work on distributed mobility management is driven mainly by IETF's DMM Working Group [16], chartered in March 2012, towards standardization in this domain. We will cover DMM in greater depth in Sect. 4.

3 Mobility Management Challenges in User-Centric Networks

The user-centric networking paradigm that we introduced earlier envisions a fully decentralized control of the network, which should be operational without relying on any dedicated network entity. Wireless access points opened up by end users to the UCN community may act as gateways for several services, including Internet access and mobility support, but they should not be deemed as robust or reliable entities in general since they are not dedicated to perform those functions only and their availability may change over time.

Since the access network elements are provided by users, there is a twofold requirement that applies in general to protocol design for UCNs. On one hand there is a need to let users have control of who is using their resources; on the other hand, the solution should take into account the dynamic nature of the network. The structure of a user-provided network would have a high level of dynamicity and mobile nodes should be able to use ANs provided by different operators or by other users, and to dynamically switch among them when needed. In user-centric scenarios, users should be aware of the operating context and be able to take decisions based on their own preferences, which may drive the requirement that handover procedures are MN based. The MNs need to measure the performance of different networks/ANs and accordingly take proper handover decisions.

We can highlight the main challenges in the design of mobility support mechanisms for user-centric network environments as follows.

- *Coping with highly dynamic environment*—unreliable nodes and links, mobility anchors possibly coming up and going down frequently.
- *Mobility anchor selection*—nodes in the user-centric network having different user and device behavior, e.g. trust level, available resources, mobility pattern, etc. that should affect the selection.
- *Handover decision*—unnecessary handovers and ping pong effects are more likely and should be avoided. Several sources of information are of potential interest in order to take the handover decision: geographical mobility estimation, context info, social mobility aspects, etc.

Those intrinsic characteristics of the UCN environment naturally call for distributed and dynamic approaches also for the mobility management. On the other hand, with the evolution of commercial operator networks towards a flattened all-IP model and the increasing traffic from mobile users, there has already been a strong interest and great deal of research on more dynamic and distributed mobility management approaches, as discussed in [17]. In the next section we will go over the DMM literature and then present a distributed, per-application mobility management solution in detail.

In the dynamic environment of UCNs, the selection of a reliable mobility anchor may be even more crucial than the decision of when to perform a handover, since the anchor selection may directly affect the reliability of connectivity and session continuity. Condeixa et al. introduce a range of scenarios for user-centric networks in [18] and affirm that a proper approach should consider a dynamic and optimized mobility control point distribution according to mobility models and considering network changes. They also identify two main blocks that present major issues to be considered for a user-centric mobility management approach:

- *Binding*: Users already have several devices, each with multi-access capabilities, and this will be even more the case in the future. IP address should not be related to user identification, and instead, used only for location procedures. The binding process should support the association of one user identification to several IP addresses. With this new association, the binding update/maintenance process would also need a reformulation.
- *Forwarding*: The mobility control element in its current form, performing both data plane and control plane functions, should be refactored considering the splitting of these functionalities. Such split would make data forwarding more flexible, since several data plane elements can be placed in different places in the network, providing the possibility of dynamically choosing the best data forwarding point for each MN.

Nascimento et al. further extend these concepts and identify the functional building blocks of mobility management in user-centric networks as device identification, binding mechanism, routing or forwarding, handover negotiation, resource management, and mobility estimation [19].

4 Distributed and Dynamic Mobility Management

Although the term *DMM* often refers to *Distributed Mobility Management*, mainly driven by the IETF working group on DMM [16]; it is usually used, implicitly or explicitly, also to capture the concept of *Dynamic Mobility Management*. So we first clarify our interpretation of these two complementary concepts, in line with the IETF definition of DMM. In general, *Distributed* MM contrasts with centralized MM by using a multitude of mobility anchors dispersed in the network, and removing the reliance on centrally deployed anchors to manage IP mobility sessions. Mobility anchors can still be assumed to be mostly fixed and robust in nature within this concept. *Dynamic* MM involves the additional concept of dynamic activation/deactivation of mobility protocol support (i.e. giving mobility services only to users or applications that need it) [20]. In user-centric networks, *Dynamic MM* should also capture the ability to cope with the more dynamic environment characteristics, such as frequently changing topology of anchor nodes as well as their changing resource availability and other contextual properties.

The requirements for distributed mobility management has been recently given in the Internet-Draft [15] by Chan as follows.

1. *Distributed processing*
   DMM solutions must enable distributed processing to avoid traffic traversing single mobility anchor.

2. *Transparency to upper layers when needed*
   Not every application needs a stable IP address, i.e. mobility support. DMM solutions must provide such transparency above the IP layer as needed.
3. *IPv6 deployment*
   DMM solutions should primarily consider IPv6, and not just IPv4, as the target environment.
4. *Existing mobility protocols*
   DMM solutions should consider reusing and extending IETF-standardized protocols before specifying new ones.
5. *Co-existence with deployed networks and hosts*
   DMM solutions must be able to co-exist with existing network deployments and end hosts.
6. *Security considerations*
   DMM solutions should not create new or amplified security risks.
7. *Multicast considerations*
   DMM solutions should enable multicast solutions to be developed to avoid network inefficiency in multicast traffic delivery.

In [21] Bertin et al. propose a dynamic approach based on IPv6 to provide mobility support while keeping traffic as close as possible to the user in the access network. Traffic of a moving user is managed with a tunnel, but the Mobility Agent only takes part in the procedures during the handover. Access points or base stations, termed as *access nodes* in this work, support two distinct functions in this proposal: *AAN* (Anchor Access Node) performs the anchor functionality for MN's IP address traffic on the access point (or base station) to which it has been currently associated and *VAN* (Visited Access Node) is used for delivering MN's traffic sessions with IP addresses not anchored to the current access node. This distinction between AAN and VAN functions allows supporting simple, dynamic and distributed mobility management. Figure 2 illustrates an example where MN's current access node serves both a flow anchored to it (from correspondent node 2) and a flow previously anchored to another access node (from correspondent node 1).

Chan proposes a modified PMIPv6 architecture to distribute mobility anchors over different networks in [22]. The proposal is based on the observation that traffic deriving from signaling is several orders of magnitude lower than that of data traffic. So it is proposed to distribute only the latter one, by replicating the re-routing functionalities on the access network, while keeping the control plane centralized or hierarchical (using some duplicated servers to avoid single points of failure).

A distributed mobility management scheme is presented in [23] based on a flat network architecture, replacing the hierarchical network architecture elements and reducing the signaling cost. Following a clean-slate approach, the locator/id split approach is employed for mobility support by relieving the IP address overload problem, which consists in the fact that in the IP architecture the IP address is used both for identifying an end-point and for locating it.

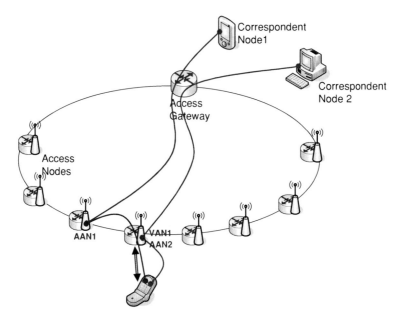

Fig. 2 Illustration of distributed mobility management components proposed in [21]

Xu et al. introduce a new Local Mobility Anchor (LMA) selection algorithm, called Mobile Controlled Movement Tracking (MCMT), for distributed mobility management in IP networks by taking the mobility pattern into consideration for LMA selection [24]. The main idea is to constantly monitor the node mobility and attempt to locate an LMA that is stable and closest to the mobile node, with the objective of providing low latency handovers and load balancing through the selected LMA. The changing mobility characteristics are detected by the algorithm and adapted in discovering new LMAs that are more suitable for the new mobility pattern.

A nice overview of the distributed mobility management literature is given in the recent work [25], which also includes the standardization activities from both IETF and 3GPP perspectives. It is clear that DMM solutions can be exploited well in the UCN context due to the removal of centralized mobility anchor requirement; however, no existing solution is ready to be applied directly as a standalone solution, since there's still the involvement of some fixed entities for the coordination and selection of distributed mobility anchors. Nevertheless, the concepts presented here—the separation of control and data plane, replacement of the hierarchical architecture with a flat one for distributed mobility, keeping traffic closer to the user in the access network, and the LMA selection mechanisms—are all important building blocks towards truly decentralized and dynamic mobility support in user centric networks.

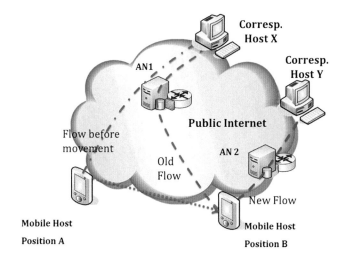

Fig. 3 UPMT-DAM, multi-AN case

5 UPTM-DAM: A Distributed, Per-Application Mobility Management Solution

In this section we present a distributed and dynamic mobility management solution, called UPMT-DAM, which extends the basic UMPT (Universal Per-application Mobility Management using Tunnels) solution introduced earlier in [26]. Our extension of the UPMT solution is mainly towards removing its reliance on a centralized *Anchor Node* (AN), which is usually offered by a provider.

UPMT is based on "IP in UDP" tunneling and provides per-application flow management, i.e. traffic flows of different applications can be independently routed on different access networks. UPMT is well suited for Always Best Connected scenarios [27], whose basic idea relies on the automatic selection "at any time" of the "best" interface for sending and receiving data. It allows performing vertical handovers over different access technologies without session disruption. UPMT acts at application level, without changes on the TCP/IP stacks in the mobile host and is fully compatible with existing network infrastructures. Using UPMT, a host can manage its network flows separately for each application. A set of policies dynamically select the best interface to use for each flow, basing the decision on the availability/quality/cost of the different interfaces. The UPMT mechanism is seen by the Mobile Host (and by the applications therein) exactly as a NAT service. In principle, all applications that can be run behind NAT boxes can also run using UPMT.

In UPMT-DAM, we consider a distributed approach by replicating the centralized anchor point into multiple ANs, potentially at the edges of the network, and allowing the users to select the "best" one for their purposes. Figure 3 depicts the UPMT-DAM multi-AN scenario.

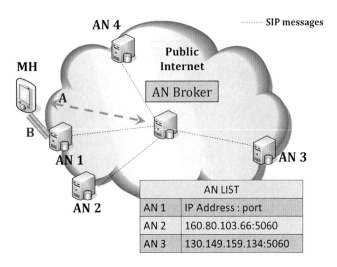

Fig. 4 The AN brokering function

Having multiple ANs creates new issues to face: how to signal their presence in the network, how to select them and how to switch the applications from one to another.

5.1 AN Brokering

In order to support multiple ANs, which are supposed to join and leave the UPMT overlay network over time, we introduce a new entity called "AN Broker". This can be a special AN or a dedicated node to which the other ANs signal their presence and capability as they activate UPMT functionalities (see Fig. 4).

The AN Broker keeps a constantly updated list of ANs, and provides it to the UPMT clients that request it. When a new AN becomes available, it sends a registration message to the Broker, signaling its presence and features. A Keep Alive function is called to refresh the parameters at regular intervals, allowing to update the AN List. When a Mobile Host (MH) activates the UPMT functionalities, it is unaware of the presence and address of the ANs. The client first connects to an AN Broker (arrow A in Fig. 4), sends an "AN List Request" message to the AN Broker and receives the updated list of ANs. From the received list of ANs, the MH can select the best AN (or ANs) to connect to (arrow B in Fig. 4) and start creating tunnels.

Note that this approach has the same effect of control and data plane separation as in [22] presented in the earlier section; the control plane employs the centralized approach by using the AN Broker, while the data traffic is dispersed over different anchor nodes. The AN Broker functionality can also be decentralized for better applicability in the UCN context, e.g. by a peer-to-peer overlay approach, which we will revisit in a more general context in Sect. 6.

5.2 Interface and AN Selection Policies

The selection of the interface and of the AN to be used is made through policies. A *policy* in UPMT-DAM includes a set of interfaces (or ANs) and the criteria to make the selection among them. The selection of interfaces and ANs is a continuous process, driven by events like interface connections and disconnections or by the updates of performance metrics gathered by the MH. The MH uses a default interface policy for the selection of the interface and a default AN policy for the selection of the AN. Moreover, independently for every application, an interface policy and/or AN policy that overrides the default behavior can be configured. A policy can be based only on information about the availability of a given interface or of an Anchor Node or it can use performance metrics dynamically gathered, for example related to packet delay, packet loss rate, and estimates of available bandwidth. When the MH is equipped with several access technologies, it will establish for each interface a tunnel toward the AN that it may want to use for sending and receiving data. Starting from this moment, the MH can select, independently for each application, which tunnel (i.e. which interface) to use to send packets, using the policy definition. For what concerns the interface selection, the following policies can be independently associated to each application:

- *Block*: The packets of the selected application will not be forwarded in any tunnel.
- *Static*: An interface is indicated and will be used for every packet of the selected application. If not available, the policy will be set as Block.
- *PriorityList*: The user gives a set of interfaces in order of preference. The first interface available on the list will be chosen. If none of them is available, the Block policy is selected.
- *Random*: A random interface between the ones available is selected and used for the application.
- *PerfThreshold*: The user provides two thresholds with the maximum allowed value of RTT and PL and a list of interfaces. The first interface among them that fulfills the requirements will be selected. If no interface fulfills the requirements, the policy will be read as a normal PriorityList.
- *VoIP*: This is a special performance policy that will select the best available combination of RTT and PL for real time applications, based on the Mean Opinion Score (MOS).

According to our previous definition, the first four are "availability based" policies, while the other two are "performance based" policies.

Once the client is connected with more than one AN, we need a mechanism to manage its flows through this multiplicity of ANs. The main objectives we aim at are to balance the load and keep traffic local. If more ANs are available, the MH should select in every moment the best AN, for example considering link-level/IP-level performances.

Currently, there are four kinds of implemented policies for the Anchor Node selection.

- *PriorityList* takes a list of IP addresses given by the user and looks for a match with the IP of the associated ANs. The first one that matches is selected as AN for the given application.
- *Any* takes whatever anchor node is available. In the current implementation, there is a default active anchor node that will be selected by the Any Policy.
- *Static* defines the IP address of an anchor node to connect to. If this address is found among the associated ANs, it will be used.
- *Random* selects randomly between the available anchor nodes.

The system reacts to events like the disconnection of interface or the update of the performance metrics according to the active policies. As far as the interface policies are concerned, the system may re-route the active flows on other tunnels. For example, if a PriorityList interface policy is active, the available interface with the highest priority will be chosen. When such a handover occurs the communication can keep on seamlessly, thanks to the presence of the AN, that hides the change of interface to the correspondent host. In this case we refer to a "flow-level handover". As far as the AN policies are concerned, the change of the selected AN does not impact existing flows, only applies to the new flows that will be originated by the applications.

5.3 Load Balancing and System Reliability

Since the user-centric networking environment may typically involve ANs with limited capacity, we introduce a mechanism to make sure that this limit is not reached, giving ANs the ability to refuse connections. To do this, each AN can set a maximum number of MHs. If the number of associated MHs reaches this value, the AN will refuse all the following associations, until some associated MH leaves.

Refusing clients can also be based on AAA considerations, for example refusing "bad" users on a black list, or allowing only users from a given subset to use an AN (e.g. only from Operator A and not from Operator B). In the current implementation, the black-list is simply read from a configuration file, as a list of users SIP IDs to refuse. As shown in Fig. 5, if the MH receives a negative reply, it skips the association to that specific AN and tries the next one on the list, if present.

5.4 Handover Management

In the existence of multiple ANs, we also need to take in consideration how to manage handovers, i.e. the change of selected AN for a specific application. We note that under our architecture it is not possible to reroute an active flow, anchored at a given AN, to a different one without session disruption for our Correspondent Hosts (CH) that are UPMT-unaware. In fact, by changing the AN, the "virtual NAT"

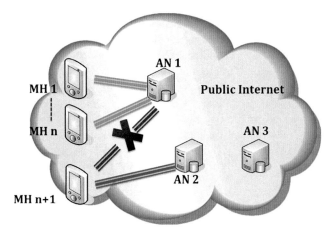

Fig. 5 Depiction of a MH refused by an AN

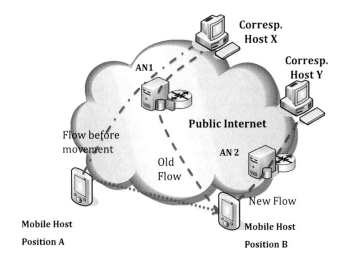

Fig. 6 A MH starts using a new AN

operation is made using a different public IP address, and the applications running on the CH will see a different source sending packets. While obviously we can perform seamless handovers across tunnels connected with the same AN, we are not able to perform flow-level AN handovers. The AN handover is a different type of handover that we may refer to as "application-level AN handover". It simply means that an application will change the selected AN to be used by new flows, starting from a given time. As shown in Fig. 6, if the old AN remains available, the pre-existing flows will keep going through it while the new ones will rely on to the other AN.

There are several events that could lead to performing an application-level AN handover, and these are processed by the AN policies of the application, which will select the new AN if needed. For example, an application-level AN handover could be recommended when some operating conditions change (e.g. the MH moves or the used AN gets overloaded) and another AN could grant better performance. In this case the AN policies should trigger the application-level AN handovers as defined above. At the current status of our implementation, the AN policy can react to the failure of an AN, or to the occurrence of a new AN higher in the priority list. In order to change the selected AN among two active ones, the defined policy for a given application can also be manually edited. If that happens, the new flows will be routed via the new AN.

5.5 UPMT-DAM Deployment

The proposed solution has been fully implemented and the developed components are available under the GPL license [28]. We have deployed UPMT-DAM in a testbed and performed a set of experiments to verify the functionality of the system. As for the mobile hosts we have implemented a Linux version and an Android version of the UPMT-DAM MH. The Linux version is based on Linux kernel 2.6.35.4-upmt, but porting to later kernels should not be a problem. The Android version has been developed for an HTC Desire HD (ARM Snapdragon S2 processor) with Android 2.3.7 based on CyanogenMod 7.1.0. As physical hosts, we use laptops equipped with built-in Wi-Fi card and 3G USB stick. As for the Anchor Nodes and AN Broker nodes, they are desktop Linux PCs running the same Linux kernel as the MHs. They can be physical Linux hosts or guest Linux virtual machines (VMs) running in a host server. In the experiments we considered two ANs running in two guest virtual machines on a server at Tor Vergata University in Rome and two ANs running on guest virtual machines on a server at TU Berlin. We used as access networks the campus WiFi networks at Tor Vergata, and at TU Berlin the wired campus Ethernet and a public 3G network provided by a network operator. In the testbed, we use netem in order to add delay and loss over given interfaces (or even selectively for given flows) and synthetically recreate impairments that can happen in real networks. For example we can add some delay on a given interface of the MH to simulate the delay on the access network even if the real delay between the MH and the AN is limited as we are performing a simple lab experiment.

We have verified the MH to AN broker interface and the ability of the MH to choose an AN according to the AN policy, to associate with it and setup the tunnels over all its physical interfaces. Then we have verified the flow-level handovers (among tunnels towards the same AN), both considering "make before break" handovers, where the new tunnels are available before the old one becomes unusable, and "break before make" handovers in which the connectivity over a tunnel is lost before a new tunnel is available toward the AN.

6 P2P for Fully Decentralized Mobility

The last step in the distribution of mobility management further to the edge of the network gets its main inspiration from a full peer-to-peer (P2P) networking approach. There is growing interest in P2P overlay networks, which are already in use for many purposes as file sharing, gaming, storage and processing applications [29]. Lua et al. [30] made a comprehensive survey on the most used P2P solutions, which are divided into two categories: structured and unstructured. Structured P2P overlay networks do not grow randomly, but follow a controlled pattern, and the contents are placed in an efficient way. The most known structured topologies are Chord [31], CAN [32], Pastry [33] and Tapestry [34]. Structured P2P overlay networks use Distributed Hash Table (DHT) so that every content can be found in a small number of logical hops. This does not ensure that the delay will also be low, as the physical distance between peers could be not so small as the overlay distance. Moreover, since the load is equally shared among all the peers, if some of them are resource-limited, there is the risk of creating some bottlenecks. The network has to monitor and maintain the state of its peers' presence with background signaling. If the joining/leaving rate is too high, lookups may fail due to the fact that topology and availability information gets outdated quickly. All these solutions have good reliability and fault-resistance features. A drawback of DHT-based overlay systems is that they can suffer from security issues, in particular when malicious peers participate in the network [35, 36].

In unstructured P2P overlay networks the peers organize themselves in a random topology, in a flat or hierarchical manner without any control. The most famous unstructured overlay networks are Freenet [37], Gnutella [38], FastTrack [39], KaZaA [40] and BitTorrent [41]. The unstructured approach is less efficient as it relies on flooding, random-walks or expanding-ring search, and could reach the time to live before finding a rare content, but provides shorter lookup time for widely replicated contents. Furthermore, flooding search provides great resistance to the changes of the network when a peer joins or leaves, but generates a high load on the network. In order to avoid this drawback, other solutions employ a hybrid structure [42] or rely on the presence of "super-peers" with more bandwidth and processing-power that make search more efficient on behalf of other peers, as in FastTrack.

Mobility Management could take advantage of the distributed, self-organizing and scalable nature of P2P overlay networks, in order to share the workload over the peers avoiding the problems of a centralized approach, as depicted in Fig. 7. Farha et al. [43] propose such a P2P approach that could provide robustness, scalability and availability to the system. The peers virtualize the MA and FA functionalities of Mobile IP and form a mesh network, with a structured topology based on the Chord ring. All mobile nodes connect to the MAs in the Chord ring following a parent/child relationship. In this solution the peers are fixed nodes with enough resources to support more MNs at once. MNs do not participate in the Mobility Management, as they are thought to be resource limited (i.e. battery, bandwidth) but, if needed, they could be promoted to the ring if the capacity of the fixed nodes is exceeded.

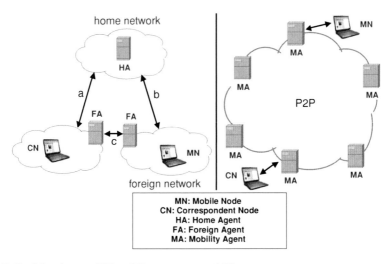

Fig. 7 Traditional versus P2P mobility management [43]

Every MN has two identifiers: one permanent (i.e. hash of MAC address), given by a permanent MA (pMA), and one temporary (i.e. hash of new IP address) given by a temporary MA (tMA). As the MN moves, his temporary identifier changes, while the permanent remains constant. A permanent MA keeps track of the visited MAs. In order to avoid longest routing of data packets due to the overlay proximity, there is also a bootstrap MA, chosen on physical proximity criterion. The strong point of this solution is that mobility management is transparent to MNs and CHs, there is no need for tunnels, no triangular routing and no single point of failure. As a drawback, there is the high signaling associated to the lookup of the MN, and to the joining/leaving of MAs.

A similar solution is given by Lo in [44], which employs a P2P network overlay for the organization of mobility anchors (HAs in Mobile IP). Users' address bindings are hashed in the P2P network, which can be queried using P2P lookup mechanisms. The authors rightly argue that it would be impractical to provide a single P2P network for worldwide usage. Therefore they opt for a multi-operator model, where each operator constructs its own P2P network. In this multi-operator environment, depicted in Fig. 8, each MN belongs to a specific P2P network, which can be queried by correspondent nodes through the DNS. Once the *home P2P network* of the MN is identified through the DNS query, the MN is located through the P2P lookup as in a single operator (i.e. single P2P network) case.

Mobility anchors are organized in a two-level hierarchy, and the signaling remains local unless the MN changes its domain. Domain-level mobility anchors are represented in the figure as GFA (*Gateway Foreign Agent*). A GFA that currently has MN in its service range is selected as a temporary HA of the MN. In order to balance traffic, every HA has the same chances to become the permanent HA of a node, and if it is overloaded, it can refuse the association, which will then be handled by another

Fig. 8 P2P-based mobile IP in
multi-operator environments
[44]

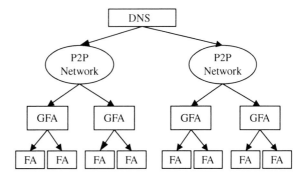

HA. As peers in the P2P network (HAs) are not supposed to join and leave frequently
in this solution, the signaling of binding renewals between MNs and temporary HAs,
and between temporary and permanent HAs are performed with different intervals.
A caching strategy with a hot list is proposed to speed up the search time of the
permanent HA associated to the MN.

In [45] by Gonen et al. mobility is managed by the MNs (the peers) in a transparent way for the network. The authors change the perspective on the MNs, usually
seen as resource limited devices, as they consider that recent smart phones have sufficient resources to manage different functionalities, so that they could rely on their
own resources for mobility. The MN is supposed to have several access technologies
and interfaces with a separate MAC address and protocol stack for each. A distributed lookup server gives information on the position of the peers. Soft handover
is performed informing the CN on the change of interface and by multicasting the
packets on both interfaces until it is concluded. If the CNs are not provided with such
technology, backward compatibility for moving peers is provided by P2P mobility
proxies that act as the HA in MIP. As more MNs have these functionalities, the role
of proxies decreases. The location management is performed relying on a Lookup
Server. This solution has good scalability properties since it relies only on mobile
nodes and thus the network does not need an upgrade of certain entities as the number
of users grows.

Peer-to-peer approaches may represent an indispensable component for the design
of dynamic and decentralized mobility management solutions in user-centric networks. The arrangement of distributed mobility anchors in a P2P fashion, as opposed
to their coordination through central or hierarchical controlling entities, better captures the UCN scenario and characteristics. When used in combination with the DMM
approaches, P2P-based solutions may provide full decentralization of mobility support entities and functions. An important aspect to consider here is the performance of
the solution, since mobility management has typically much more stringent latency
requirements than other traditional uses of P2P architectures.

7 Conclusion and Future Research Directions

The user centric network vision poses many new challenges to existing networking protocols, including mobility management. We have reviewed those challenges and presented some background work from the relevant literature on mobility management, with a focus on distributed, dynamic, and peer-to-peer based fully decentralized solutions. We have also presented a new solution proposal that can be utilized as a component for mobility support in user-centric networks.

Going over the mobility related state-of-the-art and the specific challenges in user-centric networks, we observe that there are many available partial solutions that can collectively be employed and integrated as components of a truly decentralized and user-centric mobility management solution that can cope with the dynamic characteristics of UCNs. In fact, the next paper of this book [46] presents such solution that uses the PMIP solution as a baseline and introduces additional mobility support elements based on distributed mobility management and peer-to-peer networking concepts, in order to deal with the changing topology and availability of user-provided mobility anchors. However, a through performance analysis, analytical or experimental, of such distributed and peer-to-peer solutions in the UCN environment is still an open research question.

On the other hand, there are new research opportunities in improving the reliability, robustness and performance of mobility management solutions in user-centric networks. Designation of which user-provided network entities should perform anchor point functionality, and which entities should be involved in the decentralized coordination of those anchors are some crucial design considerations with potential for further research. Another important aspect is the dynamic selection and update of the mobility anchor for each flow. Both of these issues become even more interesting and challenging due to the trust and incentive mechanisms that have to be taken into account in UCN protocol design and operation.

The UCN paradigm mainly focuses on the automatic proliferation of user-provided network resources and operation, but the concept does not exclude the involvement of existing centralized network architectures, especially those of commercial telecom operators. For example, the EU Project ULOOP [47] includes, from the beginning, the role of network operators in its original concept and scenarios. This creates a nice mixture of robust and fixed network entities as well as many dynamic, less-reliable user-provided entities, with commercial interests of operators in addition to the trust and incentive aspects in the UCN community. The search for a mobility solution in such scenarios should not only consider the design of a technical solution, but even more the socioeconomic dimension and interoperability aspects of the problem.

Finally, since users are the core elements of the UCN concept, the research on user mobility patterns also becomes a more important dimension for mobility support in UCNs than in legacy network architectures. Mobility estimation with high accuracy could make a big difference for arranging in advance the user-provided network entities and resources for a more seamless mobility support. There is already substantial research in mobility modeling and their implications on various network functions, including mobility management, but the user-centric paradigm provides a different setting where any network entity can potentially be mobile with a variety of mobility patterns. More on mobility estimation aspects in the context of DMM and UCNs can be found in [48].

References

1. Sofia R, Mendes P, Moreira W, Ribeiro A, Queiroz S, Carvalho L, Jamal T, Chama N, Junior A (2011) User-provided networks: living-examples, challenges, opportunities. SITI technical report SITI-TR-11-03, Mar 2011
2. Akyildiz IF, Xie J, Mohanty S (2004) A survey of mobility management in next-generation all-IP-based wireless systems. IEEE Wirel Commun 11(4):16–28
3. Saha D, Mukherjee A, Misra IS, Chakraborty M (2004) Mobility support in IP: a survey of related protocols. IEEE Netw 18(6):34–40 Nov–Dec 2004
4. Bolla R, Repetto M (2013) A comprehensive tutorial for mobility management in data networks. IEEE Commun Surv Tutorials PP(99):1.22
5. Chan HA, Yokota H, Xie J, Seite P, Liu D (2011) Distributed and dynamic mobility management in mobile internet: current approaches and issues. J Commun 6(1):4–15
6. Li T (2011) Recommendation for a routing architecture. RFC 6115, Feb 2011
7. Perkins CE (1998) Mobile networking through mobile ip. IEEE Internet Comput 2(1):58–69
8. Soliman H, Castelluccia C, ElMalki K, Bellier L (2008) Hierarchical mobile IPv6 mobility management (HMIPv6), RFC 5380 (Proposed standard) Oct 2008
9. Kong KS, Lee W, Han YH, Shin MK, You H (2008) Mobility management for all-ip mobile networks: mobile ipv6 versus proxy mobile ipv6. Wirel Commun 15(2):3645
10. Wakikawa R, Devarapalli V, Tsirtsis G, Ernst T, Nagami K (2009) Multiple care-of addresses registration. Internet-draft, work in progress, Oct 2009
11. Ernst T, Montavont N, Wakikawa R, Kulandinithi K, Ng C (2007) Motivations and scenarios for using multiple interfaces and global addresses. Internet draft, work in progress, July 2007
12. Castelluccia C (2000) Hmipv6: a hierarchical mobile ipv6 proposal. SIGMOBILE Mob Comput Commun Rev 4(1):4859
13. Gundavelli S (2008) RFC 5213, Proxy mobile IPv6, Aug 2008
14. Kuntz R, Sudhakar D, Wakikawa R, Zhang L (2011) A summary of distributed mobility management. IETF internet draft, work in progress, May 2011
15. Chan H (ed) (2013) Requirements for distributed mobility management. IETF internet draft, work in progress, Nov 2013
16. IETF charter for working group DMM. https://ietf.org/wg/dmm/charter/
17. Bertin P, Bonjour S, Bonnin J (2009) Distributed or centralized mobility? IEEE global telecommunications conference (GLOBECOM 2009), Nov 30–Dec 4 2009
18. Condeixa T, Matos R, Matos A, Sargento A, Sofia R (2010) A new perspective on mobility management: scenarios and approaches. Second international ICST conference on mobile networks and management, Sept 2010
19. Nascimento A, Sofia RC, Condeixa T, Sargento S (2011) A characterization of mobile management in user-centric networks. 11th international conference on next generation wired/wireless advanced networking, New2AN 2011
20. Chan AH (ed) (2011) Problem statement for distributed and dynamic mobility management. IETF internet draft, work in progress, Mar 2011
21. Philippe B, Bonjour S, Bonnin J-M (2008) A distributed dynamic mobility management scheme designed for flat IP architectures. In: IEEE new technologies, mobility and security (NTMS'08)

22. Chan HA (2010) Proxy mobile IP with distributed mobility anchors. In: IEEE GLOBECOM workshops (GC Wkshps)
23. Yu L, Zhijun Z, Tao L, Hui T (2010) Distributed mobility management based on flat network architecture. In: Proceedings of wireless internet conference (WICON'10) pp 1–6
24. Xu Y, Lee HCJ, Thing VLL (2003) A local mobility agent selection algorithm for mobile networks. IEEE international conference on communications, ICC'03, vol 2, pp 1074–1079, May 2003
25. Zuniga JC, Bernardos CJ, De La Oliva A, Melia T, Costa R, Reznik A (2013) Distributed mobility management: a standards landscape. IEEE Commun Mag 51(3):80–87
26. Salsano S, Bonola M (2009) UPMT universal per-application mobility management using tunnels. IEEE GLOBECOM 2009, Honolulu, Hawaii, 30 Nov–4 Dec 2009
27. Gustafsson E, Jonsson A (2003) Always best connected. IEEE Wirel Commun 10(1):49–55
28. UPMT home page. http://netgroup.uniroma2.it/UPMT
29. David L-N, Balakrishnan H, Karger D (2002) Analysis of the evolution of peer-to-peer systems. In: Proceedings of the twenty-first annual symposium on principles of distributed computing, ACM 2002
30. Lua EK et al (2005) A survey and comparison of peer-to-peer overlay network schemes. IEEE Commun Surv Tutorials 7(1–4):72–93
31. Stoica I et al (2003) Chord: a scalable peer-to-peer lookup protocol for internet applications. IEEE/ACM Trans Netw 11(1):17–32
32. Ratnasamy S et al (2001) A scalable content-addressable network. In: ACM vol 31 No. 4
33. Rowstron A, Druschel P (2001) Pastry: scalable, decentralized object location, and routing for large-scale peer-to-peer systems. Middleware 2001. Springer, Berlin
34. Zhao BY et al (2004) Tapestry: A resilient global-scale overlay for service deployment. IEEE J Sel Areas Commun 22(1):41–53
35. Castro M et al (2002) Secure routing for structured peer-to-peer overlay networks. ACM SIGOPS Oper Syst Rev 36(SI):299–314
36. Wallach DS (2003) A survey of peer-to-peer security issues. Software security theories and systems. Springer, Berlin, pp 42–57
37. Clarke I et al (2001) Freenet: a distributed anonymous information storage and retrieval system. Designing privacy enhancing technologies. Springer, Berlin
38. Gnutella. http://en.wikipedia.org/wiki/Gnutella
39. FastTrack. http://en.wikipedia.org/wiki/FastTrack
40. KaZaa. http://en.wikipedia.org/wiki/Kazaa
41. BitTorrent. http://en.wikipedia.org/wiki/BitTorrent
42. Loo BT et al (2005) The case for a hybrid P2P search infrastructure. Peer-to-peer systems III. Springer, Berlin, pp 141–150
43. Farha R et al (2006) Peer-to-peer mobility management for all-IP networks. In: IEEE international conference on communications, ICC'06, vol 5, pp 1946–1952, June 2006
44. Lo S-C (2007) Mobility management using P2P techniques in wireless networks. J Inf Sci Eng 23(2): 421–439
45. Gonen EK, Xu H, Pranav J (2005) A peer-to-peer architecture for mobile communications. 2nd international symposium on IEEE wireless communication systems 2005
46. Peters S, Pardo DP, Zhou Q (2014) Mobility management in ULOOP. In: User-centric networking future perspectives. Springer LNSN Book, London
47. User-centric wireless local-loop (ULOOP). EU FP7 project. Available http://uloop.eu
48. Sofia R, Sargento A, Condeixa T (2014) Mobility estimation in the context of distributed mobility management. In: User-centric networking future perspectives. Springer LNSN Book, London

Mobility Estimation in the Context of Distributed Mobility Management

Rute Sofia, Tiago Condeixa and Susana Sargento

Abstract This paper addresses mobility management in the context of user-centri networking, alerting to the need to consider new paradigms to adapt mobility management solutions to future Internet architectures. The paper provides notions concerning distributed mobility management aspects, as well as what is mobility estimation and how it can be applied to current or to future mobility management solutions, based on an existing proof-of-concept.

Keywords Mobility estimation · Dynamic mobility management · User-centric networks · Mobility tracker · Social mobility analyzer · Handover predictor

1 Introduction

Internet services and models have been going through a paradigm shift, product of three main factors: (i) widespread wireless technologies; (ii) increasing variety of user-friendly and multimedia-enabled terminals; (iii) availability of open-source tools for content generation. Together, these three factors are changing the way that Internet services are delivered and consumed as the end-user has a particular role in controlling content as well as connectivity, based upon cooperation. Specifically focusing upon Internet access, Internet connectivity models that rely upon cooperation (*User-centric Networking*, UCNs) [12, 41] are already being commercially

R. Sofia (✉)
Copelabs, University Lusófona, Building U First Floor, Campo Grande 388, 1749-024,
Lisbon, Portugal
e-mail: rute.sofia@ulusofona.pt

T. Condeixa · S. Sargento
University of Aveiro/Telecommunications Institute, Aveiro, Portugal
e-mail: tscondeixa@gmail.com

S. Sargento
e-mail: susana@ua.pt

A. Aldini and A. Bogliolo (eds.), *User-Centric Networking,*
Lecture Notes in Social Networks, DOI: 10.1007/978-3-319-05218-2_14,
© Springer International Publishing Switzerland 2014

adopted in some networks (e.g., FON, OpenSpark), from a nomadic perspective only. Nomadism can be seen as a property of global mobility management, property which relates to permanence of a subscribed environment, i.e., the possibility for an end-user to access his/her set(s) of subscribed services anytime, anywhere. In addition to nomadism, global mobility management also incorporates session continuity. Session continuity requires functionality capable of transparently and seamlessly diverting active sessions to whichever access location and whichever terminal (and interface) the end-user activates at an instant in time. In other words, while an end-user is on the move, the active sessions (independently of the type of application in use) are kept running without noticeable interruptions. Both support for session continuity as well as for nomadism should be contemplated in future Internet architectures and are also essential from a user-provided model perspective. The reason is that being user-centric and based on wireless technologies, these models rely on end-user mobility patterns to self-organize.

Future Internet models have to integrate properties that allow nomadic end-user experience for any application across multi-access or single-access networks, assuming that one or more operators are involved. Session continuity must also be considered in the new models, given that micro-providers (e.g. a user or a community of users) are also moving nodes. Depending upon micro-operator mobility patterns, it is likely that end-users receiving shared connectivity may experience disruptive connectivity breaks, which can be avoided if adequate session continuity support is provided.

Currently, the most popular solutions for global mobility management have in common a model where a centralized and static mobility anchor point is in charge of keeping some form of association between previous and current identities for a mobile node that roams across different networks. In UCN, these models should rely on a distributed architecture which raises the need to develop efficient anchor selection mechanisms based on reputation models, as well as models providing incentives to be a mobility anchor point. Moreover, time availability of mobility anchor points may be short in the presence of mobile connectivity access points. This poses an extra stress on seamless mobility mechanisms, which may need to perform handovers more often. Nevertheless, the impact on seamless mobility depends considerable on mobility patterns of mobile users (e.g. mobile users may follow the same movement of their current mobility anchor.

Moreover, in UCNs, strangers are expected to interact passively or actively for the sake of robust data transmission, several wireless access points and additional user equipment is controlled by the end-user, in addition to the regular model of control by operators. From a mobility management perspective, this implies that *Mobility Anchor Points* (*MAPs*) will often be physically located in *Customer Premises* (*CP*).

The term MAP used in the following description refers to a functional entity that takes care of all of the processes required to sustain movement of nodes that cross different networks, with reliability. For instance, in *Mobile IP* (*MIP*) [33] the MAP is the entity known as *Home Agent* (*HA*). In the *Session Initiation Protocol* (*SIP*) [34], it corresponds to the SIP server and proxy entities.

As several of these devices are controlled and/or owned by the end-user, the availability of MAPs cannot be easily controlled from a centralized perspective, as these elements they may disappear and appear in a dynamic and self-organizing way, thus disrupting the operation of mobility management solutions available in the network. Hence, for the case of UCN, the Internet end-user is also a network stakeholder. Assuming, for instance, that a MAP resides on an end-user device or in any element of the CP, then the period of time a mobility anchor point is available may vary frequently. This poses extra stress on seamless and centralized mobility mechanisms, which have to manage handovers more often.

The paper is provided in the context of this debate, giving insight to two new directions which are being addressed in the context of future Internet architectures, namely, distributed mobility management solutions, the impact and differences towards centralized solutions; mobility estimation as a handover optimization add-on to mobility management solutions.

This paper is organized as follows. Section 2 goes over related work, while Sect. 3 covers notions and current *distributed mobility management* (*DMM*) solutions. Section 4 explains the relevancy of DMM in the context of UCNs, covering also the main directions that are being developed. Then, Sect. 5 goes over mobility estimation, providing operational examples on how estimation can assist mobility management, in improving handover performance. The paper concludes with Sect. 6 where some guidelines concerning future research are also provided.

2 Related Work

Attempting to understand potential mobility patterns related to spontaneous environments, a study on an urban landscape based on the Google Wi-Fi mesh network [1] provides a good basis for the analysis of wireless usage. In particular, the authors show that such usage is split into three classes mostly based on user devices, namely, traditional mobile computers (notebooks), APs, and smartphones. The authors also show that urban mobility patterns exhibit the property of geographic locality. Specifically regarding accounting of mobile users in wireless environments a solution considers the application of agents that track node mobility, the *Mobile Agent* (*MA*) middleware. Such solution is based on having agents sent on demand to administer nodes. The central block works on the control plane only, in contrast to centralized mobility management solutions of today. A few proposals [2, 3] consider the application of overlays to deal with mobility from a global perspective. This gives the means to consider mobility management from a distributed perspective, where the mobility anchor point may be placed within the user premises. However, these solutions do not consider de-centralization or decoupling of mobility functionality.

A proposal for spontaneous environment mobility architecture based on the definition of more adequate addressing schemes and hence of more adequate routing [35] combines the notions of geographical routing based on ballistic trajectories with a location service based on *Distributed Hash Tables* (*DHT*) to achieve seamless mobility management in a k-neighborhood. Mobility management is based on the definition of an identifier that identifies the node on its constructed pseudo-geographical space and which associates the node with a k-neighborhood thus providing an identifier to its mesh area.

In what concerns optimizing the user roaming experience, roaming has been up until recently considered an integrated part (commodity) of a communication service provided to the user. In other words, mobility management in the form of roaming was not, up until recently, considered to be a service per se. This was mostly due to the fact that roaming was an integrated part of a subscribed plan in mobile networks. With the rise of wireless technologies and in particular of user-centric wireless architectures, roaming becomes a value-add service which is worth to be analyzed and optimized. Related work has analyzed a few aspects related to the optimization of such experience. From a user-centricity perspective, where user expectations drive the network selection Sofia et al. [5] propose an utility function (*Satisfaction Degree Function*) which based on specific criteria (user expectations) assist in dynamically switching between available networks. The exploitation of satisfaction degrees and algorithms from a user expectation (QoE) perspective has been pursued [21, 32] with the common goal of attempting to optimize the network selection, from a user-centric perspective.

New Internet paradigms and the need for scalability have been driving mobility management proposals which include the potential need to analyze mobility functional blocks and a potential better positioning from a networking perspective. Specifically considering mobility management in packet-based environments, Sofia et al. [40] analyze the possibility to separate control and data functionality in a way to provide a more flexible mobility management framework and to assist in developing non-centralized (e.g. distributed or hierarchical) mobility architectures. Following this work, Nascimento et al. validate, via simulations, a potential instantiation of such model [31], showing improvements in terms of node reachability time, and end-to-end delay.

Having in mind the recent trend of flatter mobile network architectures, Seite [36] addressed the concept of "flattening" by confining mobility support in the access network e.g. only confining it to access routers through a specific implementation of the application of *Proxy Mobile IP*. Following the same line of though, i.e. IP mobility management in flatter mobile networks, Liu et al. [26] debate on the idea of IP mobility management in such flatter environments.

Mobility estimation as a way to improve mobility management has been considered in the context of cellular works, context in which there are several studies dedicated to movement prediction. Several techniques have been considered, for instance, prediction based on *Signal-to-Noise* (*SNR*) ratio levels [9, 38]. Improvements have been considered, e.g. by adding a probabilistic selection based on user location (GPS). Such related work fell short in terms of adequately estimating movement, partially as it there was not a solid understanding on users' roaming behavior. In the most recent years, the availability of large-scale data sets, such as mobile-phone records and global-positioning-system (GPS) data, has offered researchers from various disciplines access to detailed patterns of human roaming behavior, greatly enhancing our understanding of human mobility. Even though this is a recent effort, we highlight that attempting to capture social movement behavior is an old field of work. One of the first works in this field relates to human mobility modeling concerning diffusion (epidemics), and was based on the diffusion analysis of over one million dollar bills [4], attempt which lead the authors to derive universal properties concerning human mobility, which gave rise to one of the most popular diffusion theories.

US2013040644 [25] provides a method for movement prediction, considering one device and its neighborhood, in regards to different cells, by predicting population movement in each cell. While their solution is based on traffic volume assessment and on a collective behavior (of the mass of nodes roaming between cells), our proposed solution is based on passive collecting of available network information (without recurring to any traffic volume assessment) and a priority is provided based on specific user parameters (that are provided by the user, passively, or actively, e.g. preference of a network). EP1903828 [11] explores a technique for preferred network selection based on the notion of roaming units. Their solution relies on feedback provided in the form of call admission control of requests concerning roaming requests (e.g. rejected requests) to assist in influencing home and roaming network selection. While our solution provides a unique way to rank preferred network selection in a completely passive way and based on information that visited networks today already send around to nodes involved in visits. WO2012080305 [27] describes a method for controlling connections between multiple user devices and a network entity via a communication network, where in this part a controlling entity gets information from a social system to retrieve data from users, in order to police the network.

Our work shares with the related literature the concept that decoupling of mobility functionality may assist in better understanding how and where to manage mobility. As described, most of today's attempts of flattening mobility management are being applied in the evolved packet core being the sole reason the urgent need to simplify mobility management. Albeit such urge is not as significant in today's wireless networks we believe that there are two aspects that are relevant to undertake: i) understanding on how distributed mobility management functionality can be optimized across the different network regions and what is more beneficial concerning all of the stakeholders involved; ii) applying mobility estimation as a way to assist mobility management in making handover decisions.

3 Distributed Mobility Management

Mobility management embodies different perspectives depending on the object that it refers to. *Personal mobility* refers to the ability of a user to reach any of its services based on a personal identifier. This identifier allows the network to bind the user to one (or several) reachability profiles, anywhere, anytime. *Device* or *Terminal mobility* relates with the capability of a device in motion to access the user services independently of the location. Terminal mobility ensures that packets continue to be delivered to a terminal as it moves through the network and its point of attachment to the network changes. In *Service mobility* the same service (look and feel) is provided independently of access type and of the terminal in use.

Central to the topic of mobility are two additional definitions, namely, *nomadism* [13] (or *nomadicity*) and *session continuity*. Nomadism relates to the ability of a subscriber to obtain the same set of services independently of the location or device used. Nomadism implies physical reconnection (performed either manually or automatically), while session continuity refers to the property of keeping connectivity while on the move.

3.1 Nomadism/Nomadicity

Within the global concept of mobility, nomadism relates to usage models where session continuity is not required, e.g., the use of a *Virtual Private Network* (*VPN*) service across different networks. When the user moves between networks covered by the same VPN, session continuity is not a requirement of this service. Behind the nomadic notion is usually the underlying assumption that the user will be able to recover its service(s) profile(s) independently of location and device. This is in fact the model for the current 3GPP roaming service, which allows a user subscribed to a specific service from a specific operator, to obtain that same service (look and feel) by means of another access operator. Therefore, the nomadic perspective relies on a policy-based architecture. MIPv6 is therefore key to achieve nomadicity, given that it allows any IP host to be reachable independently of its access location. In this case, a subscriber (and not a device) is assigned to a personal identifier, which would be the key to access the subscriber's profile(s). Whenever the subscriber moves to a new location, a re-connection is triggered and the subscribed set of services can be accessed independently of the access media. However, additionally to the global addressing requirement of nomadism, there is the need to rely on a global *Authentication, Authorization, and Accounting* (*AAA*) architecture, capable of providing the service profiles with the lesser disruption.

3.2 Session Continuity

Session continuity is provided both in the form of (automatic) IP continuity and of session continuity. In other words, session continuity is kept despite the change of access point, or UE. While this characteristic is crucial within the mobile environment, such is not necessarily the case within fixed network environments, given that in the majority of cases, service continuity for fixed environments implies physical re-connection. However, there are cases where session continuity may be required, e.g., between the fixed and the wireless networks. For instance, if a device holds more than one network interface service continuity should be possible. A specific example for this scenario is a user located in its office, and using its smartphone via fixed access. It then travels home on a train. While in the train, he might wish to keep his previous service session. These are issues currently being discussed in IEEE 802.21 [18], which explores the creation of *Media Independent Handovers*. Within the context of IEEE 802.21, session continuity covers adaptation to new link both on L2 and L3 (address adaptation), as well as session continuity at the application layer. 802.21 does not provide a standard in terms of which mobility stack (MIPv4 or MIPv6) to use, even though the discussion is currently following the direction of MIPv6.

From the 3GPP IMS perspective, currently SIP provides some session mobility management to (multimedia) services, when coupled with MIPv6. In this case the SIP server is used as a HA, and handover notification messages are traded via regular SIP messages to the HA (register) and to the CN (re-invite). The problem with the SIP approach is latency (it inherits all the delays of the transport and of the application layers). This aspect is not however crucial to the fixed network mobility, given that the underlying technology always requires the user to physically reconnect.

3.3 Mobility Management Roles and Functional Blocks

In terms of mobility management, one can isolate today three main roles: the *Mobile node* (*MN*), which usually is incorporated in portable device with limited energy capacity and also normally limited to short-range wireless or some form of cellular technology (e.g. LTE) through which the end-user connects to the Internet. Today, these devices also hold significant storage support (e.g. a tablet).

The MN is the element for which the mobility management service is offered. It has an identification (usually an IP address, or a username), that is associated to the IP address of the device's interface currently connected. Moreover, this role exists both on the control plane and data plane of current mobility management solutions.

The *MAP* is a role which refers to control functionality that may reside in the network (e.g. router or access element) or in a server. This role has as usual main functions: to be capable of providing the up to date location of MNs; to perform the translation between old and up to date identifiers, and to dictate the rules that may influence the forwarding packets to the real destination of a MN active communication.

The third role is the *Corresponding Node* (*CN*). The term CN is here applied with a broader meaning than the one of MIP and in fact being applicable to any existing centralized mobility management solution as of today, being an "active partner" to the MN and therefore requiring mobility management support. Relevant to the definition of mobility management is understanding the "proximity" (physical or based on a community perspective) of the CN to the MN and vice-verse.

In what concerns mobility management functionality, based on a thorough analysis of existing solutions which we summarize in Fig. 1, we consider ten possible functional blocks [30]:

- *Device Identification*: corresponds to the network identification for the MN. Usually the main mechanism for a location management is the association between the device's known-address and the device's real-address. In MIP, known-address and real-address are IP addresses; in SIP, the known-address is a URI, and the real-address is a IP address. In MIP the Identification control is the Home Agent/Mobility Anchor Point/Correspondent Node cache binding. In SIP it is the user database used by the Proxy server.
- *Identification database control*: corresponds to the mechanism that is applied to control the database identification. This is normally a block relevant from an access perspective, and which today follows a centralized approach.
- *Mobility anchor point location*: corresponds to the mechanism applied to support all the processes that assist in identifying the MN at any instant in time, as well as processes that support communication to and from the MN.
- *Binding registration*: it is the signaling related to the first time a device registers to the mobility system. It creates a record in the Identification control binding cache, associating the known-address to the real-address. In MIP it is the first Binding Update message sent to a HA/MAP/CN. In SIP it is the REGISTER message sent to the Registrar server.
- *Binding update*: it is the signaling to update a record in the Identification control. Binding update are used when the real-address of a device changes, or periodically to maintain the register in the Identification control. In MIP, it is the Binding Update sent from the MN to the HA/MAP/CN.
- *Routing or forwarding*: it is the process of intercepting the packets destined to the known-address, encapsulating them with the real-address, and forwarding them. In MIP this is performed by the HA/MAP, in SIP this process is performed by an element named RTP translator (when it is used).
- *Handover negotiation*: the process taken when the device has its real-address changed. It involves negotiation and signaling. The main objective is to guarantee that the user will keep active all its sessions during the handover process. In MIP, the handover negotiation may be anticipated with the Fast Handover extension, and the SIP does not implement any anticipation, performing a re-negotiation after the connection between the peers is lost.
- *Resource management*: the resource management is a necessary procedure for the mobility management to guarantee the quality of the connection when the MN changes its point of attachment to the network. However, it is not provided by most of the mobility management approaches. The 802.21 MIH standard is focused on the handover process based on a resource management aware negotiation for vertical handovers.
- *Mobility estimation*: it is the procedure of changing the MN point of attachment to the network before its current connection breaks. The extension Fast Handovers for MIP, and the 802.21 MIH provide this functionality.
- *Security*: it refers to every security mechanism to assure the integrity of the elements and signaling of the mobility management system.

Parameter	2G/3G	WiMAX	SIP	MIP	HIP	M-SCTP	IEEE 802.21
Device Identification	SIMCard (unique, tied to subscription)	IP address (unique, tied to interface)	SIP URI (unique, tied to user)	IP	Locator ID	IP and Port	MIHF ID
Identification database control	Centralized, controlled by the provider; access through the mobility anchor point	Centralized, controlled by the provider; access through the mobility anchor point	Centralized, controlled by the provider. access through the mobility anchor point	Mobility anchor point	Mobility anchor point	Does not integrate an identifier database	Does not implement an identifier database
Mobility anchorpoint location	Centralized solution, located in the provider premises	Centralized solution, located in the provider premises (in the ASN-GW element)	Centralized solution, located in the provider premises (Proxy SIP (server)	Centralized solution, located in the provider premises (HA, access router)	Centralized solution, located in the provider premises	Centralized solution, located in the provider premises	Does not implement a location management element
Binding registration	REGISTER message, MN to Registrar Server	MIP based and only for vertical handovers	REGISTER message, MN to Registrar Server	Binding Update message, MN to HA, MAP or CN	-	-	Does not implement location management
Binding update	REGISTER message, MN to Registrar Server or Outbound Proxy	MIP based and only for vertical handovers	REGISTER message, MN to Registrar Server or Outbound Proxy	Periodic Binding Update message, MN to HA, MAP or CN	-	-	Does not implement location management
Routing / Forwarding	Proxy or RTP translator	Point-to-point, Proxy based	Proxy or RTP translator	IP based, regular routing	Dual, based on locator and on IP	IP based, regular routing	Does not implement location management
Handover negotiation	Break-before-make, Register required	Break-before-make, register required	Break-before-make, RE-INVITE message, MN to CN	Make-before-break, with FMIP access routers negotiation	Make before break	Break before make, requires set up of new TCP connection	Make-before-break, attachment points negotiation
Resource management	None	None	None	None	None	None	Yes. The Standard describes functions and parameters to obtain information on network resources
Security/privacy	Yes, based on 3GPP system requirements	Yes, for the channel established between UE and BS	None	None	Yes, inherent to HIP	None	Yes, in a security related extension (P802.21a)
Mobility estimation	Static roaming agreement prevents the need to anticipate movement (pre-established tunnel)	No	No	Yes (e.g. FMIP)	No	No	Yes

Fig. 1 Mobility management functional blocks, characterization across different existing solutions

Out of the analysis provided it is clear that today's mobility management solutions lack a few aspects in particular in the face of dynamic environments such as UCN, being the major gaps concerned with resource management as well as security aspects. Both aspects are crucial in wireless networks, and in particular in networking architectures where there is a highly variable roaming behavior. Database control is normally centralized, an aspect which may not be compatible with the notion of communities that UCN scenarios are based upon. Routing and forwarding is also based on mechanisms (e.g. proxy mechanisms) which may not be completely compatible with the fact that users in our scenarios are expected to roam frequently. This is an aspect that can be improved by integrating mobility estimation mechanisms as explained in Sect. 5.

4 Distributed Mobility Management in the Context of UCNs

4.1 UCN Mobility Management Characterization

Mobility management aspects in UCN contemplate an end-to-end Internet perspective, emphasizing the nature of spontaneous wireless environments on the fringes of the Internet, the last hop towards the end-user. This implies that when considering movement of people and of devices, one must take into consideration aspects that relate to spontaneous wireless environments.

UCNs are based on the notion of users carrying (or owning) low-cost and limited capacity portable devices which are cooperative in nature and which extend the network in a user-centric way, not necessarily implying the support for networking services such as multihop routing. For instance, in UCNs transmission may simply be relayed based on simple mechanisms already existing in end-user devices. These emerging architectures therefore represent networks where the nodes that integrate the network are in fact end-user devices which may have additional storage capability and which may or may not sustain networking services. Such nodes, being carried by end-users exhibit a highly dynamic behavior. Nodes move frequently following social patterns and based on their carriers interests; inter-contact exchange is the basis for the definition of connectivity models as well as data transmission. The network is also expected to frequently change (and even to experience frequent partitions) due to the fact that such nodes, being portable, are limited in terms of energy resources. From a mobility perspective UCNs therefore exhibit a highly dynamic behavior where the selection of the "best" mobility anchor points requires the pursuit of two main aspects: adequate selection and redundancy. This has to be achieved by always weighting user expectations and the support each user is willing to give as well as the network support (access sharing) each user can in fact provide to its counter peers in the network. Mobility anchor point location and selection optimization is therefore a crucial requirement of UCNs.

Table 1 Mobility management assumptions and requirements in UCNs

Assumptions	• Users willing to share resources may be stakeholders of networking management functions, including mobility management
	• MAPs may reside in CP
	• Users (and carried devices) roam frequently—devices carried or owned by humans
	• Node movement follows human movement patterns
Requirements	• Mobility anchor point redundancy
	• Optimal mobility anchor point selection
	• Flexible mobility management architecture (most likely, decentralized)

Table 1 summarizes both assumptions and requirements for mobility management in UCNs.

4.2 Problems

Current approaches to distribute mobility anchors focus on a light distribution of all mobility functionality, as a unique block (such as the Home Agent in MIP), through some network elements. Thus, most of the approaches do not decouple the mobility management functionality, which in our vision is crucial to achieve a better performance of mobility management when integrated with user-centric scenarios. Most of the proposals do not implement any kind of decoupling of functionality, and consequently they cannot have different degrees of distribution for different mobility functionality. Besides, most of the solutions do not follow flexible architectural principles, which allow the integration of selection mechanisms to optimize the mobility management [7]. Furthermore, DMM approaches do not consider the case of an anchor failure, so they do not implement mechanisms to support fault-tolerance, which can be really important when considering that some anchors can be placed closer to the user in user-centric environments.

From a mobility management perspective, the major implications faced by those scenarios is the fact that if we consider the possibility of placing some mobility management functionality in the customer premises, the MAPs may appear and disappear suddenly, thus exhibiting a high degree of variability which impacts the network operation as well as the user *Quality of Experience* (*QoE*). The QoE perception of two users that use the same mobility management mechanism could be completely different, depending on the user behavior and required services. Thus, QoE metrics and mechanisms should give an important contribution in the selection of mobility management. Another aspect relates to the fact that the roaming environment on user-centric scenarios is strongly dependent on human behavior, which, from a networking perspective, relates to social networking aspects. Thus, daily user's movements usually present mobility patterns, by repeating the same actions and visiting the same places. Furthermore, the behavior of each user can show us his/her type of mobility, such as the pause times and crossed distances.

An aspect that may be critical to mobility management in the face of UCN is that due to the bet on centralized approaches, mobility management solutions are today available at OSI Layers 2, 3 or 6, or even between layers (as is the case of the *Host Initiation Protocol, HIP* [28]). Focusing on the network layer, solutions are normally based on a centralized approach, client/server paradigm. Part of the functionality resides on the MN, but the control-plane intelligence resides on a centralized MAP. The mobility management protocol is typically based on the principle of distinguishing between node identifier and routing address and maintaining a mapping between them, called binding. For instance, in MIPv6 the *Home Address* (*HoA*) serves as the identifier of the device whereas the *Care-of Address* (*CoA*) takes the role of routing address; the binding between them is maintained at the MAP, in this case, the *Home Agent* (*HA*). MIPv6 is a centralized mobility management approach; therefore, the mapping information between the stable node identifier and the changing IP address of an MN is kept at a centralized HA. Besides centralized binding placement, routing/forwarding is also centralized, since the HA acts also as a data central anchor, routing packets destined to an MN whenever is necessary. In other words, such mobility management systems are centralized in what concerns control and data plane and hence not the most suitable to be directly integrated into scenarios that, such as UCNs, where nodes roam frequently, even if following a specific movement pattern, tied to a specific routine which can be statistically defined.

Several other existing mobility management deployments make use of centralized mobility anchoring in hierarchical network architectures. Another example of such centralized mobility element is the *Local Mobility Anchor* (*LMA*) in *Proxy Mobile IP* (*PMIPv6*) [16]. Moreover, current mobile networks such as the *Third Generation Partnership Project* (*3GPP*) UMTS networks, CDMA networks, and 3GPP *Evolved Packet System* (*EPS*) networks also employ centralized mobility management, with the *Gateway GPRS Support Node* (*GGSN*) and the *Serving GPRS Support Node* (*SGSN*) in the 3GPP UMTS hierarchical network, and with the *Packet data network Gateway* (*P-GW*) and the *Serving Gateway* (*S-GW*) in the 3GPP EPS network [24].

The main problems of current centralized approaches for mobility management in flat networks can be summarized as follows:

- Signaling via a centralized anchor is often longer, so that those mobility protocol deployments that lack optimization extensions results in non-optimal routes, affecting performance; whereas forwarding optimization may be an integral part of a distributed design.
- As a mobile network becomes less hierarchical, centralized mobility management can become more non-optimal. In contrast, distributed mobility management can support both hierarchical networks and flat networks. Furthermore, the recent trend in network flattening, with connectivity sharing among users in the same geographical area and direct communications among them, reinforce centralized architectures weaknesses.

- Centralized route maintenance and context maintenance for a large number of mobile hosts is more difficult to scale. Scalability may worsen if there is no mechanism to determine whether mobility support is needed; dynamic mobility management (i.e., selectively providing mobility support) may be better implemented with distributed mobility management.
- Excessive signaling overhead should be avoided when end nodes are able to communicate end-to-end; capability to selectively turn off signaling not needed by the end hosts would reduce the handover delay.

4.3 Splitting Mobility Management Functionality

As mentioned before, a first step to address in terms of understanding up to which point and how to develop de-centralized mobility management architecture relates to being able to better compose the functional sub-blocks required by such architecture. Therefore, thinking about mobility management functioning in a fine-grained way, we have identified a group of functional blocks as addressed in Sect. 3. Based on the dynamics of UCNs, the first step towards a more suitable mobility management approach is by understanding and further analyzing the basic tasks a mobility management system should provide.

Towards the idea of making mobility management more flexible (being the aim a reduced operational cost), Chan et al. suggest to position the mobility anchors closer to the mobile nodes, ideally in the first element visible on the path from a MN perspective [8]. Sofia et al. proposed and validated in comparison to centralized approaches the separation of management functionality into two elements, attempting to decouple data plane and control plane [31, 40]. In the proposed architecture, the HA-C (control plane element) is located in a server, and HA-Ds (data plane elements) are positioned in the access nodes, close to mobile nodes. Chan relies on the PMIPv6 solution, and also splits the mobility anchor functionality into three logical blocks [26]. Although the author states that those blocks of functionality are placed in the home network, they do not need to be placed in the same physical entity. Those works can be considered as a first step towards an architecture where the management functionality is split and distributed in different places in the network.

Condeixa et al. [10] provide a performance evaluation of a generic model of centralized vs. distributed mobility management approaches, showing significant improvement when distributed mobility management solutions are applied to environments such as UCNs.

Such approaches, the positioning of the MAP as well as the definition of interactions between the different roles of mobility management have been object of heavy analysis. Still, today there is not truly consensus in where MAP and additional functionality should reside. Such positioning depends on the network architecture and requirements; on the OSI Layer being tackled, as well as on the overall complexity from a technical and policing perspective. Considering that user-centric networks present particular characteristics (e.g. there is no clear splitting between network elements and end-devices), the current centralized standards may not be suitable. Thus, a novel mobility management approach should be designed for such networks, considering all its particularities and following this trend of rethinking the mobility anchor point element.

4.4 Solutions to Optimize the Distribution of Mobility Anchor Points

As the first step towards the distribution of mobility anchors is the analysis of the currently available proposals. A first approach relates with distributing the mobility management control in a way that becomes more functional from an end-to-end perspective. Several solutions today realize the need to address such control, by providing a better distribution of mobility anchor points, e.g. *Hierarchical MIPv6* (*HMIPv6*) [42]. PMIPv6 splits the mobility management functionality that is in MIPv6 integrated into the MN, between this node and the network element *Mobile Access Gateway* (*MAG*). The MAG and the LMA manage the binding between the HoA and CoA, perform encapsulation and de-encapsulation, and are the tunneling endpoints for the traffic between the MN and the CN. The *Flat Access and Mobility Architecture* (*FAMA*) [14, 15] suggests moving the functionality of the HA closer to the edge of the network and placing it in the default gateways that provide IP connectivity to the MNs. Thus the access routers, called *Distributed Access Routers* (*DARs*), provide not only the access to the Internet for these MNs but also perform mobility management. However, FAMA does not specify when and under what conditions an MN would want to retain use of its old IP address. FAMA also does not specify whether the MN is associated with a permanent address that can be used to reach it by default. The use of multiple anchored addresses mandates a mechanism (such as DNS) on the correspondent node side to retrieve a proper and valid destination address for the MN. Care should also be taken to avoid routing loops between DARs and routing dead ends whenever the MN mutually binds a new and old address to two different DARs.

A second category of work takes a look at reliability and availability of MAPs by considering semi distributed architectures, i.e., some form of redundancy of elements. For instance, the *Home Agent Redundancy Protocol* (*HARP*) [17] applies the notion of HA cluster to provide reliability, where one HA from the group becomes the active HA and receives binding requests and updates from the MNs. Following the same line of though, *Dynamic Mobility Anchoring* (*DMA*) [37] proposes to distribute mobility traffic management with dynamic user's traffic anchoring in access network nodes. The solution relies on a flat architecture, where the *Mobility capable Access Router* (*MAR*), implemented in access nodes supports both traffic anchoring and MN's location management functionality. This is also the context of the *Signal-driven Distributed PMIP* (*SD-PMIP*) [22] which considers redundancy of all elements, control and data plane.

Closer to our belief that the functional blocks of mobility management themselves must be also distributed (and not only the physical elements that compose the mobility management solution), the *Proxy Mobile IP with Distributed Mobility Anchors* [6] propose to split the logical functions of a mobility anchor into three functional blocks: (1) allocation of home network prefix or HoA to a MN that registers with the network. (2) inter-network location management (LM) function; managing and keeping track of the inter-network location of MN, which include a mapping of the HoA to the mobility anchoring point that the MN is anchored to. (3) mobility routing (MR) function; intercepting packets to/from a MN's HoA and forwarding the packets, based on the inter-network location information, either to the destination or to some other network element that knows how to forward to the destination. The logical functions of the *Home Mobility Anchor* (*H-MA*) do not need to be in one single physical entity or even co-locate. It is possible to have multiple instances of the LM and MR functions, and they do not need to be in one-to-one relationship.

5 Predicting Movement to Optimize Mobility Management

Mobility estimation, when applied to mobility management, can assist in reducing signaling overhead derived from the roaming behavior of MNs in self-organized environments such as UCN. Using the identifiers of visited networks together with the collection of network parameters for ranking visited networks enables the estimation of user roaming behavior in an easy way with no signaling overhead based on existing technologies which requires only small modifications on existing systems. Hence, based on intelligent and local algorithms, mobility estimation can be used to rank preferred network selection in a passive way, without any additional signaling overhead, but instead based on information that controllers in visited networks already send around automatically to nodes involved in visits. A resulting list of visited networks can reflect user and network preferences/policies. It can also incorporate some properties learned with the roaming behavior of the user in regards to visited networks.

Figure 2 illustrates three wireless visited networks served respectively by AP1, AP2, and AP3. MN periodically visits the three networks. Moreover, each visited network is served by a specific MAP which can be co-located to the AP, or placed somewhere else on the network, as occurs today.

MN follows a regular routine trajectory e.g. during a day, where it crosses the three different visited networks. Following the regular IEEE 802.11 operation MN is set to perform *passive scan*ning, i.e., while roaming it passively receives *Beacon* frames sent by the surrounding APs. It can therefore get a list not only of APs that it regularly attaches to, but also of neighboring APs that it did not visit. We highlight that there is no relation whatsoever with GPS location or tracking of the nodes. MN integrates also some mobility management solution, e.g. MIPv6 or PMIPv6. On its list of visited locations, it keeps track of the different SSIDs, MACs of the APs, as well as a list of relevant parameters, for which an example is provided later

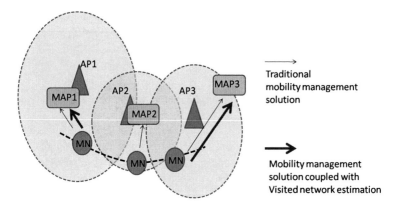

Fig. 2 Example of Impact of Mobility Estimation

in Sect. 5.1. Periodically and based on the collected input MN ranks the different visited networks and provides a time estimate for a move. Based on such estimate, MN can selectively trigger binding updates to the respective MAPs, as illustrated. In the next section we describe a tool that we have developed and that is available as open-source software, to incorporate mobility estimation into current or future mobility management solutions.

5.1 The MTracker Solution

We have developed an instantiation of the proposed estimator in the context of the European project *ULOOP - user-centric Wireless Local Loop* [12], a tool which is named MTracker. The *Mobility Tracker (or MTracker)* [39] is an application that passively tracks anonymous properties of a user's roaming behavior, and ranks each visited network based on a specific algorithm which takes into consideration aspects such as number of visits to a given access point and the average duration of such visits. Our solution is available as software licensed as LGPLv3.0. It relies on an algorithm which shall provide as outcome an estimate of time for the next move, and potential targets (list of ULOOP gateways with a priority) for the move. This outcome is then fed to the handover support module of ULOOP.[1]

The MTracker application then tries to predict how much time the node will change the network connection, and which will be the next visited network. Figure 3 illustrates the main components of the MTracker, which has as input parameters that are passively collected by a node. This functionality is expected to reside on the end-user device, and as shown, has as input parameters that are passively collected by a node. The input is expected to be passively collected and to be obtained via

---

[1] Refer to the other papers in this chapter, concerning the global ULOOP mobility mechanism.

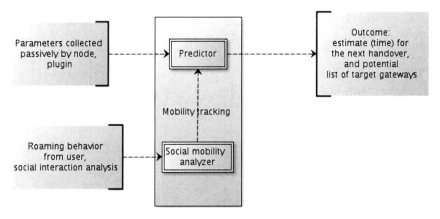

Fig. 3 High-level architecture for social mobility tracking and estimation

background processes. The roaming behavior is analyzed via the *Social Mobility Analyzer (SMA)* entity which then periodically feeds the outcome to the *Predictor* entity.

A few examples for parameters that can be collected both directly via a software plugin (based on passive listening) in regards to visited networks preference characterization:

- List of the MAC addresses for the most visited gateways.
- Average node speed.
- Number of times the node has seen a MAC attachment requested redirected
- Average stationary association time.
- Aggregate perspective on the roaming behavior of a node, e.g. daily, weekly, monthly values.

The Predictor is the entity that provides the final estimate for a potential handover. The Predictor's core is an algorithm which shall provide as outcome an estimate of time for the next move, and potential targets (list of ULOOP gateways with a priority) for the move. This outcome can then be fed to a mobility management process, which can then decide whether or not to activate a handover.

The Predictor entity relies on the time T estimate for a potential move to target G. The selection of target G is based on a ranking utility function [39] which receives as input the parameters collected, and provides as output a rank for the gateway. The time estimate T relates to the average duration of a visit. The way this is computed in our tool is based on the average time a node connects to a gateway. In other words, the MTracker definition of stationary time is: the period between the instant a node becomes connected to a ULOOP gateway, until this connection is first broken, provided in seconds.

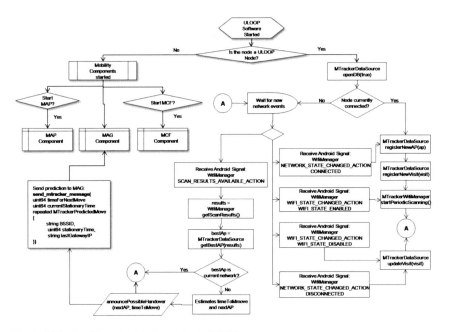

Fig. 4 MTracker Flow-chart in the context of UCNs

5.1.1 Operation

Figure 4 provides a flow-chart for the MTracker operation in the context of the ULOOP project, which assumes, in the context of mobility management, two key aspects: (i) there is a process, the MCF, which optimizes the availability of anchor points; (ii) in its proof-of-concept, ULOOP considered PMIPv6 as a potential mobility management solution.

The MTracker runs in the background. ULOOP assumes two different node roles: a regular node represents, for the context of mobility management, a role in the MN. While a gateway corresponds to the server side, where the control intelligence resides.

5.1.2 The MN Side

MTracker operates on the client side, as previously explained - refer to the right-hand side of the flow-chart. There are two main processes that the MTracker performs. One relates with tracking (collecting) data concerning visited networks. The other relates with estimating the next potential handover.

The first step after opening the MTracker list of visited networks (which is kept locally or remotely on another server). Assuming the node has Wi-Fi access it goes through the regular 802.11 operation (scanning) and periodically updates the data concerning the list of visited networks, including the computation of the ranking for each visited network.

The second part, also processed in background, relates with the time estimate and MAP target for a next handover. We highlight that the MTracker only notifies an entity (a user, some entity on the network, or even some other process in the local device) that a potential move may occur, so that a decision may assist the device in reaching some form of reliability in terms of active communication flow. Hence, it is up to the mobility management solution to perform such a move, or not, based on the information provided by the MTracker.

The estimate is based both on the computed ranking cost, as well as on the average duration of a visit. To provide a concrete example based on Fig. 2 we assume that MN has recorded an average visit duration of 15 min to AP1. On the current visit, 6 min have elapsed. Periodically, MN analysis its list of visited networks and checks whether or not the average duration visit is being reached. From a computational perspective, this means that MN has a time threshold which in this example we consider to be e.g. 1 minute, to reach and eventually send a notification to an entity in the current visited network (e.g., AP, MAP, etc). So, in our example, after 6 min, MN realizes that there is still a gap of 9 min and therefore does not send any information. When MN1 realizes that the current visit has reached 14 min, it sends a notification about the best possible visited network which, in our example, is the visited network served by AP3. In that notification, it therefore sends information to MAP1 about the best next MAP–MAP3, and also about how much time in average is left for a move. MN does not perform, however, any decision concerning moving (handover).

5.1.3 The MAP Side

In the context of PMIPv6, that information is provided to the MAG, which does not require any modification to handle such information. MTracker provides a server-side abstraction which takes care of collecting the notification provided by the MN, and sending it to the intended server-side component—the MAG, for our example.

6 Summary and Future Work

In this paper we describe our investigation concerning a better application of mobility management solutions in the context of environments such as UCNs, i.e., environments where networking elements are also subject to the Internet end-user control and willingness to share resources.

We have explained the main problems with current solutions, in particular, resource management, security, and predicting some characteristics of movement as a way to reduce the cost of dealing with mobility management. The work explains why distributed mobility management is more suitable for environments in which Internet connectivity is also provided and controlled by the end-user, by tackling aspects like the optimization of the distribution of mobility management elements in the network, the de-centralization of functionality in a way to achieve scalability and alleviate the load, mechanisms for automatic detection of such decentralized elements, and the signaling mechanism required by those elements to perform all the mobility management functions.

Based on a specific proof-of-concept developed so far in the context of the project ULOOP and also based on validation of solutions that we have developed in related work, we can state that the decoupling of mobility management functionality and the distribution of mobility management functionality in different points in the network are beneficial and improve network aspects, representing an adequate path towards mobility support in future networks architectures. Moreover, we have also exemplified the application of mobility estimation to mobility management, showing that it does not impose any significant additional cost, and explaining which benefits may arise from integrating mobility estimation even in the context of current centralized solutions.

As ongoing work, we are currently evaluating the proposed estimation in the context of today's most popular mobility management approaches, namely, PMIPv6, and understanding how to better integrate such solution in the context of distributed mobility solutions, such as the DMA proposal.

Acknowledgments We thank all of the elements involved in the *User-centric Mobility Management* (UMM, http://copelabs.ulusofona.pt/~umm) project, which involved COPELABS and University of Aveiro/Instituto de Telecomunicações, Portugal. Project funded by Fundação para a Ci?ncia e Tecnologia, reference PTDC/EEA-TEL/105709/2008. Acknowledgement goes also to Jonnahtan Saltarin, whom implemented the second version of the MTracker solution in the context of the European project ULOOP.

References

1. Afanasyev M, Chen T, Voelker GM, Snoeren AC (2010) Usage patterns in an urban WiFi network. IEEE/ACM Trans Netw 18(5):1359–1372
2. Bolla R, Ranieri A, Rapuzzi R, Repetto M (2009) Moving towards user-centric paradigms for internet mobility. In: Proceedings of the international intermedia summer school 2009, pp 37–44
3. Bolla R, Ranieri A, Repetto N (2009) A user-centric mobility framework for multimedia interactive applications. In: Proceedings of the 6th international conference on Symposium on Wireless Communication Systems (ISWCS'09), pp 293–297. ISBN: 978-1-4244-3583-8
4. Brockmann D, Hufnagel L, Geisel T (2006) The scaling laws of human travel. Nature 439:462–465
5. Cai X, Chen L, Sofia R (2007) A dynamic and user-centric network selection in heterogeneous wireless networks. In: IEEE international performance, computing, and communications conference (IPCCC 2007)
6. Chan H (2010) Proxy mobile IP with distributed mobility anchors. In: Proceedings of GLOBECOM workshop on seamless wireless mobility, pp 16–20
7. Chan H (2011) Problem statement for distributed and dynamic mobility managemen. Draft-chan-distributed-mobility-ps-05 (Work in Progress), Oct 2011

8. Chan HA, Yokota H, Xie J, Seite P, Liu D (2011) Distributed and dynamic mobility management in mobile internet: current approaches and issues. J Commun, 6(1):4–15
9. Choi CH, Kim MO (2000) Adaptive bandwidth reservation mechanism using mobility probability in mobile multimedia computing environment. In: Proceedings of the 25th annual IEEE conference on local computer networks, 2000
10. Condeixa T, Sargento S, Nascimento A, Sofia R (2012) Decoupling and distribution of mobility management. In: 4th international workshop on mobility management in the networks of the future world (MobiWorld)/Globecom, 2012
11. Elkarat S, Florence S, Wolfman S, Ophir S, Fester K (2008) Preferred network selection. IPR EP 1903828 A1, Star Home GmbH, March 2008
12. EU IST FP7 User-centric wireless local loop project. 2010–2013, grant number 257418, http://uloop.eu
13. European project (2004–2007) MUSE—Multi-service access anywhere. http://www.ist-muse.org/
14. Giust F, de la Olivia A, Bernardos CJ (2011) Flat access and mobility architecture: An IPv6 distributed client mobility management solution. In: IEEE computer communications workshop, April 2011
15. Giust F, Olivia A, Bernardos CJ (2011) Flat access and mobility architecture: an IPv6 distributed client mobility management solution. In: IEEE computer communications workshop, April 2011
16. Gundavelli S, Leung K, Devarapalli V, Chowdhury K, Patil B (2008) RFC 5213 - Proxy Mobile IPv6. IETF http://tools.ietf.org/html/rfc5213
17. Gundavelli S, Leung K, Devarapalli V, Chowdhury K, Patil B (2008) RFC 5213—Proxy Mobile IPv6. IETF, 2008
18. IEEE 802.21 Media independent handover
19. Johnson D, Perkins C, Arkko J (2004) Mobility support in IPv6. IETF RFC 3775, Standards track, June 2004
20. Kassi-Lahlou M (2003) Dynamic mobile IP (DMI) IETF draft, Draft-kassi-mobileip-dmi-01.txt, Jan 2003
21. Kim J-H, Yong Kim Y Analysis of user utility function in the combined cellular/WLAN environments. In: Centre for Intell. Dynamic Commun., University of Strathclyde, Glasgow, U
22. Kim J, Koh S, Jung H, Han Y (2012) Use of proxy mobile Ipv6 for distributed mobility management. IETF draft. Draft-jikim-dmm-pmip-00.txt, Work in Progress, Mar 2012
23. Koodli R (ed) (2005) Fast handovers for mobile IPv6. IETF RFC 4068, Experimental, July 2005
24. Korhonen J, Soininen J, Patil B, Savolainen T, Bajko G, Iisakkila K (2012) RFC 6459 – Ipv6 in 3rd Generation Partnership Project (3GPP) Evolved Packet System (EPS). IETF, Jan 2012. http://tools.ietf.org/html/rfc6459
25. Lin Y-B, Huang-Fu C-C (2013) Movement predicting method. IPR US 2013004064A1, Feb 2013
26. Liu D (China Mobile), Cao Z (China Mobile), Seite P (France Telecom—Orange), Chan H (Huawei Technologies) (2010) Distributed mobility management problem statement. IETF Networking Group Informational Draft v3, Sept 2010
27. Marcé O, Maknavicius L (2013) Control of connection between devices, WO 2012080305 A3, Alcatel-Lucent, Jan 2013
28. Moskowitz R, Nikander P, Jokela P, Henderson T (2008) Host identity protocol, RFC 5201, April 2008
29. Moskowiz R, Nikkander P (2006) Host identity protocol (HIP) architecture. IETF RFC 4423, Informational, May 2006
30. Nascimento A, Sofia R, Condeixa T, Sargento S (2011) A characterization of mobile management in user-centric networks. In: 11th international conference on next generation wired/wireless advanced networking (New2AN 2011)
31. Nascimento A, Sofia R, Condeixa T, Sargento S, Matos R (2012) A decoupling approach for distributed mobility management. In: The second workshop on cooperative heterogeneous networks (coHetNet), July 2012

32. Nguyen-Vonga QT, Agoulmine N, Ghamri-Doudane Y (2008) A user-centric and context-aware solution to interface management and access network selection in heterogeneous wireless environments. Comput Netw 52(18):3358–3372
33. Perkins C (ed) (1996) IP mobility support. IETF RFC 2002, Standards track, Oct 1996
34. Rosenberg J (2012) RFC 3261—SIP: session initiation protocol. IETF, 2002, http://www.ietf.org/rfc/rfc3261.txt
35. Rousseau F, Heusse M, Grunenberger Y, Untz V, Schiller E, Starzetz P, Theoleyre F, Alphand O, Duda A (2007) An architecture for seamless mobility in spontaneous wireless mesh networks. In: MobiArch, 2007
36. Seite P (France Telecom) (2010) Dynamic mobility anchoring. IETF informational draft (expired), May 2010
37. Seite P, Bertin P (2010) Dynamic mobility anchoring. IETF Draft May 2010. draft-seite-netext-dma-00.txt
38. Shen X, Mark JW (2000) Mobility profile prediction using fuzzy inference in cellular networks. In: Pedrycz W, Vasilakos AV (eds) Computational intelligence in telecommunications networks, CRC Press, pp 449–477
39. Sofia R (2013) Mobility management method and apparatus, Copelabs, EP 13186562.9 (under application)
40. Sofia R, Hof A, Wevering S (2008) Method for packet-based data transmission in a network having mobility functionality. In: Siemens AG, EP1883196, 2008
41. Sofia R, Mendes P (2008) UCN: consumer as provider. In: IEEE communication magazines, feature topic on consumer communications and networking—gaming and entertainment. vol 46. Issue: 12 pp. 86–91, December 2008
42. Solliman H, Castellucia C, El Malki K, Bellier L (2005) Hierarchical mobile IPv6 mobility management (HMIPv6). IETF RFC 4140, Experimental, Aug 2005
43. Wakikawa R, Valadon G, Murai J (2006) Migrating home agents towards internet-scale mobility deployments. In: Proceedings of the 2006 ACM CoNext, 2006

Mobility Management in ULOOP

Sebastian Peters, Denis Pozo Pardo and Qing Zhou

Abstract This paper provides an overview of the mobility management (MM) solution designed and developed in the context of the User-centric Wireless Local Loop Project (ULOOP).

Keywords Mobility management · Peer-to-peer · PMIPv6 · User-centric networks · Distributed mobility management · Mobility anchor coordination

1 Introduction

The concept of user centric networking empowers the end-users as new Internet stakeholders with their own network resources, rather than considering them as just consumers and producers of content [1]. User Centric Networks (UCNs) allow end-users to cooperate by sharing their network resources and services. It is crucial for the success of UCN vision to provide a certain level of quality of experience to the members of the UCN community, comparable to those expected from legacy network architectures. The dynamic environment of UCNs and the diversity of user entities (UE), acting either as micro-providers or as clients, are certain factors responsible for the increased complexity in designing such enabling mechanisms for UCNs. The EU IST FP7 User-centric Wireless Local Loop Project (ULOOP) [2] tries to address those complexities, in particular by designing and implementing the foundations for

S. Peters (✉) · D. P. Pardo
DAI-Labor, Technische Universität Berlin, Berlin, Germany
e-mail: sebastian.peters@dai-labor.de

D. P. Pardo
e-mail: denis.pozo@dai-labor.de

Q. Zhou
Huawei Technologies Düsseldorf GmbH, Düsseldorf, Germany
e-mail: zhouqing@huawei.com

A. Aldini and A. Bogliolo (eds.), *User-Centric Networking,*
Lecture Notes in Social Networks, DOI: 10.1007/978-3-319-05218-2_15,
© Springer International Publishing Switzerland 2014

trust and incentive mechanisms, resource management, and mobility management (MM) in UCNs. This paper aims to provide a detailed overview of the MM aspects in the context of the ULOOP Project.

The structure of the paper is as follows: Section 2 provides a general introduction on the MM challenges tackled in ULOOP, which includes the background and overall objectives. Section 3 describes the mobility coordination aspects in ULOOP for a single mobility domain, including the technical challenges, how they were tackled, and provides operational examples in form of use cases. In Sect. 4 we extend the perspective to multiple MM domains and explain how the mobility coordination works in this scope. Finally, Sect. 5 concludes the paper.

2 ULOOP Environment and Challenges from a Mobility Management Perspective

As mentioned earlier, the ULOOP project is based on the notion of UCNs, in which users cooperate by sharing wireless resources as well as Internet services. The UCN environment envisioned in ULOOP is a highly dynamic one where user-provided entities, possibly providing various network support functions, may frequently appear and disappear. To deal with this dynamism, the ULOOP concept handles mobility of users in a way that is graceful towards disappearing MM entities such as mobility anchor points (MAPs).

Any ULOOP user, whether providing a part of the UCN infrastructure or just taking advantage of it, is seen as a member of the *ULOOP community*. Conversely, ULOOP communities are based on nodes made available based on users willingness, which means they are not provided and controlled by network operators. Within a ULOOP community, any ULOOP node assuming the role of *gateway*, can also be a potential MAP. Therefore it is relevant also to assist in an adequate coordination of the MAPs made available. Instead of defining a new MM solution, the definition of a signaling architecture that can assist in a better coordination of available MAPs in the dynamic UCN environment is the design methodology adopted in ULOOP. In other words, ULOOP networking architectures are expected to rely on existing mobility solutions (e.g. Proxy Mobile IPv6–PMIPv6 [3], Mobile IPv6–MIPv6 [4], Session Initiation Protocol (SIP) [5]) and to address aspects that are crucial to sustain the least disruption when nodes roam across the dynamic infrastructure. PMIPv6 fits well into the design concept of UCNs as it does not involve the mobile node (MN) in the MM methodology. However, it is necessary to extend PMIPv6 to overcome the challenges posed by the UCN environment (i.e. an environment where the infrastructure is not provided by a network operator as it is usually the case with PMIPv6). In the remainder of this paper we will show novel MM concepts for the scope of UCNs, using PMIPv6 as a baseline solution.

In a common PMIPv6 environment the Local Mobility Domain (LMD) only supports a single mobility anchor. In UCNs this is problematic since a single-MAP environment poses a single point of failure. On account of this multi-MAP environments have been discussed, most notably in the PMIPv6 extension by Korhonen et al. called "Runtime Local Mobility Anchor (LMA) Assignment Support for Proxy Mobile IPv6" [6]. The runtime LMA assignment of [6] is designed for the purpose of load balancing in a multi-LMA environment with a static number of mobility anchors. However, in a UCN environment the number of available LMAs is not known at the time of deployment, as is implicitly assumed in [6]. In this paper the MAPs that are currently offering their anchoring service to MNs, and which are thus available for a dynamic selection are coordinated in the PMIPv6 domain by a broker-like entity that is described in detail in Sect. 3. Furthermore, PMIPv6 only provides MM for MNs of a single LMD, lacking support for roaming across multiple LMDs. A concept for inter-domain mobility based on P2P is detailed in Sect. 4.

To summarize, we identify the following set of technical challenges to be overcome with regard to MM aspects in ULOOP:

- ULOOP communities are based on nodes owned by Internet users and are out of the control of the operator. Their dynamic behavior which may even be erratic, can cause unreliabilities from a MM perspective. The Mobility Coordination Function (MCF) is expected to assist in a better coordination of MAPs, in the face of this dynamic behavior.
- ULOOP communities interoperability to other systems requires a way to deal with the dynamic behavior of UCNs. A MCF domain shall coordinate with other MCF domains to provide for roaming of MNs across multiple MCF domains.

3 Mobility Coordination in ULOOP

As described in Sect. 2, UCNs create the need for a MM mechanism that is able to cope with the dynamic behaviour in order to still provide the best possible continuous network connection to a ULOOP Node. Another characteristic apart from the dynamic infrastructure is that the user provided infrastructures only have limited resources in terms of the available network bandwidth and computational potency. The mechanism that has been developed within ULOOP to meet these requirements has the following major components:

- MCF.
- MAPs.
- Mobility Access Gateway (MAG).

The MCF component serves as a broker of MAPs that helps in managing the limited resources of the ULOOP Gateways (UGWs) that offer MAP services by selecting MAPs on behalf of the users. The MAP component is located in a UGW and provides mobility anchor functionality to ULOOP nodes that are associated with another UGW. This way ULOOP Nodes are less affected if the associated UGW disappears. It should be noted that the MAP component is based on the LMA functionality of PMIPv6 which is modified to provide a loose coupling between

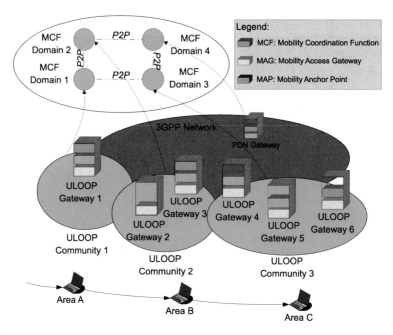

Fig. 1 Overview on mobility coordination concept

MAG and LMA. We refer to this modified version of the LMA as MAP. The MAG component also corresponds to the entity in PMIPv6, however, it is modified to be able to request an anchor node (re-)selection from the MCF.

Figure 1 provides an example of how these functional entities can be distributed within the user provided Gateways, thus making up the user provided MM infrastructure. It can be seen that each UGW runs a selection of the three ULOOP MM components. While the MAG is obligatory, users can choose wether they would like to provide a MAP or offer their Gateway to make the MCF service available. If multiple MCFs are available in a MCF domain, coordinator election algorithms could be used to retrieve the most suitable Gateway to provide the MCF functionality for the ULOOP community, however, this is not within the scope of this paper.

In Fig. 1 we have depicted three ULOOP communities as an example. ULOOP community 1 covers area A; ULOOP community 2 covers area B, and ULOOP community 3 covers area C. In area A the ULOOP node associates to UGW1 in ULOOP community 1, which also provides a MAP and a MCF service. When the ULOOP node moves to area B, UGW2 in ULOOP community 2 is selected as the UGW and the MAG component obtains the mobility context and updates the MAP in UGW1 with a Proxy Binding Update (PBU). That way the ongoing session of the user continues, however, since the MAP is from a remote ULOOP community selecting a new MAP from the current community for newly started sessions is beneficial to the performance. As UGW2 does not provide a MAP functionality itself, a query is sent to the local MCF to select a new MAP for this user. The MAP in UGW3 is selected, being currently the only active MAP in this community. Finally, in area C the MCF has multiple MAPs registered and can select the one with the most free resources in terms of available network bandwidth or based on stability metrics (uptime).

While the ULOOP MM infrastructure is intended to be mainly provided by users, the MCF entity and MAPs can also be located in the access network of an operator to support the mobility of ULOOP users (as indicated by the 3GPP coverage area).

3.1 Mobility Coordination Function Architecture

On the highest abstraction level the MCF architecture is composed of a Mobility Decision (MD) sub-block and a Mobility Cooperation (MC) sub-block, as depicted in Fig. 2. The MD sub-block groups the functionality to decide on a suitable MAP for a ULOOP node based on certain triggers. The MC sub-block groups the functionality to provide context on available MAPs to the local, as well as remote instances of the MD sub-block. As we proceed to detail the design specification of the sub-blocks we now further decompose them into the smaller components of the respective blocks in Fig. 2. Inside the MD sub-block the Mobility Trigger & Information Collection is responsible for receiving triggers from other blocks and triggers the MD. Subsequently the Mobility Mapping performs the MD and selects a suitable MAP for the ULOOP node that caused the trigger. Within the MC sub-block the Mobility Context Register is responsible for providing a constantly updated registry of MAPs that is sorted either by the id or the current score of the MAP. Finally, the Inter-MCF Coordination component indicates the inclusion of information from other MCF domains, which is described in Sect. 4 of this paper. Situated at the left side of the figure the interface to the Mobility Tracking component (located in a ULOOP node, optional) provides further input to help selecting a suitable MAP. It does so by tracking some properties of user mobility, and by providing an estimate of potential target gateways and time to move. For more information, please refer to the paper on Mobility Tracking in this book [7].

On the top left side of Fig. 2 the interfaces to other ULOOP blocks are indicated. In ULOOP device-internal socket communication is used to share information that is relevant to MM functionality. Most notably the Resource Management block provides information on the current load and the overall available bandwidth, which the MAP then provides to the MCF to allow for load-balancing mechanisms.

Summing up, all the described components are working together to form what we call the MCF to coordinate the mobility of ULOOP nodes. Having described the architectural perspective we now explore how the MAG, MAP and MCF entities interoperate to achieve this architecture.

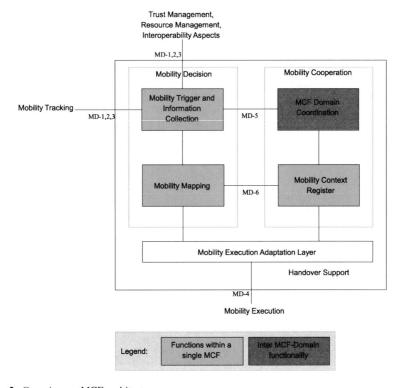

Fig. 2 Overview on MCF architecture

3.2 The Mobility Coordination Function (MCF)

Figure 3 provides an overview of the basic MCF functionality and the messages other MM entities exchange with the MCF. The UGW with MAG is depicted by the router symbol to the left, the MAP to the right and MCF in between them. The functionality of each component is implied by boxes, beginning with the initialization of the respective component on top to operational parts further down the column. The events at the MCF are as follows:

1. The MCF service initializes itself and prepares to receive incoming messages.
2. MAPs register at the MCF when they are ready to accept ULOOP Nodes.
3. The information obtained from a MAP registration message is inserted into the Mobility Context database.
4. MAG sends a message to trigger the MCF when a new user in need of a MAP connects to it.
5. The MCF then performs Mobility Mapping to select a suitable MAP for this user.
6. The decision taken in this process is then sent to the MAG.

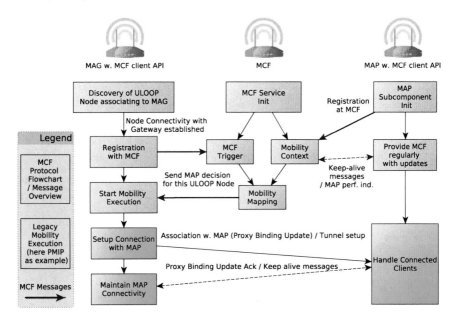

Fig. 3 Overview on MCF with MCF protocol and exchanged messages (PMIP example)

The message protocol that is used between MCF and MAG and between MCF and MAP is described in more detail in the following paragraph. In addition to this the figure also shows how the Mobility Execution part relates to the Handover Support architecture.

The MCF Protocol

This part of the paper describes the messages that are exchanged among the MAG, MAP and MCF entities. The purpose of these messages is the exchange of information between MCF–MAP and MCF–MAG to enable decision making at the MCF. The specification of these messages is tailored to the specific needs of ULOOP mobility aspects and therefore implements a custom protocol. In Fig. 3 the respective messages are indicated by blue arrows. The transmission of these messages is handled by the TCP protocol. *Messages from MAP to MCF.* Messages from MAP to the MCF are sent on three different occasions:

- MAP registration
- MAP update / keep-alive
- MAP deregistration

The involved parameters in the message are:

- MAP Message Reason—The reason code for MCF sending this message.

- MAP ID—The IP address of the MAP (IPv6).
- MAP Upload Bandwidth—Overall upload bandwidth of the MAP specified in 100kbit/s unit.
- MAP User Provided Flag—Indicates whether the MAP is provided by a user or by a network operator.
- MAP Registration Flag—Indicates whether the message is a registration or a deregistration message.
- MAP Congestion status—Indicates the congestion status of the MAP, interpretable as a percentage value of used resources.

Message from MAG to MCF. A *Request MAP for MN* trigger message is sent by the MAG whenever a MN requires a new MAP.
The involved parameters in the message are:

- MN Message Reason—The reason code for sending this message
- MN ID—The IP address of the MN (IPv6)
- MN Registration Flag—Indicates whether the message is a registration or a deregistration message
- MAG ID—The IP address of the MAG (IPv6)

Message from MCF to MAG. The *Selected MAP for MN* response message is sent by the MCF to the MAG to answer the request message with a suitable MAP for this user. The involved parameters in the message are:

- MAP ID—The IP address of the MAP
- MN ID—The IP address of the MN
- MCF Message Reason—The reason code for sending this message

This protocol description concludes the single MCF domain perspective and we will now shift our attention to the interworking of MCF domains. In Sect. 3 we have already hinted at this in the top left part of Fig. 1 showing the P2P interconnection of MCF domains, however, so far we just explained that the mobility context information for a specific user is retrieved in some way. The following section explains the involved concepts in detail.

4 Distributing the Mobility Coordination Function

The MCF entity that has been introduced in the previous chapter of this paper is responsible for the coordination of anchor nodes within a so called MCF domain. Such a MCF domain corresponds to a PMIPv6 domain in which the MCF manages anchor nodes.

Since the MCF is playing the decision maker role, the number of addressed requests depends directly on two factors: the number of units to be managed and the proper movement of them. So, we may foresee the first constraint that this element introduces in the scenario; i.e. the amount of users managed by one MCF has to be limited. Therefore, in terms of scalability, the extension from one to multiple domains seems to be the logical solution to overcome this issue. Apart from scalability, the MCF entity itself introduces the well-known drawbacks associated to centralized systems because of its way of working. The proposed system deals with these inconveniences, such as bottlenecks, congestion and single point of failure. On the other hand, there is the stronger constraint of meeting the design concept of UCNs. MM is one potential service provided within a UCN, and it is of utter importance as it greatly enhances the user experience of UCN users. Having said that, the challenge of MM in UCNs is that the user who offers the connectivity shall not be burdened with complicated administrative or technical tasks in order to provide this service to other users. While the MCF solves the problem of anchor node discovery, selection and assignment to users within the boundaries of a single MCF domain, the challenge of a hassle-free, reliable MM service that spawns across multiple UPN communities (i.e. multiple MCF domains) is tackled here.

In the following we will introduce User-Centric and Distributed Mobility Management (UDMM), a unified MM system for UCNs that is based on P2P mechanisms, PMIPv6 and the MCF. Before going into the technical details we will first illustrate why a P2P approach was chosen over other solutions, present related papers from the area of P2P based MM, and conclude with a description of the specific use cases.

4.1 Distributed Mobility Management with P2P

Along the bibliography related to distributed MM we may find a great diversity of contributions. For instance, in [8, 9] authors propose a session establishment approach, providing seamless connectivity for the MN while moving along different LMDs. Since both are based on PMIPv6, network entities such as LMAs and MAGs handle mobility on behalf of MNs. The main idea behind these drafts is to establish a session between the MN and the first reached LMA. For example, in [9] the first LMA becomes the Session Mobility Anchor (SMA) for this MN. Then, when the MN moves to another domain, the visited LMA and the SMA have to setup a tunnel between them, aiming to keep the roaming transparent from outside these domains. What is proposed here is to replicate the PMIPv6 tunnelling based performance to manage inter-domain mobility, but held by different entities: LMA and SMA instead of MAG and LMA. In these interesting works the scalability problem is thus avoided since tunnelling facilitates the connectivity between different domains. However, proper session maintenance plays an important role against UCN requirements in terms of flexibility. The intrinsic features of the UCN, in terms of the random availability of devices, do not allow us to rely on the uninterrupted performance of any device.

On the other hand, there is a trend within the related work that plays an important role in our solution. In [10], for example, the authors propose a translation from Mobile IP (MIP) to an innovative architecture based on a peer-to-peer overlay, with the aim to overcome all those inconveniences of the centralised systems, previously mentioned. One of the key points here is how they break the fixed and inflexible architecture of home networks and foreign networks by using a P2P overlay. All the entities, here called Mobility Agents (MA), start to implement both functionalities, assuming not only home agent responsibilities but also foreign agent ones depending on which MN they are managing. In this work they use three MAs to track this movement. The key point here is the virtualisation of the MM performance to distribute the system using the features provided by P2P techniques. With them, specifically by using the Data Hash Table (DHT) properties, they assign automatically the responsibilities between the MAs to manage a certain node, increasing the level of abstraction to a virtual one rather than a physical. Other work pointing to P2P networking as a basis for MM is described in [11]. Authors provide here a MM solution based on grouping the Home Agents onto P2P networks. Then they let the network operators create their own P2P domains, forming communities. Moreover, users' mobile phone numbers are translated by using the Domain Name Service (DNS), with the purpose of identifying MNs in the P2P network with a unique identifier (UID). Once the HA joins, the P2P overlay may accept users. These users are usually selected according to proximity criteria, in order to reduce the registration updates while moving. Then necessary user's information, as the binding address, will be accessible by the HAs thanks to the P2P overlay. Thus, in [10, 11] we have two examples of how useful the P2P mechanisms can be for distributing the MM protocol, providing a solution where the well known weaknesses of centralised systems; i.e., single point of failure, bottlenecks or low scalability, are overcome.

In summary, the DHT features provide us, on the one hand with the automatic distribution of the mobility context information and, on the other, with the proper abstraction level to make it feasible. It should be noted that the discussed approaches are based on the MIP-related standard instead of PMIPv6. However, the P2P concept can be transformed to be used with PMIPv6, as we will show in the following. Using a P2P based architecture has the following advantages:

• Decoupling of MM infrastructure entities
• Accessibility of mobility context from arbitrary MM nodes in order to allow for non-operator-provided MM infrastructure
• Self-management capabilities for user-provided networks
• Resilient infrastructure that can cope with parts of the MM infrastructure while nodes appearing/disappearing
• No need to rely on operator MM service

Applying P2P networking in the solutions however has some inconveniences as well. For instance:

• Messaging overhead, as P2P information, needs to be exchanged among the participating MM entities.
• Delay for queries, MNs can experience a longer delay when the DHT table is queried upon association with MAG.

In the paper at hand we have chosen an approach based on the well known Kademlia protocol, which was introduced in [12]. It is a structured P2P overlay based on DHT, where the nodes, here peers, are distributed among a virtual overlay regardless

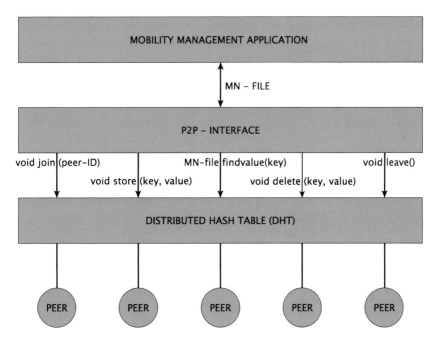

Fig. 4 Services provided by the P2P network

of the physical location. The virtual shape of the network depends directly on the implemented protocol, nevertheless all of DHT based protocols have in common the same principles:

- We have an N-bits virtual space with 2^N addresses that may point both nodes or data objects.
- A hash function maps UID into N-bit keys, so that any identifier corresponds to a different address.
- After the hashing process of a node's UID, each one is allocated into the virtual overlay and it receives an amount of keys to be responsible for in case they are fulfilled by data objects. The number of keys one receives may vary and depends directly on the number of nodes there are in the network to divide the entire space and their own keys as well.
- This particular manner of distributing the nodes and the information among them ensures that every search returns successfully the requested object.

Since the persistence of the data is ensured by the P2P protocols even when nodes leave the network, the entities performing MM relay on the P2P layer to manage the necessary information. In our solution we propose to promote only the MAGs as peers because they are the MM entities who are closer to the MNs and therefore they have the current information per node on real-time. To achieve this functionality the MM layer makes use of the API provided by the Kademlia implementation, as shown in Fig. 4. Then by using this API every MAG may perform the following actions in a transparent manner: join, leave, store, update and remove. According to the use of a

P2P overlay there is still an issue left. This aspect is related to the stored information. Intuitively, if we wonder what information we would need to manage the movement of a mobile object, the position sounds as a reasonable answer. Applying the logic we can obtain the answer for our specific case. It is important to know the position of the MN to manage it while moving. However here, in a world controlled by the IP rules, the physical one is not as important as the IP addresses of the entities that keep the MN attached to the network. Hence, each MN will be represented in the system by an information unit containing the IPs of its current MAP and MAG, as well as the MCF domain where they are located. Finally, this information unit will be stored with the key obtained by hashing a UID representing the MN, in this case the MAC address.

In the following section we will show how the MM layer and the P2P layer cooperate to manage MNs roaming among communities.

4.2 UDMM: User-Centric and Distributed Mobility Management

In UDMM we distinguish between two operation modes. The first one is called intra-domain mode and is related to the MCF performance explained in Sect. 3. The second one is referred to as the inter-domain mode and distributes the first mode performance among communities by using the P2P overlay. As a design decision the MAGs are the only entities with P2P capabilities, as explained earlier. Since MAGs have a new role in this architecture, they become the key-entity here, where both operation modes converge. In Fig. 5 we may observe the system performance when a node is being managed while moving. In the upper side of the figure we have arrows representing intra-domain operations, while in the lower one the arrows indicate the inter-domain mode:

1. The MN reaches the network (both diagrams).
2. The MAG sends a request for context information about this concrete node. The result may be either unsuccessful or successful search. As the node is new in the MM system the result has been the first one, which means no context information (inter-domain).
3. Starting from scratch: MAG sends a MAP request to the MCF (intra-domain).
4. The MCF answers with the selected MAP (intra-domain).
5. MAG and selected MAP establish the binding with the messages PBU and PBA introduced in the PMIPv6 protocol (intra-domain).
6. Then the MAG creates/updates the information related to this MN in the DHT. The information fields may be extended but so far they are current MAG, MAP and MCF-domain (inter-domain).
7. The MN moves to other domain (both diagrams).
8. It tries to attach to a new MAG there (both diagrams).
9. The new MAG queries the DHT for the context information. In this case the answer is successful search. Some time later (represented by the label 9'), the

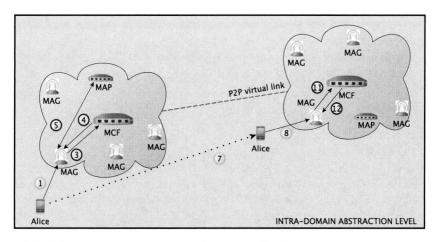

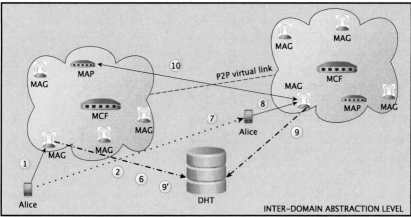

Fig. 5 System performance splited in abstraction layers

old MAG sends another request, querying for the MN context information. When there is no change in the MAG field of the resulting mobility context information (it is still its own identifier), it removes the entry for the MN from the DHT, assuming that this MN has left the MM system. Otherwise it does not need to do anything because this MN now is under the management area of another MAG. With this mechanism we ensure that the system is only storing the mobility context information about the nodes that are active in the global MM domain.

10. In order to maintain the session, the MAG establishes a binding with the previous MAP, by using again PBU and PBA[1] messages (inter-domain).

11. Afterwards the MAG sends another request to the MCF for a possible MAP placed in their domain (intra-domain). This request may be sent in some other

---

[1] Denotes standard PMIPv6 PBU and Acknowledgement messages

cases as well. For example when the MAG notices that the current MAP has disappeared (intra-domain).

12. Where appropriate the MCF answers with another MAP, which brings us to the 5th step (intra-domain).

5 Conclusion

In this paper we have presented a MM solution to meet the specific requirements of UCNs. The MCF provides a brokering entity for the user provided MAPs and applies a simple loadbalancing mechanism. A prototype has been implemented and tested under lab conditions, however an extensive field testing remains to be done within the future work. As an ongoing work we are also preparing an Internet Draft that proposes the dynamic mobility anchor coordination presented in this paper, using PMIPv6 and the Runtime LMA assignment as a basis for our efforts.

The extension of the MCF-based MM domain with the help of P2P technology enables a simple to use infrastructure that interconnects the different ULOOP communities. A prototype has been implemented for this part as well, further work includes testing QoS parameters and analyzing the delay in establishing new mobility bindings as well as the examination of inter-domain roaming performance.

References

1. Sofia R, Mendes P (2008), User-provided networks: consumer as provider. IEEE Commun Mag 46(12):86–91
2. Uloop Consortium. Uloop project (2010)
3. Gundavelli S, Leung K, Devarapalli V, Chowdhury K, Patil B (2008) Proxy Mobile IPv6. RFC 5213 (proposed standard). Updated by RFC 6543
4. Perkins C, Johnson D, Arkko J (2011) Mobility support in IPv6. RFC 6275 (proposed standard)
5. Rosenberg J, Schulzrinne H, Camarillo G, Johnston A, Peterson J, Sparks R, Handley M, Schooler E (2002) SIP: Session initiation protocol. RFC 3261 (proposed standard). Updated by RFCs 3265, 3853, 4320, 4916, 5393, 5621, 5626, 5630, 5922, 5954, 6026, 6141, 6665, 6878
6. Korhonen J, Gundavelli S, Yokota H, Cui X (2012) Runtime local mobility anchor (LMA) assignment support for Proxy Mobile IPv6. RFC 6463 (proposed standard)
7. Sofia R, Sargento A, Condeixa T (2014) Mobility estimation in the context of distributed mobility management. In: User-centric networking-future perspectives. Lecture Notes in Social Networks. Springer (2014)
8. Na JH, Park S (2008) Roaming mechanism between PMIPv6 domains (IETF Internet-draft <draft-park-netlmm-pmipv6-roaming-01.txt>). Available http://tools.ietf.org/id/draft-park-netlmm-pmipv6-roaming-01.txt (expired 12 Jan 2009)
9. Neumann N, Fu X, Lei J (2009) Inter-domain handover and data forwarding between proxy mobile IPv6 domains (IETF internet-draft <draft-neumann-netlmm-inter-domain-02.txt>). Available http://tools.ietf.org/id/draft-neumann-netlmm-inter-domain-02.txt (expired 10 Sept 2009)
10. Farha R, Khavari K, Abji N, Leon-Garcia A (2006) Peer-to-peer mobility management for all-ip networks. In: IEEE international conference on communications, ICC'06, vol 5, pp 1946–1952

11. Lo SC et al (2007) Mobility management using p2p techniques in wireless networks. J Inf Sci Eng 23(2):421–439
12. Maymounkov P, Mazires D (2002) Kademlia: a peer-to-peer information system based on the xor metric. In: Druschel P, Kaashoek F, Rowstron A (eds) Peer-to-peer systems, vol 2429. Lecture Notes in Computer Science. Springer, Berlin, pp 53–65

Part V
Market Perspective

Market Analysis and Exploitation

David Valerdi, Imanol Fuidio, Luis Gómez, Ricardo Mota, Alfredo Matos, Olivier Marce, Paolo Di Francesco and Qing Zhou

Abstract This paper goes through different business exploitation opportunities that may arise around User-centric Wireless Local Loop—ULOOP-project [1] concepts. This market analysis takes as a starting point the initial approach covered by "D2.2: ULOOP socio-economic sustainability report" [2], a public document released during the first year of the project. On top of this deliverable, a deeper analysis on

D. Valerdi (✉) · I. Fuidio
Fon Labs S.L., Getxo, Spain
e-mail: david.valerdi@fon.com

I. Fuidio
e-mail: Imanol.fuidio@fon.com

L. Gómez
Fon Technology S.L., Madrid, Spain
e-mail: luissimon.gomez@fon.com

R. Mota
ZON TV Cabo Portugal SA, Lisboa, Portugal
e-mail: ricardo.a.mota@parceiros.zonoptimus.pt

A. Matos
Caixa Magica Software LDA, Lisboa, Portugal
e-mail: alfredo.matos@caixamagica.pt

O. Marce
Alcatel-Lucent Bell Labs France, Centre de Villarceaux, Nozay, France
e-mail: olivier.marce@alcatel-lucent.com

P. Di Francesco
Level7 s.r.l., Palermo, Italy
e-mail: paolo.difrancesco@level7.it

Q. Zhou
Huawei Technologies Duesseldorf GmbH, Dusseldorf, Germany
e-mail: zhouqing@huawei.com

A. Aldini and A. Bogliolo (eds.), *User-Centric Networking*,
Lecture Notes in Social Networks, DOI: 10.1007/978-3-319-05218-2_16,
© Springer International Publishing Switzerland 2014

operators' (incl. service providers) market opportunities is performed. Additionally, other telco market stakeholders' perspective, like infrastructure and device vendors', is considered.

Keywords Exploitation · Operator · Infrastructure vendor · Device vendor · End-user · Subscriber

1 Introduction

Starting from the early days of the project, the consortium put special emphasis on the project results exploitation, by addressing the practical issues of usability and sustainability as part of the overall framework specification activities. More specifically, socio-economic sustainability of user-centric networking and the impact of regulation, business models, and public policies were assessed in [2], laying down a significant basis for the commercial exploitation of ULOOP results. Moreover, by explicitly considering the interoperability and integration of ULOOP functionality with legacy systems, exploitation of project results in the current telecommunications market is an inherent aspect of the research and development activities of the project.

In [2], a first approach is performed on how key features of ULOOP can be perceived as added values by main players involved in ULOOP ecosystem. That analysis identifies "a priori" most valued ULOOP features and main value chain players. The market analysis covered in this paper goes beyond this first analysis by extending the value chain with the inclusion of additional stakeholders as shown in Fig. 1. Device vendors and infrastructure vendors are added to the picture as equipment and technology providers who sell both to and through operators or service providers or directly to end users and consumers.

The exploitation and new potential business opportunities and the targeted markets are derived from the relevant features identified in [2] and, particularly, from the potential of dynamical grow of wireless local loop networks. By relying on existing and low-cost infrastructures, ULOOP will allow operators and manufacturers to offer a new, disruptive, cost and resources efficient way to provide wireless access, based on existing resources and aligned with community expectations. The market analysis begins by describing these opportunities for each of those stakeholders in the value chain that can implement business models around ULOOP concepts: device, infrastructure vendors and operators and service providers.

This paper also covers a market assessment from a technology perspective. Existing and future technologies may represent potential enablers or barriers for ULOOP concepts exploitation. In that section, main related technologies are identified and it is assessed whether these technologies can benefit ULOOP potential market opportunities or not.

Finally, SWOT analysis is performed as a summary of the market analysis.

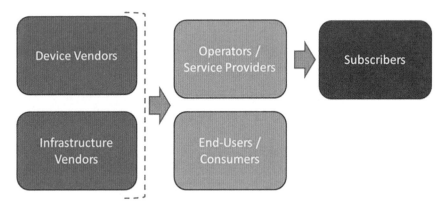

Fig. 1 Basic ULOOP value chain and main players

2 Operators/Service Providers in the ULOOP Value Chain

In general terms, an operator exploits a network infrastructure and offers services on top of the networks. The traditional services are telephony and internet broadband (double-play), but in the last years TV (triple-play) has been added to the offering and recently the mobile services (quadruple-play), in the case of integrated operators (fixed and (virtual) mobile). With ULOOP, the network of the operator will coexist with user-centric networks. Deliverable 2.2 [2] already highlights that operators can benefit from ULOOP as long as they are directly involved in the value chain, while they can penalize or even block user centric networks, if they are not involved. With that in mind, this section explores how operators can fit into this value chain by describing a bundle of business opportunities and benefits for them. In addition to "traditional" operators, other services providers including "overt the top" are considered in this section. Depending on the type of service provider (e.g. owns infrastructure, OTT, etc.), they may show a different fit in the value chain and motivations with regards to traditional operators.

The market of networks is demanding more speed, flexibility and value added services. The most relevant events occurred lately in the network providers market seem to show the interest in the implementation of new broadband technologies like FTTH/FTTC/FTTS, LTE (4G) and Wi-Fi based solutions to allow higher transmission rates, for both fixed and mobile, in order to enable more and faster services.

Operators seem to be under a period of big changes. On one hand, telecommunications operators are attempting to implement measures against the trend of decline in profitability of their transport markets (e.g. internet access, IP transit, etc). On the other hand, they are driving value added services, for example, in the mobile market, which is identified as the true source of future business performance.

Furthermore, competition among operators is increasing day by day, and implementation of new fixed and mobile networks (fiber and 4G), seems to expand over time because of the need of high investments. Therefore, this sector is seeking for ways to differentiate from their competitors by providing new services to their clients.

ULOOP will enable operators to provide their customers with new services that users can help and contribute to design and develop (e.g. community-based services) and that will differentiate the operators from their competitors (e.g. to offer internet access subscriptions with worldwide wireless roaming included). In other words, operators provide the technology and the framework that enable users to create their customized ULOOP community ecosystem and experience. Within the ULOOP community, the end user controls the access as s/he has to decide the trust on other ULOOP users and the number of credits in order to gain some services access [3]. Within the ULOOP community, the QoS offered among users is also controlled because the allocated bandwidth changes dynamically based on trust and credits [3, 4]. The most evident service defined in ULOOP is the internet connection but also other services can be provided such as, network printing, file sharing, local RSS, and messaging. Thus, ULOOP opens opportunities to create new business models around the ULOOP communities and the exchange of services between the operator and the community.

Current main business model of an operator is to sell (broadband) internet access to their subscribers. With ULOOP, additional opportunities may come from new services, higher customer retention and new subscriptions motivated by the ULOOP derived services and the Wi-Fi offloading towards ULOOP communities through Wi-Fi access points.

Collaboration between operators and end-users in terms of user-centric network-ing services is crucial and opens up new possibilities in terms of business models. In order to succeed, these business models should be defined in a way that revenues and other benefits are shared between operators and users. For instance, an operator may propose to ULOOP communities to enable advertising when certain resources are exchanged (i.e. it can even be a target advertising service based on monitoring information provided by ULOOP). Revenues obtained from this advertising service can be shared with the users providing the resources. Alternatively, revenues can be indirectly shared with users through discounts on subscriber invoice, increase of data tariff cap, etc.

This new range of opportunities can strengthen relation between the end-user and the operator. This will be resulting in a higher competitiveness and differentiation, a higher customer satisfaction and client attraction, and a reduction of cost for extend-ing local loops by relying on communication opportunities. First approaches to user centric networks like the ones proposed by FON have been proved very effective in that sense. When FON partners with a fixed operator to deploy a community Wi-Fi (i.e. a Wi-Fi network formed by the aggregation of residential Wi-Fi access points), the operator provides the service (i.e. user roaming across the community Wi-Fi) at no cost for the subscriber, as a value added service. This have resulted in a significant reduction of churn (i.e. the proportion of contractual customers or subscribers who leave a supplier during a given time period), what means savings in customer reten-tion campaigns, revenues increase coming from new and more loyal subscribers, etc.

As another approach, operator may be involved in the value chain by offering resources within a ULOOP community. In other words, ULOOP Community can be a distribution channel for operators to offer their existing and new services. These resources can range from additional internet bandwidth, VoIP services, etc. to access to cloud services (e.g. storage). Users might be paying for these resources under certain conditions or even free as a valued added service for the subscribers and members of ULOOP community. In this sense, for instance, ZON will potentially assess the opportunity of offering VoIP type services for ULOOP communities.

Another opportunity opened for mobile operators (both traditional and virtual) is that they will be capable to expand capillarity and capacity in a low-cost/low-investment way. This is particularly useful to enable mobile data offloading towards ULOOP communities. The benefits of data offloading for operators are well known. Basically, it allows mobile operators to cope with data usage exponential increase and to defer capital expenditure for network capacity enhancement (i.e. financial benefit). ULOOP can enable a community driven offload. In addition, by deploying user-centric wireless local-loops, operators can keep traffic local and have as a consequence a reduction in the access OPEX as well as an increase in spectrum and energy efficiency in managing wireless communications. ULOOP will be a perfect solution for operators who look for higher density at limited cost, letting them rely on created communities, in order to provide the required resources to demanding users at specific periods of time. This will offer an energy-efficient and cost optimized solution to increase density of the operators' networks. As an example of beneficiary of community driven offloading, ZON is currently a virtual mobile operator. That means ZON is charged for any mobile traffic exchanged by its mobile subscribers. Therefore, any new solution, which enhances the mobile offloading experience and encourages subscribers offloading practices, is beneficial for them, primarily, in the shape of OPEX savings.

ULOOP can also open new markets for OTT like service providers. New and existing companies may arise to develop and provide ULOOP related services regardless the network operator. For instance, the advertising service previously mentioned can be offered to ULOOP communities by an OTT service provider. Alternatively, "traditional" operators may also participate in this scenario if collaboration with the OTT service provider is set up, as it happens with FON. For example, the operator can offer a different configuration of QoS over those services.

Tables 1, 2, 3 and 4 summarize these conclusions for fixed, mobile and other service providers (e.g. Wi-Fi) operators.

3 Infrastructure Vendors in the ULOOP Value Chain

Infrastructure vendors provide the equipment and services to build networks under different shapes: hardware, software, maintenance services, etc. These organizations are experts in telecommunications equipment (both software and hardware) able to manage a big amount of users with high availability. This infrastructure is usually

Table 1 Exploitation opportunities and business impact for fixed operators

ULOOP impact	Business impact

New business models around ULOOP communities, enabled by:
• New services where end users and operators cooperate and share benefits
   – Internet sharing over Wi-Fi for mobile offloading
   – Other services (advertising, collaboration with OTT service providers)
• Operators offering/making available resources through ULOOP communities as a distribution channel for existing and new services, etc. (e.g. voice services for mobile users)
	• New revenue possibilities
	• Higher customer retention
	• Higher customer attraction
	• Differentiation towards competitors

Table 2 Exploitation opportunities and business impact for mobile network operators

ULOOP impact	Business impact
Promote the deployment of ULOOP communities regardless the fixed network providing the broadband access. Implement new business models based on: • Deployment of new services where end users and operators cooperate and share benefits • Offering resources using ULOOP communities as a distribution channel for existing and new services, etc Additionally, these mobile operators can benefit from: • Mobile data offloading (e.g. community driven offload) • Network capillarity expansion	• New revenue possibilities • Higher customer retention • Higher customer attraction • Differentiation towards competition • Increase capacity and coverage with low OPEX and deferred CAPEX • Energy efficiency improvement thanks to mobile data offloading

built by following standardization bodies such as the IEEE, IETF and, 3GPP. ULOOP contributions will impact the infrastructure vendors market, since network equipment will be needed to make ULOOP available and expand user centric networks.

The telecom infrastructure market has undergone dramatic transformation over the years. Successful vendors and system integrators are driven to provide more complex solutions with a consultative and seamless approach. Product offerings now tend to integrate or interconnect fixed and mobile platforms; bundled products, cloud solutions and other emerging technologies, together with a more demanding customer, all make this a fast-developing sector.

The service provider network has been under nearly constant pressure to evolve over the past decade as enterprise and consumer users, alike, have come to demand

Table 3 Exploitation opportunities and business impact for integrated operators

ULOOP impact	Business impact
Similar case than a mobile operator with the difference that an integrated operator may only promote the creation of ULOOP communities within their own infrastructure	• New revenue possibilities • Higher customer retention • Higher customer attraction • Differentiation towards competition • Increase mobile network capacity and coverage with low OPEX and deferred CAPEX • OPEX reduction (virtual operator reducing expenditure by avoiding paying for mobile data roaming) • Energy efficiency improvement thanks to mobile data offloading

Table 4 Exploitation opportunities and business impact for (wireless) service providers (incl. OTT)

ULOOP impact	Business impact
Service providers may promote over the top services making use of own infrastructure (Level7), by a partnership with fixed operators (FON) or by their own regardless the fixed operator Business models and opportunities are similar to operators': • Deployment of new services where end users and service providers cooperate and share benefits • Offering resources using ULOOP communities as a distribution channel for existing and new services, etc • Offer mobile data offloading to third operators • Network capillarity expansion	• Increase service offering portfolio • New revenue possibilities • Higher customer retention • Higher customer attraction • Differentiation towards competition

more from their connectivity options and operators have looked for new ways to drive revenues and monetize existing services. This evolution continues today and the impact on the infrastructure deployed by telecom operators is well understood. New fixed and mobile services demand ever more advanced network solutions from the access layer through the transport, application and service delivery layers. Network vendors, in turn, are forced to deliver increasingly efficient R&D and innovative product roadmaps simply to meet customer requirements and remain competitive.

The ULOOP project [1] shows direct impact over this sector and organizations since it relies on deployed wireless infrastructure. Additionally, although they may directly offer some products to end consumers, infrastructure vendors are mainly targeting operators. Thus, market success is again driven by operators. In other words, these vendors will implement and develop those products and services that operators require and may consume.

The ULOOP project [1] demonstrates how networks can evolve from a fully centralized (as in wireless realm) or a partially distributed (i.e. internet world) model towards a user-centric approach that changes the way that the network and infrastructures are considered. It is expected that ULOOP results will have an impact on network architectures similar to the impact on application architecture the P2P model did have.

Table 5 ULOOP impact as per infrastructure solution type

Type of solution	ULOOP impact	Business impact
Wireless access	Aggregation of users' WLAN resources	Additional wireless capacity at low cost
	Community driven off-load	Service differentiation and personalization
Customer care system	Banking and trust system	Refined user management
Payment system	Banking and trust system	Disruptive plan

For Huawei, for example, ULOOP concepts open opportunities for the Huawei SingleRAN solution, to provide mobile broadband services to the end user via wireless network from another user. The ULOOP concepts are beneficial for Huawei SingleRAN solution on several aspects; first, by integration of the user-centric network into the current mobile broadband network, the wireless coverage can be extended; then, the mobile signaling between end user and mobile network can reduced, as all mobile signaling from the user-centric network are integrated into one mobile node; in the end, the ULOOP concepts can also be used to provide emergency services. The Mobility Management Element (MME) and PDN GW (Packet Data Network Gateway) functions in 3GPP Evolved Packet Core (EPC) should be extended to support ULOOP concepts. Similarly, ULOOP results will potentially impact Alcatel Lucent "Light Radio" family. The "Light Radio" concept defines the conjunction of a reduced size for wireless base station, together with an optimized IP backhaul network. In addition to the benefits of a better deployment of cellular technologies (3G or 4G), the approach also consider aggregation of various resources, including WLAN (Wi-Fi) ones. Thus, "Light Radio" is a potential equipment to implement ULOOP functionalities.

Table 5 gives some potential impacts for infrastructure vendor solutions, which are closely linked to impact for operators, as main customers.

4 Device Vendors in the ULOOP Value Chain

All terminals, which the user interacts directly with, are included in the analysis. The type of mobility is fundamental to differ between mobile devices versus those who remain more static or nomadic at home or office. In this sector the main characters are PCs, laptops, video game consoles, tablets and smartphones. Nevertheless, tablets and phones are the ones that will be more impacted by ULOOP.

The mobile revolution really started in 2007 with Apple's introduction of the iPhone. The iPhone radically and irreversibly redefined mobile devices with relatively fast and simple web access, an innovative and intuitive touch screen, and the creation/promotion of mobile applications and a mobile app marketplace.

The iPhone ushered in a new mobile era and growing unit sales continue to generate incredible financial returns for Apple. It also generated an immediate competitive response from Google.

In 2008, Google launched Android as an alternative mobile operating system through a no-fee, open-source licensing model, providing device manufacturers and mobile carriers significant freedom and flexibility to design products while relying on manufacturers to build and promote Android devices and the carriers and other retailers to sell them to consumers.

Google's no-fee licensing model has resulted in rapid adoption in the smartphone market such that Android is now the leading mobile operating system globally.

In fact, the Android OS accounted for 72 % of all smartphone shipments in the third quarter of 2012. Samsung is the leading Android smartphone manufacturer by a large margin, accounting for 22.9 % of global smartphone shipments in the third quarter of 2012 [5].

At the close of 2011, there were approximately 6 billion mobile subscribers on a global basis. This represents a penetration rate of approximately 87 % based on a current global population of 7 billion. Developed countries represent 25 % of these subscriptions with 122 % penetration, while developing countries represent 75 % of the subscriptions and 78 % penetration [6].

The rapid proliferation of smartphones and tablets is deeply changing the way that we behave, consume content and conduct commerce. The ubiquity of wireless connectivity combined with increasing functionality and speed of connected devices and mobile networks will further drive consumer demand for media, content and advertising while monetization models continue to evolve.

Established and emerging mobile advertising technology companies are also developing and providing solutions to enable marketers and advertisers to reach cost-effectively this vast mobile audience on a targeted basis at scale.

The mobile market is complex, fragmented and both growing and evolving rapidly. Therefore, it is considered that we are in the early innings of a mobile revolution engendered by more affordable mobile computing and internet access on a global basis.

ULOOP will have a direct impact on this market. Opportunities from new classes of applications taking advantage of convergence will appear. The user provided network proposed by ULOOP is community centric. The social networks, which are application facet of the community centric paradigms, showed their value in impacting the way people behaves on internet. It is expected that a similar revolution will hit the wireless network area, by allowing a community based connection that opens the doors towards new classes of applications leveraging on the user controlled network. A device manufacturer (e.g. Huawei) can take an active role implementing ULOOP functionalities in their devices. They can even create communities related to a specific device vendor, becoming an OTT service provider. This approach will allow differentiation, which is a challenge in current market scenario.

Future devices with ULOOP capabilities will have a set of features needed in the OS of the devices and distributed with new ROMs versions. These new capabilities will enable phones the fast deploy of user centric networks (Table 6).

Table 6 ULOOP impact as per device type

Device type	ULOOP impact	Business impact
PCs	Internet access without the need of broadband service subscription	User satisfaction and device utility increase
	Social applications	Market opportunities
	Advertising	User satisfaction increase
	ULOOP installation for community based connection	User satisfaction and device utility increase Enhance the user experience
	Internet access without the need of broadband service subscription	User satisfaction and device utility increase
	Social applications	Market opportunities
	Advertising	User satisfaction increase
Laptops	Internet access without the need of broadband service subscription	User satisfaction and device utility increase
	Social applications	Market opportunities
	Advertising	User satisfaction increase
	ULOOP installation for community based connection	User satisfaction and device utility increase Enhance the user experience
Video game consoles	Internet access without the need of broadband service subscription.	User satisfaction and device utility increase
	Social applications	Market opportunities
	Advertising	User satisfaction increase
Tablets/Smartphones	Expand capillarity of regular operator coverage	User satisfaction and device utility increase
		New user profiles available

An equivalent analysis can be performed from an OS perspective. Within the scope of personal computers, above 90 % of legacy computers use Windows operating systems while the rest is shared between MAC (7.2 %) and Linux versions (1.2 %) [7]. ULOOP would show a faster implementation in Linux computers due to the flexibility and the openness this operating system provides. On the other hand, implementation in Windows and MAC would require an engagement with Microsoft and Apple from a business and technical perspective. Smartphones and tablets OS's show a similar scenario. Although it would require going beyond the user space applications, Android, as an open source OS, shows enough flexibility for ULOOP software integration, allowing faster business exploitation. However, Windows Phone and, particularly, iOS are more closed OS's. Due to the significant share of iOS devices (both tablets and phones), an engagement with Apple is highly recommended to secure successful exploitation of ULOOP functionalities. This aspect should be taken into consideration when preparing an exploitation plan, since it can mean a significant constraint.

5 Market Analysis from a Technology Perspective

The technology evolution stimulates the development of new markets around new concepts. The technology state of the art favors or disfavors the entry of these new concepts into the market. In this section, an analysis is performed from a market and technology perspective. This analysis aims to clarify whether a bundle of technologies can be benefited from ULOOP concepts and vice versa.

Because of the nature of spontaneous/emergent networks and according to the general results of the ULOOP project [1], wireless technologies are the ones selected for this analysis. Wireless means that interconnectivity among devices is performed by using the air interface. There are a wide variety of technologies, using the air as a medium, which can be classified by different criteria. Spectrum bands regulation (i.e. licensed and unlicensed) may be one of them.

The ITU is an organization that releases recommendations about the use of the spectrum by the different communication systems. However, each country manages and sets the regulation for its own spectrum through their national regulators. The result is a wide number of wireless technologies that make a different use of the unlicensed and licensed spectrum depending on the country. Nevertheless, there is one wireless technology that shows significant worldwide harmonization and is pervasive across end user devices and legacy systems. This technology is Wi-Fi, which is identified in ULOOP project [1] as the most appropriate one for ULOOP functionalities to rely on. In any case, ULOOP is not limited to Wi-Fi and there are other technologies like Bluetooth that can mean an alternative.

In the sections below, different technologies impacted by or related to ULOOP concepts are summarized. The aim is to assess the complementarity or the competition of these technologies with ULOOP from a market point of view.

5.1 LTE Femtocells (Licensed Spectrum Bands)

Femtocells increase coverage and capacity in indoor environments within, approximately, a range of 10 meters. Normally, a femtocell has two interfaces: one connected to the operator's network (as backhaul) and the other enabling 3G or LTE.

There are femtocell commercial products since some years ago. First commercial LTE femtocells were released in 2010/11. Commercial success has been jeopardized by the complexity of their management within a well-planned network like cellular ones. This situation can change with the emergence of new self-organized and management capabilities, which have been partly standardized in the 3GPP.

Femtocells' micro capillarity enables the creation of spontaneous networks. The most evident use case (but not limited) for femtocells within a ULOOP environment is to use it as a ULOOP gateway. It would be a similar case than for a Wi-Fi access point with some differences.

Firstly, ULOOP nodes would be using LTE, which uses a licensed spectrum band, so implications on this matter should be assessed. Secondly, femtocells are normally managed by operators that set up, configure and integrate the equipment into their LTE network. Both differences drive us towards the conclusion that mobile operators should be involved into the value chain. In other words, user centric networks should provide value not only to the end users but also to operators (see Sect. 2 of this paper). Once this value is perceived by operators, users can be encouraged to act as ULOOP nodes by providing incentives to them in the shape of tariff discounts, data cap increase, etc.

In summary, there is a potential complementarity between LTE and ULOOP and in particular through the use of femtocells as ULOOP gateways.

5.2 Wi-Fi: Hotspot 2.0 (Unlicensed Spectrum Bands)

Hotspot 2.0 the upcoming WFA Passpoint^{TM} certification program [8], commonly known as Hotspot 2.0, changes the way users connect to Wi-Fi hotspot networks by defining the process of seamlessly discovering, selecting and getting access to the right network. In addition to the use of 802.11u for network discovery, it also uses WPA2^{TM}-Enterprise. WPA2^{TM} provides both security (you can control who connects) and privacy (the transmissions cannot be read by others) for communications as they travel across your network. WPA2^{TM}-Enterprise uses 802.1X and EAP for authentication.

Currently, only Hotspot 2.0 Rel1 is approved. This first release mainly covers network discovery and selection use cases and security aspects. The introduction of this technology into the market is being mainly impacted by the lack of significant end user device ecosystem and by the lack of definition of end user device Hotspot 2.0 configuration provisioning. In any case, this may soon change since main device vendors like Apple (iOS7) and Samsung (Galaxy III and Galaxy IV) are supporting the technology and configuration provisioning use case is covered by Rel2.

Additionally, there is a big operator interest and some of them (e.g. AT&T) have already started Hotspot2.0 deployments. Thus, everything points out that this technology will be widely deployed and used.

ULOOP and Hotspot 2.0 are complementary technologies that show clear synergies and can be benefited from mutual cooperation.

ULOOP could take advantage from Hotspot 2.0 if user-centric networks can be discovered and selected by using Hotspot 2.0 mechanisms. There are some technological challenges since Hotspot 2.0 is quite operator centric but this could be solved at standardization or proprietary level.

Beyond technical issues, operators need to understand that user centric networks have room into operator centric networks like Hotspot 2.0 and that they can provide benefits. Operators need to understand that this approach would open new business opportunities by the introduction of new services. Thus, an evangelization in that matter is required to be done within the corresponding industry forums.

5.3 Bluetooth v4.0 (Unlicensed Spectrum Bands)

The IC industry moves towards combo ICs where Wi-Fi and Bluetooth are integrated in the same chipset solution. Thus, Bluetooth is almost as pervasive as Wi-Fi. The evolution of this technology relies on a private and non-profitable group, named Bluetooth Special Interest Group (SIG), where the companies contribute to the different Bluetooth versions and standards. Latest version is v4.0 adopted in June 2010. It includes features such as Bluetooth high speed (based on Wi-Fi) and Bluetooth low energy.

Bluetooth has been a very successful technology with a wide range of applications as for example hands free headsets, car stereo systems, connectivity with PC input and output devices (e.g. printer, mouse, etc.) sensoring, etc.

ULOOP covers the concept of user centric networks, while Bluetooth is oriented to personal area networks/devices. In any case, this should not be considered an incompatibility because both technologies can perfectly coexist.

ULOOP might be adapted towards the concept of personal devices centric networks. The device can substitute the role of the user and the incentives can vary to exploit, for example, the available information exchanged between devices and the quality of it. The devices create spontaneous networks and the relationship between them can occur while the trust and the incentives among them can take place.

In the case of using high-speed rates under Bluetooth, the market can move towards a twin wireless access method of Wi-Fi following the same ULOOP specifications.

In summary, there is a bundle of new concepts and service opportunities in a scenario of ULOOP and Bluetooth coexistence and interaction that could be market exploited.

5.4 Li-Fi (Unlicensed Spectrum Bands)

The technology is in a very early stage of development. The industry forum, which drives this technology, was recently created in 2011. It uses visible light instead of traditional radio frequencies for exchanging data through the air interface. The light frequency is located in the visible spectrum so there should be line of sight among devices and the expected coverage can vary with the scenario light conditions.

Due to novelty of this technology, its commercial success is still uncertain. However, this novelty can also be an opportunity in the sense that we are in time of introducing ULOOP concepts to the Li-Fi consortium and these might be included in the standard. Li-Fi can be used as the layer1 and layer2 technology. There might be a Li-Fi modem acting as a ULOOP gateway and/or communications between end user devices acting as ULOOP nodes can use Li-Fi.

6 SWOT Analysis

A SWOT analysis is a structured planning method used to evaluate the Strengths, Weaknesses, Opportunities and Threats related to a project, company, investment opportunity, etc. SWOT is represented by a 2×2 matrix of beneficial and unfavorable indicators from an internal (strengths and weaknesses) and an external (opportunities and threats) perspective with regards to ULOOP project [1], in this case. The market analysis, covered by previous sections, is the main reference framework to perform this analysis. The SWOT analysis provides an indication of the feasibility of exploitation activities on ULOOP beyond the funded R&D project (Tables 7 and 8).

Table 7 ULOOP strengths and weaknesses

Strengths	Weaknesses
User centric networks already positioned in the industry	Very little bargaining power in the standardization bodies
ULOOP project [1] provided an integrated solution and easy to configure	Currently, support only for Android and OpenWRT
ULOOP image as whole software suite	Still reduced end-user testing
Complete trust, resource and mobility management solution	Room for user experience enhancement. A consistent end user-experience evaluation methodology and feedback mechanisms required
Operator relations already established through consortium partners	Still lack of a representative service integration (users are driven by services)
Dedicated standardization efforts performed during ULOOP project [1]	

Table 8 ULOOP opportunities and threats

Opportunities	Threats
Identified business opportunities for operators through the deployment of new services in cooperation with end-users and by offering resources using ULOOP communities as a distribution channel for existing and new services	Multiple new user centric networks
ULOOP communities as a driver to encourage mobile data offloading	Standardization bodies trying to define network frameworks against ULOOP current approach
European Union efforts for network convergence and data offloading	Time to market
ULOOP communities as a way to enable low cost network capillarity expansion for operators	Increasing focus on service-only cloud-based solutions (SaaS)
Increased Wi-Fi support on end-user devices and role in operator networks	Commoditization of Wi-Fi network access: Wi-Fi integrated into service plans may discourage the adoption of new models like ULOOP
Business opportunities for OTT like service providers that may offer/enable ULOOP related services regardless the network operator	
Business opportunities for infrastructure vendors as ULOOP technology providers targeting mainly operators but also consumers	
Industrial needs covered by ULOOP concerning mobility solutions at wireless scenarios	
Business opportunities for device vendors implementing ULOOP functionalities and targeting both consumers and operators. Example ULOOP communities enabled and promoted directly by device vendors	

7 Conclusions

Market analysis shows the potential business opportunities that may arise for main stakeholders involved in the value chain. Although user-centric networks are the main subject, the analysis highlights that operators are crucial for the commercial success of ULOOP derived services. Thus, operators should be involved in the value chain and should benefit from ULOOP. For instance, operators may be involved as ULOOP technology distributors and end-users can help and contribute to design and develop services on top of this technology. If the condition of operators is fulfilled, other stakeholders (e.g. infrastructure and device vendors, OTT service providers) can also take part on the value generation and obtain benefits from ULOOP.

Additionally, operators' involvement is also subject to the way their subscribers perceive ULOOP. User centric networks success depends on the existence of a critical mass and this is possible if end-users perceive ULOOP derived services as value added ones.

References

1. EU FP7 ULOOP Project. http://www.uloop.eu/
2. Bogliolo A (ed) ULOOP deliverable 2.2 Socio-economic sustainability report. ULOOP Consortium. Available http://www.uloop.eu. Accessed Apr 2011
3. Bogliolo A, Polidori P, Aldini A, Moreira W, Mendes P, Yildiz M, Ballester C, Seigneur JM (2012) Virtual currency and reputation-based cooperation incentives in user-centric networks. In: 8th international wireless communications and mobile computing conference (IWCMC-2012), IEEE, Limassol, Cyprus, Aug 2012, pp 895–900
4. Aldini A, Bogliolo A (2014) Modeling and verification of cooperation incentive mechanisms in user-centric wireless communications. Security, privacy, trust, and resource management in mobile and wireless communications. IGI Global, Hershey, pp 432–461
5. Gartner Research. Market share: mobile phones by region and country, 3Q12. http://www.gartner.com/resId=2236115
6. Mobile Marketer: Smartphones, tablets and the mobile revolution. 29 Jan 2013
7. NET Applications (web analytics firm). Internet usage market share. http://www.netmarketshare.com
8. Wi-Fi Alliance (WFA). Wi-Fi certified passpoint^{TM}. http://www.wi-fi.org/discover-wi-fi/wi-fi-certified-passpoint

ULOOP Business Case Study

David Valerdi and Ricardo Mota

Abstract This paper covers a business plan exercise based on ULOOP [1] technology and derived services. It provides a business case structure that can be used by third parties to build their own ones. This business plan primarily considers the immediate benefits of making accessible ULOOP technology to fixed and mobile networks subscribers. It also shows illustrative financial figures based on experience of ULOOP industry partners with similar products.

Keywords Business case · Value proposition · Revenue structure · Cost structure · Financial indicators

1 Introduction

This paper covers a business plan based on ULOOP technology. The business proposition takes as a basis the business opportunities related to ULOOP technology described in the previous paper of this book. It is an exercise that provides a business case structure that third parties can use to build their own ones. It also provides illustrative financial figures based on the experience of ULOOP industrial partners.

This business plan primarily considers the immediate benefits of making accessible ULOOP technology to fixed and mobile networks subscribers. However, ULOOP can be an enabler for additional services and opportunities as covered in the previous paper of this book. These other opportunities may have slightly different business

D. Valerdi (✉)
Fon Labs S.L., Getxo, Spain
e-mail: david.valerdi@fon.com

R. Mota
ZON TV Cabo Portugal SA, Lisboa, Portugal
e-mail: ricardo.a.mota@parceiros.zonoptimus.pt

A. Aldini and A. Bogliolo (eds.), *User-Centric Networking*,
Lecture Notes in Social Networks, DOI: 10.1007/978-3-319-05218-2_17,
© Springer International Publishing Switzerland 2014

models and impact. However, they can be built on top of those immediate benefits and this baseline business plan.

Paper starts with a description of the roles of companies involved in the business plan; what accountabilities of each other are and how they interact (company names are not real but can be easily related to existing companies). Then, value propositions of each of them are explained. Finally, financials are built by basing on the roles and value propositions explained in the initial sections. These financials cover revenues and cost structures, results summary and key project indicators. Additionally, a sensitivity analysis is performed by considering worst, typical and best cases.

2 Roles of Companies Involved

Before starting this exercise, it is important to describe the roles of both companies to understand business interactions.

WiCom is a wireless service provider that develops its own technology. WiCom normally partners with traditional operators (mainly fixed network operators). They deploy their wireless networks and technology on top of operator networks under a partnership framework. In this business plan, they will be developing, evolving and integrating ULOOP functionalities on top of their current technology. Additionally, they will be leading the deployment, operation and maintenance of resulting ULOOP networks. WiCom will offer related ULOOP product and services to operators (mobile and fixed/integrated). This exercise assumes that ABC Telecom is WiCom's first customer of this technology. In other words, WiCom will be the technology provider and the operator of ULOOP networks on top of ABC Telecom network infrastructure. WiCom will act as an OTT service provider but with a close collaboration with the "traditional" operator, both at technical and business levels.

ABC Telecom is a fixed and virtual mobile operator. They own the broadband infrastructure where ULOOP networks will rely on. ABC Telecom will be ULOOP technology (provided by WiCom) buyer/consumers and will also have accountabilities in ULOOP networks deployment, operation and maintenance (jointly with WiCom). Additionally, ABC Telecom will be in charge of ULOOP services marketing and distribution. They will be in charge of building ULOOP networks value propositions for their subscribers and other potential users.

3 ULOOP Network Value Proposition

Value propositions are different between WiCom and ABC Telecom perspectives. This section provides a view of what value propositions are built around ULOOP networks for each of the companies.

3.1 WiCom Value Proposition

The current targeted market does not change for WiCom. WiCom will keep on targeting telco operators with ULOOP related products and services. They will enable ULOOP technologies in Wi-Fi access points and develop the required applications for end users devices. Benefits and business opportunities related to ULOOP for operators are described in the previous paper of this book. In this business plan, we will focus on those benefits related to mobile data offloading, client retention and attraction improvement that ULOOP applications can enable. These benefits are crucial for an operator of the characteristics on ABC Telecom (i.e. fixed and virtual mobile operator).

3.2 ABC Telecom Value Proposition

ABC Telecom will make available ULOOP derived products to its subscribers as a free of charge, opt-in add-on to current broadband service.

The main benefits of having ULOOP technology and the new opportunities foreseen are described in the previous paper of this book. It keeps mainly targeting the residential customer, but with the perspective of an integrated solution that not only rewards users but allows them to create their own networks. ABC Telecom also expects that this approach will bring more customers.

Additional to the user-oriented added values of the service, it will allow the implementation of a mobile data off-loading strategy, with the corresponding savings in terms of the rent to be paid, as a mobile virtual operator. For customers, this will be an additional money-saver, from the mobile data plan they will use lesser, turning over to the Wi-Fi ULOOP networks.

4 Financials

Along the following sections, financials for WiCom and ABC Telecom related to this business plan are presented. This analysis covers revenue and cost structures and key financial indicators related to this business plan. Finally, a sensitivity analysis is presented after defining worst and best cases, in addition to the typical case.

The reader must consider these figures are illustrative and based on ULOOP industrial partners experience. The aim of this exercise is to provide a structure/baseline of a business case that can be used by third parties to build their own business plans for ULOOP like applications.

4.1 WiCom Financials

This section goes over all the financials related to WiCom business plan for a period of 6 years (2013–2018). The section is split into several subsections. It starts with the definition of the business model and revenue structure and expectations (typical case). Then, cost structure is presented, showing personnel costs as the main contributor. Business case results summary and key financial indicators are derived from revenue and cost structures. The section ends up with a sensitivity analysis that adds worst and best cases by mainly playing with operators contracts pace and number of activated ULOOP gateways.

4.1.1 Business Model and Revenue Structure and Expectations

The business model of WiCom for ULOOP related products and services is based on a fixed and a variable fee to be charged to operators:

- A fixed annual service maintenance license fee per deployed platform (it is assumed that only one platform is deployed per operator).
- A variable fee composed of the following items:

  - An annual license fee per ULOOP residential gateway. Applicable when ULOOP functionality is integrated in operators CPE.
  - An annual license fee per ULOOP public gateway. Applicable when ULOOP functionality is integrated in operators CPE.
  - A one-time fee per ULOOP gateway sold. Applicable when a separate access point is sold (only thought for mobile operators not owning a fixed network infrastructure).

By basing on this model, Table 1 shows the revenue structure and expectations (typical case) for this project.

Note that main inputs to derive revenue expectations are number of contracts for mobile and fixed / integrated operators (since different fees are applicable), number of ULOOP gateways sold (individual equipment supporting ULOOP functionalities) and number of activated licenses (in operator CPE's) split in public and residential gateways (because different prices are applicable).

It is assumed that 2013 and 2014 are years where ULOOP functionalities are still being developed and productized by WiCom. Figures shown are based on WICOM experience and operator acceptation to ULOOP like features and functionalities. Table 1 covers the typical case. In Sect. 4.1.4, best and worst cases are also considered. The following graph shows a summary of market expected acceptation to ULOOP technology, based on ULOOP industrial partners experience in deploying similar features (Fig. 1).

Table 1 WICOM revenue structure (typical case)

	2013	2014	2015	2016	2017	2018
Revenues (typ)						
Total revenue improvement (k€)	–	–	1,540	4,850	9,725	12,090
Revenue fixed fee						
Number of mobile (only) operator contracts (u)	–	–	0	1	2	3
Number of fixed (inc, integrated) operator contracts (u)	–	–	1	2	3	5
Annual fixed service maintenance license (k€)	–	–	500	500	500	500
Total revenue (fixed fee) (k€)	–	–	500	1,500	2,500	4,000
Revenue variable fee (incl, ULOOP GW's)						
Number of ULOOP gateways sold (k)	–	–	0	75	200	350
Number of activated ULOOP residential gateways (k)	–	–	125	200	350	550
Number of activated ULOOP public gateways (k)	–	–	1	3	5	6
Price/ ULOOP gateway (€)	–	–	0	25	25	15
Annual license fee per ULOOP residential gateway (€)	–	–	8	7	6	5
Annual license fee per ULOOP public gateway (€)	–	–	40	30	25	15
Total revenue (variable fee) (k€)	–	–	1,040	3,350	7,225	8,090

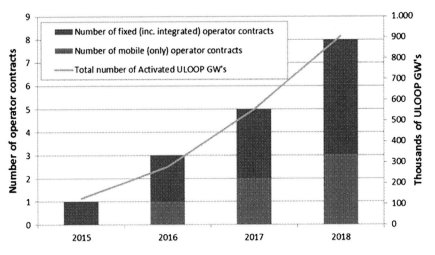

Fig. 1 WICOM market expected acceptation to ULOOP technology

4.1.2 Cost Structure

Table 2 shows the cost structure applicable to ULOOP technology development, integration and deployment and maintenance and operation. Main contribution to the overall costs is personnel. Note that this is a CAPEX free structure, since it is assumed that WiCom makes use of cloud datacentre infrastructure for their frontend and backend systems.

Table 2 WICOM cost structure

	2013	2014	2015	2016	2017	2018
Costs (typ)						
Total costs (k€)	285	283	793	3,130	6,261	9,456
Direct costs						
R&D costs (k€)	210	60	18	18	18	18
Development costs (k€)	9	158	36	36	36	36
Integration & Deployment costs (k€)	–	–	20	60	100	160
Maintenance & Operation costs (incl. Platform) (k€)	–	–	440	1,320	2,200	3,520
ULOOP GW costs (incl. CPE integration) (k€)	–	–	96	974	2,462	3,540
Total direct costs (k€)	219	218	610	2,408	4,816	7,274
Indirect costs						
General operating costs (k€)	66	65	183	722	1,445	2,182

Table 3 WiCom results summary (typical case)

Typical	2013	2014	2015	2016	2017	2018	Total
Revenues (Typ.) (M€)	–	–	1.54	4.85	9.73	12.09	**28.21**
Total revenue (fixed fee) (M€)	–	–	0.50	1.50	2.50	4.00	**8.50**
Total revenue (variable fee) (M€)	–	–	1.04	3.35	7.23	8.09	**19.71**
Costs (Typ.) (M€)	0.28	0.28	0.79	3.13	6.26	9.46	**20.21**
Total direct costs (M€)	0.22	0.22	0.61	2.41	4.82	7.27	**15.54**
Total indirect costs (M€)	0.07	0.07	0.18	0.72	1.44	2.18	**4.66**
Net impact (Typ.) (M€)	(0.3)	(0.3)	0.7	1.7	3.5	2.6	**8.00**
Cumulative net impact (Typ.) (M€)	(0.3)	(0.6)	0.2	1.9	5.4	8.0	

Table 4 WiCom key financial indicators

Key indicators	Typ.
Net present value @ 20% (M€)	3.10
Payback period (year)	2.76
ROI (%)	40

In addition to personnel costs, maintenance and operation consider costs related to data centre fees. The data centre hosts frontend and backend systems. On the other hand, ULOOP GW costs cover ULOOP gateways original manufacturer costs and ULOOP functionalities integration effort (this effort is also applicable when ULOOP functionality is integrated in a third party CPE).

4.1.3 Results Summary and Financial Indicators

Tables 3, 4 show business case results summary considering revenue and cost structures and figures covered by previous sections (typical case):

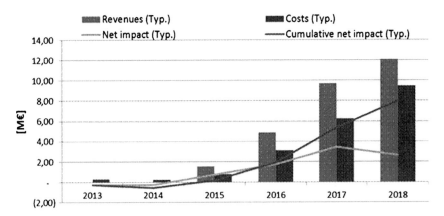

Fig. 2 WiCom results summary (typical case)

Table 5 WiCom key financial indicators worst, typical and best cases

Key indicators	Typ.	Best	Worst
Net present value @ 20% (M€)	3.10	6.92	1.48
Payback period (year)	2.76	2.36	4.03
ROI (%)	40	51	48

According to typical figures, the project shows are very positive with sustainable revenue and cost structures that result:

- An accumulated net impact of 8 MM€ that translated into net present value is 3.10 MM€.
- An acceptable payback period of less than 3 years.
- A relevant return of investment of 40%.

On the other hand, it should be noted that mayor contribution to total revenue is coming from variable fees. That means that success of the project will mainly subject to end user acceptance of ULOOP solutions.

Figure 2 depicts summary results across observation period.

4.1.4 Sensitivity Analysis: Best, Typical and Worst Cases

In addition to the typical case shown in previous version, a sensitivity analysis covering a worst and a best case is performed. Two variables are considered in this analysis: number of operator contracts and number of activated ULOOP gateways. This variables directly impact on revenue figures. In the worst case, one year delay on contracts pace and 15% less activated ULOOP gateways are considered. While in the best case, one additional contract per year (mobile and fixed/integrated operators) and 15% more activated ULOOP gateways are considered.

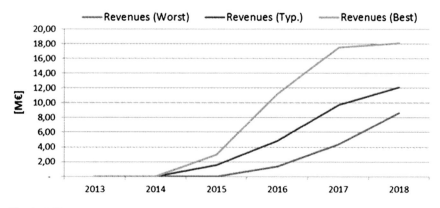

Fig. 3 WiCom revenue worst, typical and best cases

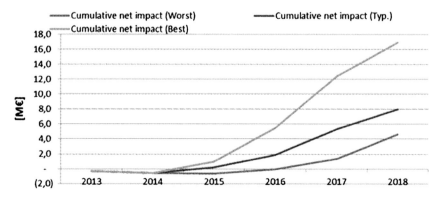

Fig. 4 WICOM cumulative net impact worst, typical and best cases

Figures 3, 4 and Table 5 summarize sensitivity assessment results. Typical case described in previous sections is also included for reference purposes .

Note that ROI is providing a better figure in the worst case than in the typical one. This is because of ROI indicator inherent error. It is simply showing a better ratio between costs and net impact. Most representative and accurate indicator is the net present value and this is the one that should be taken as the main reference.

Looking at the worst case, figures shows a relevant robustness against a pessimistic scenario where contracts are not in place (until 2016 in the case of a fixed operator and until 2017 in case of a mobile operator) and the number of activated ULOOP gateways is significantly less. On the other hand, best case proves that overall results can be considerably improved by an increase of operator contracts and ULOOP GW activation. This is mainly enabled by economies of scale, since R&D and development costs are shared among different contracts.

4.2 ABC Telecom Financials

The financial exercise considers the expected gains from the introduction of ULOOP technology, and the impacts it will represent in terms of new users and retain of present customers—sales increase and churn reduction. This exercise is done by taking into account previous experience with similar services and the feedback from users (collected at the pilot performed with real users during ULOOP project [1]). Users are expected to value the presence of such technology onto their home gateways. An important note is that the ARPU is not calculated just with Internet/broadband service, since it is always a part of a triple-play offering—as with the mobile, where the data plans are calculated with the voice plans. Nevertheless, the gains that are considered are those coming from the difference between revenues without and with ULOOP, as well as those directly and exclusively from ULOOP-derived services.

Then, it takes into account the additional costs from deploying ULOOP onto ABC Telecom gateways, which, following on their partnership with WiCom, derives directly from the fees WiCom expects to charge for the development and maintenance work, as well as the per device license fee, based on their usual business model. WiCom business model is described in Sect. 4.1.1. It is assumed that, as today's project, the deployment of ULOOP technology does not require the usage of any additional hardware, either on customer's premises or at ABC Telecom's. This means that the only costs that are considered are those related to software licensing fees and platform maintenance, which are charged by WiCom.

4.2.1 Benefits and Revenue Expectations

Table 6 summarizes the areas where it is foreseeable that there are any benefits and revenues, in a typical scenario. The exercise takes into account that 2014 will be a development and integration phase, 2015 initial deployments, so only in 2016 measurements in revenue increments are expected:

Revenue improvement is driven by an increase of customer acquisition and a reduction of churn, both in fixed and mobile business but with different rates. As mentioned before, this improvement is enabled by the expectation of an increase of value perceived by current subscribers and potential new customers. Then, total revenue improvement is derived from those figures and from ABC Telecom ARPU and subscribers' base expected evolution.

Other type of benefits is the one related to CAPEX savings provided by an increase of mobile data offloading. These are enabled by the extension of Wi-Fi network thanks to ULOOP networks.

4.2.2 Cost Structure

Table 7 shows the cost structure to be considered in ABC Telecom's case. As explained before, any new service to be deployed in ABC Telecom customer's

Table 6 ABC Telecom revenue structure (typical case)

	2014	2015	2016	2017	2018	2019
Boost in ABC Telecom business	–					
Total revenue improvement (k€)	–	–	1,063	3,705	6,234	8,166
Revenue improvement fixed						
Acquisition improvement (%)	–	–	2	1	1	1
Churn rate improvement (%)	–	–	5	5	4	4
Income w/ ULOOP (k€)	–	301,485	303,729	309,963	319,800	333,506
Income w/o ULOOP (k€)	–	301,485	303,306	308,849	318,134	331,229
Total revenue improvement (k€)	–	–	422	1,115	1,666	2,277
Revenue improvement mobile						
Acquisition improvement (%)	–	–	1	2	1	1
Churn rate improvement (%)	–	–	3	5	4	4
Income w/ ULOOP (k€)	–	732,654	756,155	790,632	834,383	883,271
Income w/o ULOOP (k€)	–	732,654	755,515	788,042	829,815	877,383
Total revenue improvement (k€)	–	–	641	2,590	4,567	5,888
Mobile business driven CAPEX savings						
Traffic offloaded/year (GB)	–	2,361	3,016	3,903	5,098	6,679
Savings/GB (€)	–	0.28	0.26	0.24	0.22	0.20
Total mobile business driven CAPEX savings (k€)	–	661	784	937	1,122	1,336

Table 7 ABC Telecom cost structure (typical)

	2014	2015	2016	2017	2018	2019
Deployment costs						
Residential footprint fees (k€)	–	1,320	1,734	1,751	1,973	2,167
Public footprint fees (k€)	–	150	138	73	66	76
Annual fixed service maintenance license (k€)	–	500	500	500	500	500
Total cost for ABC Telecom (k€)	–	1,970	2,371	2,323	2,539	2,744

premises will be done with ABC Telecom home gateways. This means that there is not a CAPEX impact due to these new services.

Thus, it only needs to cover the costs engaged by the software development, maintenance and licensing. Here, they are calculated considering a typical scenario (the reduction from 2nd to 3rd year appears due to a reduction in per-gateway fees proposed by WiCom):

4.2.3 Results Summary and Financial Indicators

As introduced before, it is foreseeable an increase in the number of subscribers, due to the positive impact of the features ULOOP will enable but also to the reduction of the churn. Experience reveals that customers value added functionalities that

Table 8 ABC Telecom results summary (typical)

Typical	2015	2016	2017	2018	2019	Total
Boost in ABC Telecom business (m€)	0.66	1.85	4.64	7.36	9.50	24.01
Total revenue improvement (m€)	–	1.06	3.70	6.23	8.17	19.17
Mobile CAPEX savings (m€)	0.66	0.78	0.94	1.12	1.34	4.84
Cost of deployment (m€)	1.97	2.37	2.32	2.54	2.74	11.95
Net impact (m€)	(1.3)	(0.5)	2.3	4.8	6.8	12.06
Cumulative net impact (Typ.) (m€)	(1.3)	(1.8)	0.5	5.3	12.1	

these networks provide to them, mainly with extra services available outside their residences. Since these new (or retained) customers will likely have triple-play packages, the impact on the revenue is quite relevant (Table 8).

From the calculated figures, it can be understood that ULOOP can represent a good improvement to ABC Telecom's business, with a payback period of less than 3 years and good net present value figure. It is clear that the greatest improvement will not come from CAPEX savings in the mobile part, but from increases in customer base and churn reductions. As mentioned before, these are enabled because it is expected that customers perceive the additional value provided by ULOOP.

4.2.4 Sensitivity Analysis: Worst, Typical and Best Cases

From the typical business case, two additional scenarios are derived: pessimistic (worst) and optimistic (best). Variables are the user-sensitive components, which reflect how the customers will react to the introduction of this kind of technology (with all the advantages already considered). The pessimistic case takes the approach of a slower adoption of ULOOP, reflected in a reduction in the number of new subscribers to ABC Telecom triple-play offering. It also considers that the deployment of user-centric networks will have less impact in reducing the churn. The optimistic approach considers just the opposite—that it will bring more new customers and that it will make more existing customers remain with ABC Telecom. Figures 5, 6 show different scenarios of subscribers' behaviors.

This variation of subscribers' evolution has also a direct impact on the amount of mobile data offloaded traffic and consequently on the CAPEX savings that offloading provides (Fig. 7).

These changes on the inputs of the business case have a direct impact on revenues and costs (at a lesser extend).

As explained above, direct costs of the ULOOP implementation come from the WiCom fees for operation, maintenance and licenses. Different scenarios do not impact much in costs, due to the fact that the biggest slice of the costs comes from a fixed maintenance fee, and the per-unit fee becomes less relevant, in terms of comparing the different scenarios. Nevertheless, it is interesting to highlight that an expected decrease in the per-unit fee (2016–2017) will, in fact, (in all scenarios) cause an overall reduction of the costs in that year (Figs. 8, 9).

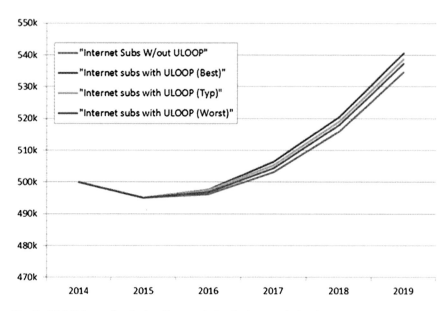

Fig. 5 ABC Telecom fixed subscribers evolution for best, typical and worst cases

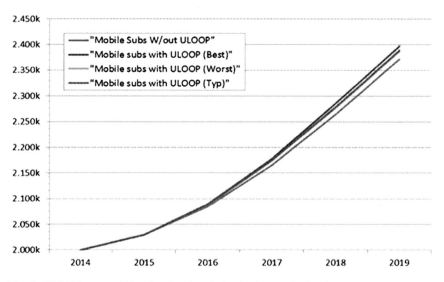

Fig. 6 ABC Telecom mobile subscribers' evolution for best, typical and worst cases

As explained in previous sections, increase in revenue is coming from monthly fees of an established base of customers, expected to increase faster with a technology like ULOOP.

From previous calculations, derived business indicators are shown in (Table 9).

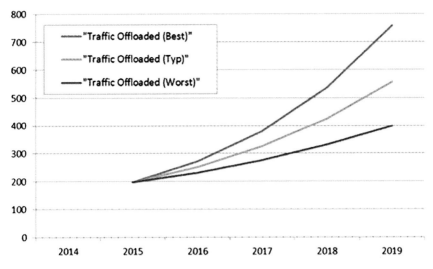

Fig. 7 ABC Telecom mobile data offloaded traffic impact for best typical and worst cases

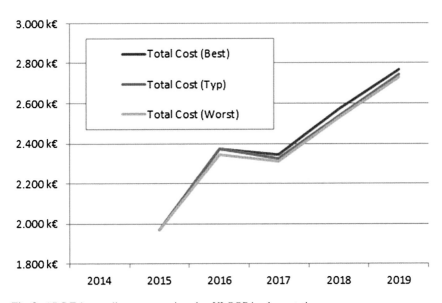

Fig. 8 ABC Telecom direct costs assigned to ULOOP implementation

It can be noted that, even in the worst scenario, business case shows very promising figures. That means that even a slow or pessimistic improvement of subscribers' base still produces good financial figures, making ULOOP deployment attractive for ABC Telecom.

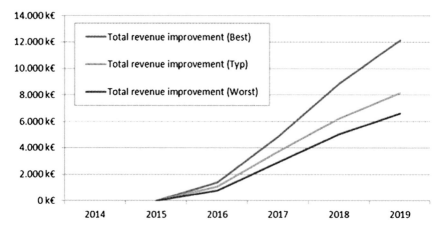

Fig. 9 ABC Telecom total revenue improvement

Table 9 ABC Telecom key financial indicators worst, typical and best cases

Key indicators	Typ.	Best	Worst
Net present value @ 15 % (m€)	6.10	11.06	3.47
Payback period (year)	2.79	2.40	3.55
ROI (%)	101	175	63

5 Conclusions

The business plan described in this paper mainly proves how two representative types of companies can collaborate and build a consistent and sustainable business proposition based on ULOOP. It shows how WiCom as an OTT service and technology provider can collaborate with an operator like ABC Telecom (fixed plus mobile virtual operator) to exploit ULOOP technologies and resulting in mutual benefits.

As in the case of other user centric network approaches, the business success depends on the existence of a critical mass. This is enabled if ULOOP is perceived by subscribers as a value added service. If these conditions are fulfilled, improved customer acquisition, reduction of churn, mobile data offloading savings, etc. justifies the purchase of ULOOP technology by operators and, consequently, the development of this technology by third parties like WiCom or infrastructure vendors. This positive potential scenario is proved across the financial calculations results provided in Sect. 4 of this paper.

Reference

1. EU FP7 ULOOP Project, www.uloop.eu

Glossary

The following list of definitions is taken from European Commission Directives.

Access [2002/19/EC Art. 2.a, as amended by 2009/140/EC]: means the making available of facilities and/or services to another undertaking, under defined conditions, on either an exclusive or non-exclusive basis, for the purpose of providing electronic communications services, including when they are used for the delivery of information society services or broadcast content services. It covers inter alia: access to network elements and associated facilities, which may involve the connection of equipment, by fixed or non-fixed means (in particular this includes access to the local loop and to facilities and services necessary to provide services over the local loop); access to physical infrastructure including buildings, ducts and masts; access to relevant software systems including operational support systems; access to information systems or databases for pre-ordering, provisioning, ordering, maintaining and repair requests, and billing; access to number translation or systems offering equivalent functionality; access to fixed and mobile networks, in particular for roaming; access to conditional access systems for digital television services and access to virtual network services.

Application Program Interface (API) [2002/21/EC Art. 2.p]: means the software interfaces between applications, made available by broadcasters or service providers, and the resources in the enhanced digital television equipment for digital television and radio services.

Associated Facilities [2002/21/EC Art. 2.e, as amended by 2009/140/EC]: means those associated services, physical infrastructures and other facilities or elements associated with an electronic communications network and/or an electronic communications service which enable and/or support the provision of services via that network and/or service or have the potential to do so, and include, inter alia, buildings or entries to buildings, building wiring, antennae, towers and other supporting constructions, ducts, conduits, masts, manholes, and cabinets.

Associated Services [2002/21/EC Art. 2.e, as inserted by 2009/140/EC]: means those services associated with an electronic communications network and/or an electronic communications service which enable and/or support the provision of services

A. Aldini and A. Bogliolo (eds.), *User-Centric Networking,*
Lecture Notes in Social Networks, DOI: 10.1007/978-3-319-05218-2,
© Springer International Publishing Switzerland 2014

via that network and/or service or have the potential to do so and include, inter alia, number translation or systems offering equivalent functionality, conditional access systems and electronic programme guides, as well as other services such as identity, location and presence service.

Call [2002/58/EC Art. 2.e]: means a connection established by means of a publicly available telephone service allowing two-way communication in real time.

Commercial Communication [2000/31/EC Art. 2.f]: any form of communication designed to promote, directly or indirectly, the goods, services or image of a company, organisation or person pursuing a commercial, industrial or craft activity or exercising a regulated profession. The following do not in themselves constitute commercial communications: information allowing direct access to the activity of the company, organisation or person, in particular a domain name or an electronic-mail address; communications relating to the goods, services or image of the company, organisation or person compiled in an independent manner, particularly when this is without financial consideration.

Communication [2002/58/EC Art. 2.d]: means any information exchanged or conveyed between a finite number of parties by means of a publicly available electronic communications service. This does not include any information conveyed as part of a broadcasting service to the public over an electronic communications network except to the extent that the information can be related to the identifiable subscriber or user receiving the information.

Conditional Access System [2002/21/EC Art. 2.f]: means any technical measure and/or arrangement whereby access to a protected radio or television broadcasting service in intelligible form is made conditional upon subscription or other form of prior individual authorisation.

Consent [2002/58/EC Art. 2.f]: by a user or subscriber corresponds to the data subject's consent in Directive 95/46/EC.

Consumer [2000/31/EC Art. 2.e]: any natural person who is acting for purposes which are outside his or her trade, business or profession; according to [2002/21/EC Art. 2.i], consumer means any natural person who uses or requests a publicly available electronic communications service for purposes which are outside his or her trade, business or profession.

Electronic Communications Network [2002/21/EC Art. 2.a, as amended by 2009/140/EC]: means transmission systems and, where applicable, switching or routing equipment and other resources, including network elements which are not active, which permit the conveyance of signals by wire, radio, optical or other electromagnetic means, including satellite networks, fixed (circuit- and packet-switched, including Internet) and mobile terrestrial networks, electricity cable systems, to the extent that they are used for the purpose of transmitting signals, networks used for radio and television broadcasting, and cable television networks, irrespective of the type of information conveyed.

Electronic Communications Service [2002/21/EC Art. 2.c and the following 2002/77/EC Art. 1.3]: means a service normally provided for remuneration which consists wholly or mainly in the conveyance of signals on electronic communications networks, including telecommunications services and transmission services in networks used for broadcasting, but exclude services providing, or exercising editorial control over, content transmitted using electronic communications networks and services; it does not include information society services, as defined in Article 1 of Directive 98/34/EC, which do not consist wholly or mainly in the conveyance of signals on electronic communications networks.

Electronic Mail [2002/58/EC Art. 2.h]: means any text, voice, sound or image message sent over a public communications network which can be stored in the network or in the recipient's terminal equipment until it is collected by the recipient.

End-User [2002/21/EC Art. 2.n]: means a user not providing public communications networks or publicly available electronic communications services.

Enhanced Digital Television Equipment [2002/21/EC Art. 2.o]: means set-top boxes intended for connection to television sets or integrated digital television sets, able to receive digital interactive television services.

Established Service Provider [2000/31/EC Art. 2.c]: a service provider who effectively pursues an economic activity using a fixed establishment for an indefinite period. The presence and use of the technical means and technologies required to provide the service do not, in themselves, constitute an establishment of the provider.

Exclusive Rights [2002/77/EC Art. 2.5]: shall mean the rights that are granted by a Member State to one undertaking through any legislative, regulatory or administrative instrument, reserving it the right to provide an electronic communications service or to undertake an electronic communications activity within a given geographical area.

General Authorisation [2002/20/EC Art. 2, as amended by 2009/140/EC]: means a legal framework established by the Member State ensuring rights for the provision of electronic communications networks or services and laying down sector specific obligations that may apply to all or to specific types of electronic communications networks and services, in accordance with this Directive.

Local Loop [2002/19/EC Art. 2.e, as amended by 2009/140/EC]: means the physical circuit connecting the network termination point to a distribution frame or equivalent facility in the fixed public electronic communications network.

Location Data [2002/58/EC Art. 2.c]: means any data processed in an electronic communications network, indicating the geographic position of the terminal equipment of a user of a publicly available electronic communications service.

Interconnection [2002/19/EC Art. 2.b]: means the physical and logical linking of public communications networks used by the same or a different undertaking in order to allow the users of one undertaking to communicate with users of the same or another undertaking, or to access services provided by another undertaking. Services may be provided by the parties involved or other parties who have access to the network. Interconnection is a specific type of access implemented between public network operators.

National Regulatory Authority [2002/21/EC Art. 2.g]: means the body or bodies charged by a Member State with any of the regulatory tasks assigned in this Directive and the Specific Directives.

Network Termination Point (NTP) [2002/21/EC Art. 2.d, as inserted by 2009/140/EC]: means the physical point at which a subscriber is provided with access to a public communications network; in the case of networks involving switching or routing, the NTP is identified by means of a specific network address, which may be linked to a subscriber number or name.

Operator [2002/19/EC Art. 2.c]: means an undertaking providing or authorised to provide a public communications network or an associated facility.

Provision of an Electronic Communications Network [2002/21/EC Art. 2.m]: means the establishment, operation, control or making available of such a network.

Public Communications Network [2002/21/EC Art. 2.d, as amended by 2009/140/EC]: means an electronic communications network used wholly or mainly for the provision of electronic communications services available to the public which support the transfer of information between network termination points.

Publicly Available Electronic Communications Services [2002/77/EC Art. 1.4]: shall mean electronic communications services available to the public.

Recipient of the Service [2000/31/EC Art. 2.d]: any natural or legal person who, for professional ends or otherwise, uses an information society service, in particular for the purposes of seeking information or making it accessible.

Service [1998/34/EC Art. 1.2 as inserted by 1998/48/EC]: any Information Society service, that is to say, any service normally provided for remuneration, at a distance, by electronic means and at the individual request of a recipient of services. For the purposes of this definition: "at a distance" means that the service is provided without the parties being simultaneously present; "by electronic means" means that the service is sent initially and received at its destination by means of electronic equipment for the processing (including digital compression) and storage of data, and entirely transmitted, conveyed and received by wire, by radio, by optical means or by other electromagnetic means; "at the individual request of a recipient of services" means that the service is provided through the transmission of data on individual request.

Service Provider [2000/31/EC Art. 2.b]: any natural or legal person providing an information society service.

Special Rights [2002/77/EC Art. 2.6]: shall mean the rights that are granted by a Member State to a limited number of undertakings through any legislative, regulatory or administrative instrument which, within a given geographical area:

(a) designates or limits to two or more the number of such undertakings authorised to provide an electronic communications service or undertake an electronic communications activity, otherwise than according to objective, proportional and nondiscriminatory criteria, or:

(b) confers on undertakings, otherwise than according to such criteria, legal or regulatory advantages which substantially affect the ability of any other undertaking to provide the same electronic communications service or to undertake the same electronic communications activity in the same geographical area under substantially equivalent conditions.

Subscriber [2002/21/EC Art. 2.k]: means any natural person or legal entity who or which is party to a contract with the provider of publicly available electronic communications services for the supply of such services.

Traffic Data [2002/58/EC Art. 2.b]: means any data processed for the purpose of the conveyance of a communication on an electronic communications network or for the billing thereof.

Universal Service [2002/21/EC Art. 2.j]: means the minimum set of services, defined in Directive 2002/22/EC (Universal Service Directive), of specified quality which is available to all users regardless of their geographical location and, in the light of specific national conditions, at an affordable price.

User [2002/21/EC Art. 2.h]: means a legal entity or natural person using or requesting a publicly available electronic communications service; according to [2002/58/EC Art. 2.a], user means any natural person using a publicly available electronic communications service, for private or business purposes, without necessarily having subscribed to this service.

Value Added Service [2002/58/EC Art. 2.g]: means any service which requires the processing of traffic data or location data other than traffic data beyond what is necessary for the transmission of a communication or the billing thereof.

The following is a list of ULOOP specific definitions.

Application: computer software design to perform a single or several specific tasks, e.g., a calendar. In ULOOP, it is an instantiation of a user service. For instance, Voice over IP is an example of a user service provided by different applications, e.g., Skype, or Gizmo.

Business Incentives: relate to incentives associated with micro-generation models. A concrete example for a business incentive could be a specific peering scheme that may assist the access operators in understanding how to obtain revenue based on ULOOP architectures.

Community: set of ULOOP nodes that hold common interests (such as sharing connectivity or resources and peripherals) at some instant in time and space. In other words, the node location exhibits a space and time correlation which is the basis to establish a robust connectivity model. The notion of community does not have any relation whatsoever to an Online Social Network (OSN), nor event to some specific OSN subset.

Cooperation Incentives: can be viewed as rewards capable of providing extrinsic motivation to be cooperative, thus realigning individuals utility to public utility. In ULOOP, a CI can be seen as a scheme that comprises a function measuring the level of cooperation (such level represents a benefit from an individual and from an aggregate perspective); a ranking related to the node reputation level (such ranking can be provided by a mechanism that can be reactive, proactive, or preventive); a payment/reward, i.e., a measure of the benefit that the node achieves for cooperating.

Crypto-ID: virtual identity generated through cryptographic means and associated to a single user, who may own more than one end-user device.

Handover: process of transferring an ongoing communication session between two networks, or two communities, from one or several ULOOP enabled devices to other devices.

Incentive: factor (economic or sociological) that motivates a particular action or a preference for a specific choice.

Interest: parameter capable of providing a measure (cost) of the "attention" of a node towards a specific location in a specific time instant. In other words, an interest is a parameter that provides a node with a measure of a specific time and space correlation.

Mobile Virtual (Network) Operator: company that provides mobile telephony services but does not have its own licensed frequency allocation, nor does it necessarily have all the infrastructure required to provide mobile telephone service.

Mobility Anchor Point (MAP): component that is located in a ULOOP Gateway and provides mobility anchor functionality to ULOOP nodes that are associated with another ULOOP Gateway.

Mobility Coordination Function (MCF): component that serves as a broker of distributed mobility anchor points (MAPs) that helps in managing the limited resources of the ULOOP Gateways that offer MAP services by selecting MAPs on behalf of the users.

Mobility Management: system providing analysis, estimation, and tracking of node movements through time, as well as handover support.

Network Infrastructure: collection of links and of networking nodes that together enable data transmission between (Internet) users. The links connect the nodes together and are themselves built upon an underlying transmission network which physically pushes the message across the link.

Network Service: system that is required to support, from a network perspective, user services. For instance, Internet connectivity is a network service. Examples of network services in ULOOP are trust management, resource management, identity disambiguation.

Owner: entity (e.g., end-user, operator, virtual operator) that is to be made responsible for any action concerning his/her device. The term "responsible" reflects liability, i.e., from an operators perspective the owner is the single responsible for the adequate/inadequate usage of the users device within a specific, trust-bounded community.

Reputation: can be seen as the "public trustworthiness" of a ULOOP user and it usually encompasses the aggregation of individual trust values computed by many ULOOP users.

Resource: physical or virtual element of a global system. For instance, bandwidth, energy, data, devices, are examples of resources in ULOOP.

Resource Management: system processing the ad-hoc, efficient, and effective deployment of the resources of wireless infrastructures. RM aspects are related to throughput maximization, congestion control, dynamic control of the community growth, which depends on the fluctuations of the network both in terms of traffic due to nodes joining and leaving frequently, as well as due to the movement of nodes.

Retailer: user who needs a monetized reward in order to share services and/or resources.

Session: permanent or transient information exchange between two or more devices and/or users.

Social Trust: trust that builds upon associations of nodes based on the notion of shared interests, or affinities between entities, like, e.g., end-users, operators, and virtual operators.

Service Provider: organization (commercial or virtual) that provides some kind of service to Internet stakeholders (users, providers). Examples of services are communication, storage, and trust management. Examples of service providers as of today are Internet service provider (ISP), application service provider (ASP), Wireless Internet Service Providers, Over the Top Service Providers (OTT).

Technical Incentives: relate to natural features of the technology that result in a win-win match when cooperation is applied.

Trust Association: unidirectional social trust association between two different nodes.

Trust Management: system processing symbolic representation of social trust. TM aspects relate to understanding how to build networks of trust on-the-fly, based on reputation mechanisms able to identify end-user misbehavior and also to reward good behavior towards communities.

ULOOP Credit: virtual currency unit used in ULOOP. Credits are used as rewards providing cooperation incentives and guarantee of reciprocity, since the credits earned by providing a service/resource can then be used to purchase other services/resources.

ULOOP Gateway: role (software functionality) that reflects an operational behavior making a ULOOP node capable of acting as a mediator between ULOOP systems and non-ULOOP systems – the outside world. The gateway role may or may not be owned and controlled by a ULOOP end-user: it may also be controlled by an access operator. The key differentiating factor of the role of gateway, in contrast to a regular ULOOP node, is the operational intelligence and mediation capability. The gateway functionality may reside in the user-equipment, in access points, or even in the access network. Hence, they exhibit a feature that is key in user-centric environments: their behavior as part of the network is expected to be highly variable.

ULOOP Node: role (software functionality) that a wireless capable device takes. Concrete examples of nodes can be specific user-equipment, access points, or event some management server.

User Service: system that fulfills a need from an Internet end-user. For instance, VoIP is an end-user service.

Value-Added Service: is historically tied to any service beyond voice calls and fax transmissions. In ULOOP, a VAS represents a new type of service that may give rise to additional benefits, from a socio-economic perspective, to the different Internet stakeholders (users, providers).

Virtual Currency: mean used to buy and sell virtual goods in a virtual persistent world without making use of real money.

Virtual Operator: entity that acts, in some aspects, as a ULOOP community coordinator. It provides services such as initial authentication or registering, and eventually, trust relationship storage. A VO is not an ISP given that it does not provide Internet Access (e.g., infrastructure, DNS). A VO is neither an ASP since it does not provide user services. The VO is therefore a new type of Service Provider emerging in today's Internet.

Volunteer: user who does not need a reward in order to share services and/or resources.

Author Index

A. Aldini and A. Bogliolo (eds.), *User-Centric Networking*,
Lecture Notes in Social Networks, DOI: 10.1007/978-3-319-05218-2,
© Springer International Publishing Switzerland 2014

V
Valerdi, D., 329, 345
van Sinderen, M., 75

W
Wang, J., 197
Wegdam, M., 75

Y
Yildiz, M., 209, 241

Z
Zhou, Q., 311, 329
Zhu, H., 135, 197